OCEANIC PROCESSES IN MARINE POLLUTION
VOLUME 1

OCEANIC PROCESSES IN MARINE POLLUTION

Editors: Iver W. Duedall
Dana R. Kester
P. Kilho Park

Volume 1: Biological Processes and Wastes in the Ocean
Editors: Judith M. Capuzzo and Dana R. Kester

Volume 2: Physicochemical Processes and Wastes in the Ocean
Editors: Thomas P. O'Connor, Wayne V. Burt, and Iver W. Duedall

Volume 3: Marine Waste Management: Science and Policy
Editors: Michael A. Champ and P. Kilho Park

Volume 4: Scientific Monitoring Strategies for Ocean Waste Disposal
Editors: Donald W. Hood, Amy Schoener, and P. Kilho Park

Volume 5: Urban Wastes in Coastal Marine Environments
Editors: Douglas A. Wolfe and Thomas P. O'Connor

Volume 6: Physical and Chemical Processes: Transport and Transformation
Editors: Donald J. Baumgartner and Iver W. Duedall

OCEANIC PROCESSES IN MARINE POLLUTION Volume 1

BIOLOGICAL PROCESSES AND WASTES IN THE OCEAN

Edited By

JUDITH M. CAPUZZO
Woods Hole Oceanographic Institution
Woods Hole, Massachusetts

DANA R. KESTER
University of Rhode Island
Kingston, Rhode Island

ROBERT E. KRIEGER PUBLISHING COMPANY
MALABAR, FLORIDA 1987

Original Edition 1987

Printed and Published by
ROBERT E. KRIEGER PUBLISHING COMPANY, INC.
KRIEGER DRIVE
MALABAR, FL 32950

Printed in the United States of America

Library of Congress Cataloging in Publication Data

Main entry under title:
Biological processes and wastes in the ocean.
(Oceanic processes in marine pollution; v. 1)

1. Marine pollution—Environmental aspects. 2. Biogeochemical cycles.
I. Capuzzo, Judith M. II. Kester, Dana R. III. Series.
QH545.W3B565 1987 574.5′2636 84-29733
ISBN 0-89874-810-0

10 9 8 7 6 5 4 3 2

CONTRIBUTORS

Arapis, Gérasime, Department of Radiobiology, Centre d'Etude de l'Energie Nucléaire–Studiecentrum voor Kernenergie, B-2400 Mol, Belgium

Bartlett, Paul D., National Research Institute for Oceanology, Council for Scientific and Industrial Research, Stellenbosch, South Africa

Bodammer, Shawn M., Environmental Concern, Inc., St. Michaels, Maryland 21663

Bonotto, Silvano, Department of Radiobiology, Centre d'Etude de l'Energie Nucléaire–Studiecentrum voor Kernenergie, B-2400 Mol, Belgium

Bossus, Albert, Department of Radiobiology, Centre d'Etude de l'Energie Nucléaire–Studiecentrum voor Kernenergie, B-2400 Mol, Belgium

Brown, David A., Southern California Coastal Water Research Project, 646 West Pacific Coast Highway, Long Beach, California 90806

Bryan, Geoffry, W., Marine Biological Association of the United Kingdom, The Laboratory, Plymouth, Devon, PL1 2PB, United Kingdom

Burridge, Leslie E., Fisheries and Environmental Sciences, Department of Fisheries and Oceans, Biological Station, St. Andrews, New Brunswick, EOG 2XO, Canada

Capone, Douglas G., Marine Sciences Research Center, State University of New York, Stony Brook, New York 11794

Capuzzo, Judith M., Biology Department, Woods Hole Oceanographic Institution, Woods Hole, Massachusetts 02543

Caspers, Hubert, Institut für Hydrobiologie und Fischereiwissenschaft, Zeiseweg 9, 2000 Hamburg 50, Federal Republic of Germany

Colard, Jacques, Department of Physics, Centre d'Etude de l'Energie Nucléaire–Studiecentrum voor Kernenergie, B-2400 Mol, Belgium

Cogneau, Michel, Laboratoire de Chimie Inorganique et Nucléaire, B-1348 Louvain-la-Neuve, Belgium

Daggett, Pierre-Marc, American Type Culture Collection, Rockville, Maryland 20852

Davis, Wayne R., U.S. Environmental Protection Agency, Environmental Research Laboratory, Narragansett, Rhode Island 02882

Donkin, Peter, Natural Environment Research Council, Institute for Marine Environmental Research, Prospect Place, The Hoe, Plymouth, Devon, PL1 3DH, United Kingdom

Farrar, Suzanne V., Natural Environment Research Council, Institute for Marine Environmental Research, Prospect Place, The Hoe, Plymouth, Devon, PL1 3DH, United Kingdom

Gerber, Georg B., Department of Radiobiology, Centre d'Etude de l'Energie Nucléaire–Studiecentrum voor Kernenergie, B-2400 Mol, Belgium

Gibbs, Peter E., Marine Biological Association of the United Kingdom, The Laboratory, Plymouth, Devon, PL1 2PB, United Kingdom

Gossett, Richard W., Southern California Coastal Water Research Project, 646 West Pacific Highway, Long Beach, California 90806

Hanstveit, Arnbjörn O., Laboratory for Applied Marine Research, Division of Technology for Society TNO, P.O. Box 57, 1780 AB Den Helder, The Netherlands

Harrison, Florence L., Environmental Sciences Division, Lawrence Livermore National Laboratory, Livermore, California 94550

Hennig, Helmke, F-K. O., National Research Institute for Oceanology, Center for Scientific and Industrial Research, Stellenbosch, South Africa

Kester, Dana R., Graduate School of Oceanography, University of Rhode Island, Kingston, Rhode Island 02881

Kirchmann, René, Department of Radiobiology, Centre d'Etude de l'Energie Nucléaire–Studiecentrum voor Kernenergie, B-2400 Mol, Belgium

Koepp, Stephen J., Biology Department, Montclair State College, Upper Montclair, New Jersey 07043

Kuiper, Jan, Laboratory for Applied Marine Research, Division of Technology for Society TNO, P.O. Box 57, 1780 AB Den Helder, The Netherlands

Lake, James L., U.S. Environmental Protection Agency, Environmental Research Laboratory, Narragansett, Rhode Island 02882

Mathot, Pierre, Department of Physics, Centre d'Etude de l'Energie Nucléaire–Studiecentrum voor Kernenergie, B-2400 Mol, Belgium

McHugh, Sophia R., Southern California Coastal Water Research Project, 646 West Pacific Coast Highway, Long Beach, California 90806

McLeese, Donald W., Fisheries and Environmental Sciences, Department of Fisheries and Oceans, Biological Station, St. Andrews, New Brunswick, EOG 2XO, Canada

Moore, Dan H., Biomedical Sciences Division, Lawrence Livermore National Laboratory, Livermore, California 94550.

Moore, Michael N., Natural Environment Research Council, Institute for Marine Environmental Research, Prospect Place, The Hoe, Plymouth, Devon, PL1 3DH, United Kingdom

Nadeau, Joseph, Geology Department, Rider College, Lawrenceville, New Jersey 08648

Nerad, Thomas A., America Type Culture Collection, Rockville, Maryland 20852

Nuyts, Gustaaf, Department of Radiobiology, Centre d'Etude de l'Energie Nucléaire–Studiecentrum voor Kernenergie, B-2400 Mol, Belgium

O'Connor, Joseph M., New York University Medical Center, Laboratory for Environmental Studies, A. J. Lanza Laboratories, Tuxedo Park, New York 10987

Pearson, Thomas H., Dunstaffnage Marine Research Laboratory, P.O. Box 3, Oban, Argyll, Scotland

Pipe, Richard K., Natural Environment Research Council, Institute for Marine Environmental Research, Prospect Place, The Hoe, Plymouth, Devon, PL1 3DH, United Kingdom

Pizza, John C., New York University Medical Center, Laboratory for Environmental Studies, A. J. Lanza Laboratories, Tuxedo Park, New York 10987

Pond, Virginia, Biology Department, Brookhaven National Laboratory, Upton, New York 11973

Pruell, Richard J., Graduate School of Oceanography, University of Rhode Island, Narragansett, Rhode Island 02882

Quinn, James G., Graduate School of Oceanography, University of Rhode Island, Narragansett, Rhode Island 02882

Renna, Mark, Department of Zoology and Physiology, Rutgers University, Newark, New Jersey 07107

Rice, David W., Jr., Environmental Sciences Division, Lawrence Livermore National Laboratory, Livermore, California 94550

Santoro, Edward D., Biology Department, Montclair State College, Upper Montclair, New Jersey 07043

Sawyer, Thomas K., National Marine Fisheries Service, Northeast Fisheries Center, Oxford, Maryland 21654

Slater, Jennifer, Marine Sciences Research Center, State University of New York, Stony Brook, New York 11794

Sunda, William G., National Marine Fisheries Service, Southeast Fisheries Center, Beaufort Laboratory, Beaufort, North Carolina 28516-9722

Thomson, Sarah, Natural Environment Research Council, Institute for Marine Environmental Research, Prospect Place, The Hoe, Plymouth, Devon, PL1 3DH, United Kingdom

Vaidya, Swati, Department of Zoology and Physiology, Rutgers University, Newark, New Jersey 07107

Varela, Marilyn, Office of Radiation Programs, U.S. Environmental Protection Agency, Washington, D.C. 20006

Weis, Judith S., Department of Zoology and Physiology, Rutgers University, Newark, New Jersey 07107

Weis, Peddrick, Department of Anatomy, University of Medicine and Dentistry of New Jersey, Newark, New Jersey 07107

Woodhead, Avril D., Biology Department, Brookhaven National Laboratory, Upton, New York 11973

Zimmer, Robin, New Jersey Marine Sciences Consortium, Sandy Hook, New Jersey 07732

Preface To
Oceanic Processes In Marine Pollution

It is becoming increasingly evident that the marine environment—oceans, seas, and estuaries comprising over 70% of the entire earth's surface—are not separate entities but are, instead, parts of a continuum. As such, the impacts dealt to one part of the marine environment must be expected to affect eventually the whole. As the ocean is used increasingly as a reservoir for wastes, it is imperative to ensure that the marine resources be used wisely and efficiently. One important approach to such a global perspective is the exchange of information vital to the health and careful management of the ocean. Many people associated with the marine scientific and technical community worldwide are now working to find solutions to problems of mutual interest. The results of their endeavors can be efficiently used through a common medium, such as *Oceanic Processes in Marine Pollution*.

The *Oceanic Processes in Marine Pollution* series is organized into individual reference books that focus on specifically related topics or themes in order to provide unity throughout. To date, six volumes are scheduled for publication:

Volume 1: *"Biological Processes and Wastes in the Ocean,"*
Volume 2: *"Physicochemical Processes and Wastes in the Ocean,"*
Volume 3: *"Marine Waste Management: Science and Policy,"*
Volume 4: *"Scientific Monitoring Strategies for Ocean Waste Disposal,"*
Volume 5: *"Urban Wastes in Coastal Marine Environments,"* and
Volume 6: *"Physical and Chemical Processes: Transport and Transformation."*

The objective of this series is to bring together selected scientific and technical information about fundamental oceanic processes, marine pollution, and marine waste management. The series will cover a broad range of topics, be interdisciplinary in scope, and be of use to scientists and researchers, environmental managers and engineers, administrative officials and legislative staffers, oceanographers, and students.

The driving force behind this series is our belief that some human activities, such as industrial and wastewater discharges to sea, ocean dumping, and inadvertent discharges from non-point sources, could lead to serious problems of marine pollution. A key element in this thought is the recognition that the combination of many dynamic biological, chemical, and physical processes in the sea sometimes reduce the effects of marine pollution; however, serious problems of marine pollution can and do occur even in the most dynamic system if it is overloaded with contaminants. The marine scientist, aware of the dynamics of this system, must, when called upon, be able to translate available technical findings into something that will be of use to the policy maker, the environmental manager, or the engineer who is given the task of regulating or setting policy for the use of the ocean. With this series we hope to provide a cohesive overview of the state-of-the-art technology and thought in a wide range of marine pollution topics.

The chapters included in this and succeeding volumes of this series have been subjected to both external and editorial reviews. We are especially grateful to the reviewers of these chapters for the time and effort that they devoted to the development of the final manuscripts.

Oceanic Processes in Marine Pollution could not have been produced without the willingness and support of the Ocean Assessments Division of the U.S. National Oceanic and Atmospheric Administration.

Iver W. Duedall
Series Editors **Dana R. Kester**
P. Kilho Park

PREFACE

The distribution, fate, and effects of wastes in the ocean are governed by the physical, chemical, and biological processes that alter the concentration, chemical form, bioavailability, or toxicological effects of contaminants. An understanding of the importance of biological processes, such as microbial reactions and biogeochemical cycling, and of biological effects of contaminants on ecologically important components of the marine ecosystem is essential in predicting the impact of waste disposal in the marine environment.

The chapters in this volume are concerned with the biological aspects of waste disposal and are directed, in particular, toward bioaccumulation and toxicological effects of contaminants in marine organisms. The volume begins with an overview of the range of biological effects of waste input to the marine environment. Part II examines the toxicity, availability, and bioaccumulation of trace metals in marine organisms. Part III presents investigations related to the occurence, degradation, bioaccumulation, and effects of organic contaminants in the marine environment. Part IV examines the uptake, distribution, and effects of radionuclides in marine organisms. In Part V, the effects of sewage-sludge disposal, specifically as it relates to the survival of pathogens and organic enrichment in marine sediments, are discussed.

We gratefully acknowledge the assistance of the scientific community in reviewing these chapters and in providing many helpful suggestions. We greatly appreciate the many laborious tasks carried out by the staff of the Center for Academic Publications; in particular, we thank Suzanne M. Ban, Theresa D. Bewig, Ellen E. Main, Mary Ann Nelson, and Diane L. Quimby for their editorial assistance and manuscript preparation. We also thank Barbara J. Struttmann for preparing the illustrations. For the photograph on the title page for Part IV we thank R. S. Dyer of the U.S. Environmental Protection Agency. This book was supported in part by a grant from the U.S. National Oceanic and Atmospheric Administration.

JUDITH M. CAPUZZO
DANA R. KESTER

Woods Hole Oceanographic Institution

CONTENTS

GLOSSARY OF ACRONYMS

AAS	Atomic Absorption Spectrophotometry
ACP	Acid Phosphatase
ANOVA	Analysis of Variance
AOML	Atlantic Oceanographic and Meterological Laboratories
BAF	Bioaccumulation Factor
BCF	Biconcentration Factor
BrdUrd	5-bromodeoxyuridine
CA	Chromosomal Aberration
CD	Chromatid Deletion
CEC	Commission of the European Communities
CFU	Colony-Forming Unit
4CP	4-chlorophenol
D_0	Mean Lethal Dose
DCA	3,4-dichloroaniline
DCP	2,4-dichlorophenol
DDA	2,2-*bis*(p-chlorophenyl)acetic acid
DDD	2,2-*bis*(chlorophenyl)-1,1-dichloroethane
DDE	1,1-dichloro-2,2-*bis*(chlorophenyl)ethylene
DDOH	2,2-*bis*(p-chlorophenyl)ethanol
DDT	1,1,1-trichloro-2,2-*bis*(chlorophenyl)ethane
DNOC	4,6-dinitroorthocreosol
EDTA	Ethylenediaminetetraacetic Acid
Eh	Redox Potential
ENZ	Enzyme-Containing
EPA	U.S. Environmental Protection Agency
FDA	U.S. Food and Drug Administration
GAE	Granulomatous Amoebic Encephalitis
GCMS	Gas Chromatography–Mass Spectrometry
GLUC	β-glucuronidase
GSH	Glutathione-Containing
HPLC	High Performance Liquid Chromatography
HFC	High-Frequency Cell
IAEA	International Atomic Energy Agency
IRE	Institut National des Radioéléments
$LD_{50/30}$	The Dose of Radiation Required to kill 50% of an Experimental Group of Organisms in 30 Days
MERL	Marine Ecosystem Research Laboratory
MT	Metallothionein-Containing
MMC	Mitomycin-C
MPN	Most-Probable-Numbers
MOP	Model Plankton Ecosytem
NAS	U.S. National Academy of Sciences
NADPH-NTR	NADPH-Neotetrazolium Reductase
NEA	Nuclear Energy Agency of the Organisation for Economic Co-operation and Development
4NP	4-nitrophenol
NRC	National Research Council
NRIO	National Research Institute for Oceanology
OECD	Organisation for Economic Co-operation and Development
PAH	Polycyclic Aromatic Hydrocarbons
PCB	Polychlorinated Biphenyl
PCBols	Oxygenated Metabolites
PCP	Pentachlorophenol
SACSA	Standing Advisory Committee for Scientific Advice of the Oslo Commission
SCE	Sister Chromatid Exchange
γ-TCCH	γ-tetrachlorocyclohexane
TLD	Thermoluminescent Dosimeter
TPBS	Tetrapropylenebenzenesulphonate
UCM	Unresolved Complex Mixture

PART I: INTRODUCTION

Chapter 1

Biological Effects of Waste Disposal: Experimental Results and Predictive Assessments

Judith M. Capuzzo

Woods Hole Oceanographic Institution
Woods Hole, Massachusetts

Dana R. Kester

Graduate School of Oceanography
University of Rhode Island
Kingston, Rhode Island

ABSTRACT

The biological effects of contaminants discharged into the marine environment as a result of waste disposal include impairment of feeding, growth, development, energetics, and recruitment that may result in alterations in reproductive and developmental success and changes in community structure and dynamics. In assessing long-term effects of waste disposal in the marine environment, it is important to understand the conditions under which contaminants persist in benthic environments and the sublethal effects that lead to loss of a population's function in marine communities. A framework for the development of monitoring programs, integrating both chemical and biological processes is addressed.

1.1. INTRODUCTION

Biological effects of waste disposal are dependent on the chemical composition of specific wastes, their persistence in the marine environment, their bioavailability and bioaccumulation, and their toxic effects. The most important biological factors to be addressed are the bioavailability and accumulation of contaminants by marine biota and long term effects of contaminant exposure on the dynamics of marine ecosystems. Understanding the bioavailability of contaminants requires detailed knowledge of the extrinsic factors that control chemical partitioning of contaminants (such as sorption–desorption reactions) and intrinsic factors that affect uptake and deposition of contaminants in biota (such as body lipids and detoxification processes). Once particulate-bound contaminants are deposited in sediments, it is important to determine how their availability is affected by sediment-transport processes, including bioturbation and resuspension. The responses of marine organisms to discharges of anthropogenic wastes can be manifested at four levels of biological organization: (1) biochemical and cellular; (2) organismal, including the integration of physiological, biochemical and behavioral responses; (3) population, including alterations in population dynamics; and (4) community, resulting in alterations in community structure and dynamics. The interrelationship of these responses is presented in Fig. 1.1. Adaptive processes are capable of countering disruptive processes until the system reaches a threshold for the toxicant, at which point the adaptive potential is completely overridden by the degeneration imposed on the system by disruptive effects.

BIOLOGICAL LEVELS OF ORGANIZATION

SUBCELLULAR	CELLULAR/TISSUE	ORGANISMIC	POPULATION	COMMUNITY
CYTOCHEMICAL AND BIOCHEMICAL RESPONSES	CELLULAR AND TISSUE RESPONSES	PHYSIOLOGICAL AND BEHAVIORAL RESPONSES	CHANGES IN POPULATION STRUCTURE AND FUNCTION	CHANGES IN COMMUNITY STRUCTURE AND FUNCTION
Lysosomal latency	Gametogenic cycle	Respiration	Biomass	Biomass
Enzyme activity	Nutrient storage	Feeding	Productivity	Species abundance
Maximum metabolic flux	Deformity	Excretion	Age and size structure	Species distribution
Energy charge	Neoplasms and tumors	Growth	Recruitment	Species diversity
Amino acid labels		Fecundity	Mortality	Trophic interactions
Blood chemistry		Reproductive effort		Energy flow
Chromosomal abnormalities		Larval viability		Spatial variability
Metallothioneins		Swimming		Temporal variability
Mixed function oxygenases				

Figure 1.1. Biological effects of contaminants at multiple levels of organization (Widdows, 1982; Capuzzo, 1981).

Biological effects of contaminants can be manifested at biochemical, cellular, and organismal levels of organization before disturbances are seen at the population level (Capuzzo, 1981). All responses are not disruptive in nature and do not necessarily result in degeneration at the next level of biological organization. Only when the compensatory or adaptive mechanisms at one level begin to fail do deleterious effects become apparent at the next level (Capuzzo, 1981). An important aspect of comparing responses at various levels of biological organization is ascertaining the degree to which adaptive responses at each of the four levels can persist with increasing concentrations of contaminants. The initial responses in each case are inductions of mechanisms to resist or reduce the impact of the toxicant; these mechanisms may include the induction of toxicant-metabolizing processes (at the biochemical level) or the selection of toxicant-resistant forms (at the population level). Adaptive processes are cap-

For predictive purposes, it is important to understand the early warning signs of stress at each level of organization before compensatory mechanisms are surpassed. From the biochemical level to the community level, the degree of system complexity, the number of compensatory mechanisms available, and the lag time to measure a response increases exponentially, thereby increasing the predictive difficulties at each level. Cairns (1983) argues that the present ability to detect toxic effects at higher levels of biological organization is limited by the lack of reliable predictive tests at population, community, and ecosystem levels. He suggests that a great deal of effort is needed in these areas before anyone can adequately address environmental hazards as a result of input of any toxicant.

Predicting the effects of contaminants on populations and communities of marine organisms requires the integration of chemical and biological monitoring; the collection of physiological and biochemical data pertinent to the growth, reproduction, and recruitment of individuals within a population; and modelling the impact both of short-term and of long-term changes in population effort and stability. Experimental studies directed at determining effects on energy metabolism or effects that influence growth and reproduction would be most appropriate at linking effects at higher levels of organization. An additional requirement is to understand the functional aspects of marine communities in order to predict the energy flow through communities and to assess resource impact. The focus of this chapter is to integrate the present understanding of the effects of contaminants on marine organisms. Using both general principles and specific examples from the extensive literature on petroleum impacts in the marine environment [U.S. National Research Council (NRC), 1985], a framework will be developed for experimental design from which predictions of effects on higher levels of organization can be made.

1.2. BIOAVAILABILITY AND BIOACCUMULATION OF CONTAMINANTS

The distribution, fate, and effects of wastes in the ocean are governed by the physical, chemical, and biological processes that generally reduce the concentration, alter the chemical form or bioavailability of contaminants, and ultimately eliminate them from the water column (U.S. NRC, 1984). In the long-term, contaminants of biological concern, such as trace metals, xenobiotics, and organic matter, will accumulate in benthic habitats, thus increasing the potential for alterations of benthic populations through organic enrichment, toxicant impacts, and disturbances of benthic–pelagic coupling. The chemical form and bioavailability of contaminants may be modified by biological processes, such as microbial degradation and bioturbation, and by chemical processes, such as sorption and desorption reactions, redox reactions, and dissolution. Sorption and desorption reactions depend primarily on the surface area and charge of a particle, competing ions for exchange, chemical speciation and ion exchange processes at the solid–solute interface, and octanol–water partition coefficients for organic contaminants (Means et al., 1979; Veith et al., 1979; Stumm and Morgan, 1981; Morel and Schiff, 1983). Even after sedimentation has occurred, contaminants may be resuspended and transported great distances by prevailing bottom currents, following remobilization through bioturbation (Aller, 1977).

For hydrocarbons (including polyaromatic hydrocarbons and chlorinated hydrocarbons), adsorption onto marine sediments is affected by the solubility of individual compounds and the grain-size distribution and organic content of sediments (Paris et al., 1978; Gearing et al., 1980; Olsen et al., 1982; Gearing and Gearing, 1983). The extent of adsorption appears to be inversely related to the solubility of the compound and the size of the sorbent particles, and it is directly related to the organic content of the sediments. Biological uptake of hydrocarbons from contaminated sediments may be attributed to desorption of hydrocarbon particles from sediment particles into interstitial waters. Thus, the bioavailability of sediment adsorbed hydrocarbons would appear to be directly related to the solubility of the compound and sediment grain size and inversely related to the organic content of contaminated sediments. Other factors such as hydrocarbon source—petrogenic versus pyrogenic—may also affect bioavailability. Farrington et al. (1983) suggested that polyaromatic hydrocarbons from petroleum sources are more available for biological uptake than are those from pyrogenic sources; this is due to the more tightly bound nature of PAHs to pyrogenic particulate matter.

Biological factors, such as deposition of organic contaminants in lipid reserves, capacity for detoxification, surface:volume ratios, feeding habits, and physiological state, may also affect the fate of chemical contaminants (Neff, 1979; Bryan, 1980). Uptake of compounds may occur through adsorption onto body surfaces, water exchange at respiratory and feeding surfaces, and ingestion of food and detritus. Removal may occur through active mechanisms such as hydrocarbon metabolism and excretion of metabolized byproducts, or through passive mechanisms such as diffusive exchange and the production of particulate products (eggs, molts, and feces). The balance of these processes for individual compounds may result in selective uptake, accumulation, and depuration.

Neff (1979) proposed a two-compartment model to partially explain the differences in accumulation and depuration of polyaromatic hydrocarbons in animals from acutely and chronically oil-contaminated sites. Uptake and storage of hydrocarbons in depot lipids, such as energy reserves or gonadal reserves, may be retained until those reserves are mobilized for nutritional or reproductive needs. Hydrocarbons associated with more labile hydrophobic compartments, such as membrane lipids and cellular macromolecules, however, may be more rapidly released when ambient levels decrease. A two-phase depuration, with an initially rapid release of hydrocarbons followed by a slower, more gradual release, could be explained by these differences in lipid compartments. Hydrocarbon metabolism, however, also plays an important role in regulating accumulation and release of polyaromatic hydrocarbons.

The relationships between body burden of contaminants and toxic effects and between reversibility of effects and release of contaminants are difficult to establish. Few studies have focused on both aspects simultaneously. Anderson (1977) concluded that there was little agreement between sublethal responses of marine organisms to petroleum hydrocarbons and the level of hydrocarbon contamination in tissues. Gilfillan et al. (1977), however, reported a negative correlation between energetics measurements, such as carbon flux, and the concentration of aromatic hydrocarbons in tissues of the soft-shelled clam *Mya arenaria*. Widdows et al. (1982) found a similar relationship between various energetics parameters and concentrations of aromatic hydrocarbons in the tissues of *Mytilus edulis*. In neither study was there a correlation observed between effects and concentration of total hydrocarbons or aliphatic compounds in the tissues.

Anderson et al. (1980) investigated the relationship between mortality and accumulated hydrocarbons in the mysid *Neomysis awatschensis* and in two species of shrimp, *Hippolyte clarkii* and *Pandalus davae,* after exposure to Prudhoe Bay crude oil. Accumulation of di- and tri-aromatic compounds in the tissues of these crustaceans was not cumulative and could not be used to explain toxic effects. The authors suggested that further analysis of mono-aromatic hydrocarbons and metabolites might provide a better clue as to the relationship between body burden and toxic effects.

Capuzzo et al. (1984) suggested that different sublethal effects might be related to exposure, uptake, and metabolism of different hydrocarbons. In studies with larval stages of the American lobster (*Homarus americanus*), suppression of respiratory activity during and after exposure to South Louisiana crude oil correlated with accumulated benzene, thiophene, toluene, and alkylbenzenes, whereas alterations in lipid metabolism occurred during exposure only and might have been related to the detoxification of aromatic hydrocarbons with higher molecular weights.

Malins and Hodgins (1981), in their review of the effects of petroleum hydrocarbons on marine fishes, concluded that although aromatic hydrocarbons are accumulated in fish tissues, subsequent metabolism and excretion of metabolized byproducts reduce body burden, so increased hydrocarbon concentrations are not always detected. They further suggested that a clearer relationship might exist between toxic effects and accumulated metabolites. Thus, it is apparent that the sublethal effects of oil exposure may be modified by the ability of the organisms to accumulate and metabolize various hydrocarbons and other lipophilic contaminants through mixed-function oxygenase reactions (Lee, 1981; Stegeman, 1981). In evaluating long-term effects of contaminant exposure, it is important to understand not only the factors that influence bioaccumulation, but also the metabolic capacity for alteration and consequences of accumulated compounds.

1.3. SUBLETHAL EFFECTS

1.3.1. Cellular and Biochemical Effects

Sublethal effects of contaminants on marine organisms may be manifested at all levels of biological organization. The level of impact will be dependent on the duration of exposure to toxic concentrations and the compensatory mechanisms available for recovery following exposure. At the subcellular level, effects can be manifested in changes in energy metabolism, alterations in cellular structure and function, and enhancement of chromosome mutation.

The mutagenic potential of metabolites of polyaromatic hydrocarbons has been well documented in mammalian systems, and they may have similar effects on fish and invertebrates. Chromosomal aberrations and increased incidence of sister chromatid exchange have been observed in fish with exposure to benzo[*a*]pyrene (Hooftman and Vink, 1981). Cod and pollock eggs collected after the *Argo Merchant* oil spill showed a wide range of developmental abnormalities, including an increase in cytological deterioration, abnormal differentiation, and incidences of mitotic abnormalities (Longwell, 1977). Other investigators have also reported a wide range of developmental abnormalities in early life stages as a result of oil exposure (Linden, 1976; Linden et al., 1980; Rabalais et al., 1981). The implications of these responses on the survival and fitness of populations remain largely unexplored.

The persistence of metabolites in the tissues of marine organisms may induce toxic effects, particularly when metabolites bind to cellular macromolecules such as DNA, RNA, or proteins. Varanasi and her co-workers demonstrated the metabolic activation of benzo[*a*]pyrene and *in vitro* binding of metabolites to DNA. Activation and binding occurred in preparations of deproteinized salmon sperm when preparations were incubated with liver homogenates from untreated, 3-methylcholanthrene- and benzo[*a*]pyrene-treated starry flounder (*Platichthys stellatus*), English sole (*Parophrys vetulus*), and coho salmon (*Oncorhynchus kisutch*) (Varanasi and Gmur, 1980; Varanasi et al., 1980). Further studies demonstrated that similar processes occurred *in vivo* in fishes exposed to benzo[*a*]pyrene (Varanasi and Gmur, 1981a, 1981b; Varanasi et al., 1981). The distribution of unmetabolized and metabolized benzo[*a*]pyrene in English sole is presented in Table 1.1.

Table 1.1. Distribution of Unmetabolized and Metabolized Benzo[*a*]pyrene in Liver and Bile of English Sole (*Parophrys vetulus*)[a]

Form of Benzo[*a*]pyrene	Amount in Liver (%)	Amount in Bile[b] (%)
Unmetabolized	2	2 ± 3
Ethyl acetate-soluble metabolites	16	14 ± 10
Glucuronides	<5	43 ± 3
Sulfates	<5	13 ± 3
Other water-soluble metabolites	45	28 ± 2
Bound to cellular constituents	34	—

[a] From Varanasi and Gmur (1981b). A dash denotes the absence of data.
[b] Mean of three replicates ± 1 standard deviation.

A wide range of morphological, cytological, and developmental abnormalities have been observed in marine animals exposed to petroleum or petroleum components (Table 1.2). The relationship between these abnormalities and metabolite production has not been clearly established although the relationship is strongly suggested (Malins, 1982). Coincident with histopathological disorders, biochemical changes indicative of altered energy metabolism and balance were observed among populations of *Pleuronectes platessa* collected from oil-contaminated estuaries following the spill from the *Amoco Cadiz* (Haensly et al., 1982; Stott et al., 1983; Neff, 1985). Malins et al. (1983) reported a high incidence of organic free radicals in liver microsomes associated with idiopathic liver lesions from English sole collected from sediments containing a high concentration of aromatic hydrocarbons. The organic free radicals appeared to be of a type derived from the metabolism of a polyaromatic hydrocarbon linked to a macromolecule such as protein. The exact relationship between free radicals derived from biotransformations and the development of lesions is not understood.

Measurement of induction of mixed function oxygenases has been proposed as a possible monitoring tool for predicting the biochemical and cellular effects of exposure to oil or other lipophilic organic contaminants on natural populations of marine animals (Payne, 1976; Kurelec et al., 1979; Stegeman, 1980). Using this parameter as a monitoring tool will require an understanding of the basis for seasonal fluctuations or sex-linked differences in mixed function oxygenase activity. The significance of chronic induction is uncertain, however, as it may imply a functional adaptation for detoxification (Spies et al., 1982) or an increased potential for pathological damage (Neff, 1985). Before this can effectively be used as a monitoring tool, an understanding of the induced changes in the mixed function oxygenase system must be linked with long-term consequences in reproduction (due to possible altered metabolism of endogenous compounds such as steroid hormones) and disease (due to possible increased production of carcinogenic and mutagenic metabolites).

1.3.2. Physiological and Behavioral Effects

The most important physiological changes associated with contaminant exposure are those that may adversely affect an organism's growth and survival and, thus, its potential ability to contribute to the population gene pool. Alterations in growth potential may take place as a result of changes in feeding behavior, respiratory metabolism, or digestive efficiencies. Animals from polar regions may be particularly sensitive to impairment of energetics and the sporadic seasonal abundance of food, because of the dependence of these polar species on long-term energy reserves, and because of slow recovery rates as a result of reduced fecundity, reduced dispersal, and the slow growth rates observed for many polar species (Dunbar, 1968; Clarke, 1979).

Reductions in physiological measurements, such as respiration rates, carbon turnover rates, and scope-for-growth indexes, have correlated with reduced growth rates measured for bivalve populations from oil-contaminated habitats (Gilfillan et al., 1976; Gilfillan and Vandermeulen, 1978). Gilfillan et al. (1977) and Gilfillan (1980) suggested that alterations in energetics and growth of bivalve populations might be related to burdens of aromatic compounds in the tissues. Widdows et al. (1982) further demonstrated that, with long-term exposure of *Mytilus* to oil concentrations as low as 30 μg liter^{-1} of total hydrocarbons, a negative correlation existed between both cellular and physiological stress indexes and concentrations of aromatic hydrocarbons in the tissues.

Behavioral responses of an organism to contaminant stress may serve as mechanisms for detecting adverse contaminant concentrations, as well as for triggering adaptive mecha-

Table 1.2. Morphological, Cytological, and Developmental Abnormalities in Marine Organisms Exposed to Petroleum Hydrocarbons[a]

Species	Hydrocarbon	Concentration and Exposure Time	Effect	Reference
Invertebrates				
Homarus americanus	crude oil	100–1000 μg liter^{-1} 15 d	Distended chromatophores	Forns (1977)
Mercenaria mercenaria	phenol	100 μg liter^{-1} 24 h	Cytological damage to the gills and digestive gland	Fries and Tripp (1977)
Mytilus edulis	aromatic	30 μg liter^{-1} 34–182 d	Functional and structural changes in lysosomes	Lowe et al. (1981) Moore and Clarke (1982)
Paracentrotus lividus	benzo[*a*]pyrene	4–5 μg liter^{-1} 3 h	Abnormal cleavage in embryos	Ceas (1974)
Fish				
Clupea harengus	benzene	0.9 μg liter^{-1} 24 h	Delayed embryonic development	Eldridge et al. (1977)
	crude oil	1 μg g^{-1} injected	Morphological abnormalities	Smith and Cameron (1979)
Fundulus heteroclitus	naphthalene	200 μg liter^{-1} 15 d	Ischemia Lateral line necrosis	DiMichele and Taylor (1978)
Hypomesus pretiosus	crude oil	54–113 μg liter^{-1} 3 h d^{-1} for 15–21 d	Necrotic neurons in embryos	Hawkes and Stehr (1982)
Menidia beryllinia	crude oil	5–100 mg liter^{-1} 5–50% WSF[b] 21–30 d	Gill and olfactory hyperplasia Pancreatic atrophy and necrosis.	Solangi and Overstreet (1982)
Menidia menidia	crude oil	140 μg liter^{-1} 168 h	Hyperplasia	Gardner (1975)
Mugil cephalus	crude oil	experimental spill (4–5 μg g^{-1}) 56 d	Fin erosion	Minchew and Yarbrough (1977)
Parophrys vetulus	crude oil	0.2–1.0% of the amount in sediments 0.75–4 months	Hepatocellular lipid vacuolation Weight loss	McCain and Malins (1982)
Pleuronectes platessa	crude oil	*Amoco Cadiz* (field sample)	Fin necrosis hyperplasia Decreased hepatocellular vacuolation	Haensly et al. (1982)
			Delayed or suppressed ovarian development	Stott et al. (1983)
Pseudopleuronectes americanus	crude oil	Amounts found in sediments (1.02 kg) 4–5 months	Liver hypertrophy	Fletcher et al. (1982)
Salmo gairdneri	crude oil	200 μl liter^{-1} OWD[c]	Gill damage Chloride cell abnormalities	Engelhardt et al. (1981)
Trinectes maculatus	crude oil	5–100 mg liter^{-1} 5–50% WSF 38–60 d	Necrosis of olfactory tissue Pancreatic atrophy and necrosis	Solangi and Overstreet (1982)

[a] Adapted from Malins (1982).
[b] Water-soluble fraction.
[c] Oil-water dispersion.

nisms, such as altered feeding behavior, or inducing an avoidance response. At an extreme level of stress, adaptive behaviors are overridden, and an organism's ability to respond to environmental stimuli may become impaired either temporarily until the stress is removed or permanently if chemosensory mechanisms are irreversibly damaged.

Blumer (1969) was the first to suggest that exposure to petroleum hydrocarbons could interfere with chemoreception and that this could result in altered behavior patterns. Behavioral aberrations associated with oil exposure include altered mating behavior (Kittredge et al., 1974), feeding behavior (Jacobson and Boylan, 1973; Atema and Stein, 1974; Berdugo et al., 1977; Berman and Heinle, 1980; Busdosh, 1981; Pearson et al., 1981; Cowles, 1983), burrowing behavior of benthic invertebrates (Prouse and Gordon, 1976; Gordon et al., 1978; Augenfeld, 1980; Olla et al., 1983), swimming activity (Berge et al., 1983), and schooling behavior of fishes (Gardner, 1975).

Neurophysiological studies with chemoreceptors in the lateral flagella of antennules of the lobster *Homarus americanus* provide evidence that oil added to mussel extracts elicit nerve-firing responses distinct from those elicited by mussel extract alone (Atema et al., 1979). Avoidance responses of fishes to low levels of petroleum hydrocarbons have also been documented (Rice, 1973; Maynard and Weber, 1981; Weber et al., 1981; Hellstrom and Doving, 1983). Behavioral responses can be observed at concentrations as low as 0.1–0.4 μg liter^{-1} of total hydrocarbons and thus may serve as extremely sensitive indicators.

Recovery from oil exposure at the level of the organism is dependent on the duration of exposure and the bioavailability of specific contaminants throughout exposure. Behavioral and physiological responses observed during short-term hydrocarbon exposure (a few hours to several days) may be restored to control levels following transfer to uncontaminated seawater (Berge et al., 1983; Cowles, 1983; Capuzzo et al., 1984), although recovery does not always occur immediately upon transfer (Baden and Hagerman, 1981; Capuzzo et al., 1984). Capuzzo et al. (1984), in their study of short-term exposure of larval stages of *Homarus americanus* to 250 μg liter^{-1} of South Louisiana crude oil, found that recovery of normal respiratory activity was not immediate upon transfer to uncontaminated seawater; values for respiration rates and ammonia excretion rates remained at a depressed level up to one week following exposure and correlated with tissue uptake of benzene, thiophenes, alkylbenzenes, and toluene. Other sublethal responses that were evident during exposure (depressed oxygen:nitrogen ratios and alterations in lipid-storage patterns) were restored to control levels soon after exposure.

Recovery from long-term contaminant exposure is somewhat more complicated. Widdows et al. (1982), in their study of long-term exposure of *Mytilus edulis* to 30 μg liter^{-1} of North Sea oil, found no evidence of gradual recovery or acclimation to exposure conditions; however, they did find a gradual deterioration in physiological condition, as evidenced by physiological and cellular stress indexes. With chronic exposure of larval and juvenile crustaceans to water-soluble (WSF) or water-accommodated fractions of crude or refined oils, increased tolerance or increased acclimation to exposure conditions has generally been observed during exposure, resulting in little difference in growth or physiological parameters between oil-exposed and control animals at the end of a long term experiment (Wells and Sprague, 1976; Laughlin et al., 1978; Cucci and Epifanio, 1979). This observation may be due in part to an increased capacity for detoxification among older, post-metamorphic animals, although few data are currently available to support this hypothesis. Benthic crustaceans and fishes exposed to oil-contaminated sediments, however, continue to exhibit sublethal responses throughout exposure (Fletcher et al., 1982; McCain and Malins, 1982), suggesting that differences in exposure conditions, particularly in the persistence and availability of specific hydrocarbons in pore waters, may elicit long-term responses.

1.3.3. Population and Community Effects

The integration of physiological and behavioral disturbances may result in alterations at the population and community levels. As illustrated in Fig. 1.2, impairment of behavioral, developmental, and physiological processes may occur at concentrations significantly lower than acutely toxic levels; such responses may alter the long-term survival of affected populations. Populations of plaice (*Pleuronectes platessa*) collected from oil-contaminated estuaries following a spill from the *Amoco Cadiz* exhibited reduced growth rates and fecundity in addition to histopathological aberrations (Conan, 1982). Hauschildt-Lillge (1982) exposed the oligochaete *Lumbricillus lineatus* through five generations (15 months) to the WSF of Arabian Light crude oil. Although this species was highly resistant to short-term oil exposure (Giere and Hauschildt, 1979), long-term exposure resulted in alterations in reproductive success. During the first generation, reduced egg fertility and hatching success were compensated by

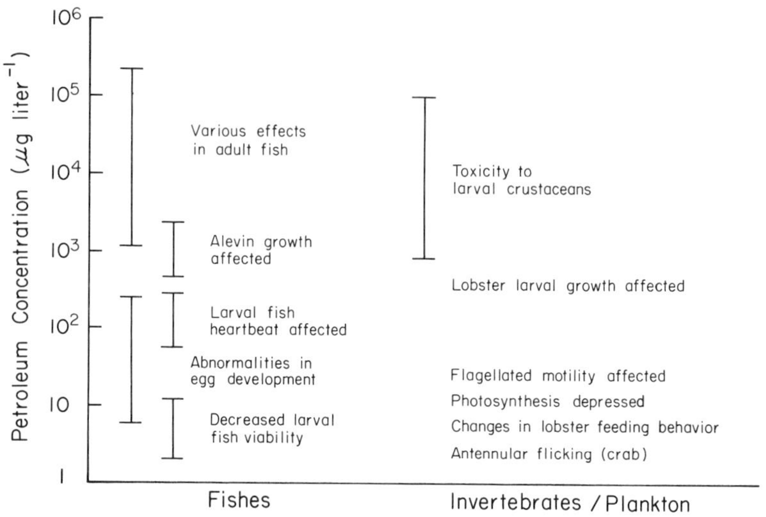

Figure 1.2. Comparison of lethal and sublethal effects of petroleum hydrocarbons on fish and invertebrates (Vandermeulen and Capuzzo, 1983).

increased coccoon production at all exposure concentrations except 100% WSF. Embryonic development was not altered by exposure, but reduced growth and prolonged maturation and generation times were evident. Subsequent generations also exhibited reduced reproductive potential, but no effects on maturation or generation times were observed. In the 20% and 100% WSF exposures, coccoon production continued to decline and ceased with the F_4 generation. In the 50% WSF exposure, however, productivity in terms of coccoon production increased among succeeding generations, but hatching success decreased and the incidence of developmental abnormalities increased with each successive generation.

Krebs and Burns (1977) studied populations of the fiddler crab *Uca pugnax* for 7 y after a spill of No. 2 fuel oil from the barge *Florida*. They observed long-term reductions in recruitment, population density, female:male ratios of adult crabs, behavioral changes, and high mortality during overwintering. Recovery of crab populations was correlated with the disappearance of naphthalenes and alkylated naphthalenes from contaminated sediments. Similar patterns of long-term changes in recruitment and density of benthic fauna have been observed with other oil spills (Gilfillan and Vandermeulen, 1978; Thomas, 1978; Cabioch et al., 1980; Sanders et al., 1980; Beslier et al., 1980; Glémarec and Hussenot, 1981; Elmgren et al., 1983). Mesocosm experiments have further documented that the major long-term changes in marine ecosystems as a result of petroleum inputs occur in the benthos (Grassle et al., 1981; Oviatt et al., 1982).

In a series of three experiments conducted in the University of Rhode Island Marine Ecosystem Research Laboratory (MERL) mesocosms from 1977 to 1979 (Grassle et al., 1981; Oviatt et al., 1982), the impacts and recovery of a controlled marine ecosystem were monitored in response to chronic inputs of No. 2 fuel oil. For the first two experiments (1977 and 1978), oil–water dispersions of No. 2 fuel oil were added on a semicontinuous basis for 24- and 17-week periods, respectively. The third experiment was designed to evaluate the recovery potential of the ecosystem, and no oil was added. In the 1977 experiment, hydrocarbon concentrations in the water column were maintained at approximately 190 μg liter^{-1} of total hydrocarbons. Toward the end of the exposure period, 151 μg of total hydrocarbons per gram of dry sediment were detected in the upper 3 cm of sediment, and concentrations as high as 527 $\mu g\ g^{-1}$ were found in the surface floccular layer (Grassle et al., 1981; Oviatt et al., 1982). Benthic macrofaunal and meiofaunal populations were reduced after the 24-week exposure; however, numbers of benthic protozoa and diatoms increased, presumably as a result of reduced predation. Time-series analyses indicate that the greatest degree of benthic impact occurred during the summer months (Oviatt et al., 1982). During the 1978 experiments, hydrocarbon concentrations in the water column were maintained at 90 μg $liter^{-1}$ of total hydrocarbons. Approximately 50% of the added oil was recovered in the surface floccular layer of the sediments. Similar effects on macrofaunal abundance were evident in the second experiment, although effects on meiofauna were less dramatic.

In the recovery experiment, benthic impacts were evident even 1 y after the cessation of oil experiments, although there were no measureable impacts on the water column. Sediment concentrations declined rapidly initially, but a residual (10–20%) remained even 1 y later. Differences in degradation rates of specific hydrocarbons were apparent throughout the exposure periods, with higher rates of loss occurring during the summer months and coincident with the highest degree of benthic impact (Grassle et al., 1981; Oviatt et al., 1982).

Elmgren and Frithsen (1982) compared the effects of oil exposure on the MERL mesocosm with changes observed in the water column and benthos following the *Tsesis* oil spill. Responses of various components of the ecosystem were similar in both situations: increased phytoplankton abundance, reduced zooplankton abundance, and reduced macrofaunal and meiofaunal abundances. Although losses of zooplankton biomass were evident, observations at the *Tsesis* spill site suggest that this reduction was transitory since the recovery time for zooplankton was relatively short. Recovery rates for the meiofaunal component of the benthos may also be rapid, but recovery of the macrofaunal component occurs slowly (Grassle et al., 1981).

Alterations in benthic biomass as a result of contaminant exposure can also result in changes in physical and chemical features of the benthos. Kalke et al. (1982) demonstrated in benthic colonization experiments that oil exposure resulted in a reduction in the depth of the oxygenated layer by approximately one-half; this could result in a reduction in subsurface benthic production and changes in processes such as nutrient regeneration. Oviatt et al. (1982) also reported reductions in benthic respiration and nutrient flux during the recovery phase of the MERL mesocosm experiments.

Although the alteration of benthic communities is an important impact of contamination inputs in many marine habitats, it is difficult to ascertain whether differential recovery rates exist among different communities (offshore versus shallow) because there is a lack of comparable data on contaminant inputs, degree of sediment contamination, and rates of recovery for macrofauna and meiofauna. Recovery rates from disturbance will be dependent on the initial degree of impact, the rate of removal and persistence of contaminants from the affected area, and the recolonization rates of indigenous fauna. More detailed characterizations of the relative sensitivities and recovery rates of different marine communities must await further field verification.

Long-term effects of contaminants on fisheries may be direct, through the loss of reproductive or recruitment success, or indirect, through the disruption in food chain dynamics. Both types of impacts are difficult to assess because not enough is known about the natural variability in reproductive effort and recruitment and the flexibility in prey selection by commercially important species of fish and shellfish. Spaulding et al. (1983) developed a multicomponent fisheries-assessment model based on a model of the fates of oil spills, a shelf-hydrodynamics model, an ichthyoplankton model, and a fishery-population model. The model is useful in predicting the spatial and temporal interactions of spill conditions with the spawning and development of early life stages. Such an approach can be modified to include the impact of chronic-exposure conditions on reproductive effort in addition to larval development and recruitment success of demersal species. Indirect effects will be more difficult to estimate and quantify; however, for alterations in food-chain dynamics to have a significant impact on demersal fishery stocks, an extensive area of the benthos would have to be degraded.

1.4. CONCLUSIONS AND RECOMMENDATIONS FOR MONITORING PROGRAMS

Biological effects of contaminants in the marine environment include impairment of feeding, growth, development, energetics, and recruitment that may result in alterations both in reproductive and in developmental success and changes in community structure and dynamics. It is difficult to ascertain, however, the relationship between chronic responses of organisms to contaminants and large-scale alterations both in the functioning of marine ecosystems and in the harvesting of fishery resources. The sensitivity of early developmental stages, the impairment of reproductive processes, and the long-term effects on populations suggest that chronic exposure to hydrocarbons may certainly alter the dynamics of benthic populations, including populations of demersal fish.

In assessing the long-term impact of waste disposal in the marine environment, it is important to understand the conditions under which contaminants persist in benthic environments and the sublethal effects that lead to reduced growth, delayed development, and reduced reproductive effort and that result in population decline and the loss of that population's function in marine communities. Such a task requires a combined laboratory, mesocosm, and field approach that addresses: (1) the physical processes (specifically, the flow characteristics) that influence the partitioning of contaminants between sediments and interstitial waters, (2) the chemical processes that influence contaminant persistence and degradation rates in sediments and interstitial waters, and (3) the long-term biological effects that alter population stability and function and the consequences of such effects on resource utilization. The first two aspects are important in establishing realistic exposure scenarios—both in time and in space—and the third is important in linking sublethal stress indexes with predictions of population and community effects.

Although a wide range of sublethal stress indexes have been proposed for monitoring the responses of organisms to contaminants (McIntyre and Pearce, 1980), few have been linked to the survival potential of the individual or the reproductive potential of a population. Particularly sensitive responses that may show a relationship with population effects include biochemical responses that relate either to energy metabolism and membrane function or to detoxication (such as induction of mixed-function oxygenase activity), and physiological responses (such as scope for growth or hormonal changes) that influence the energy available for growth and reproduction or other aspects of reproductive and developmental processes. These indexes can be integrated by the functional relationship of metabolic function and energy turnover for growth and reproduction. No single index can provide the predictive capability to evaluate population changes and future studies should be directed at defining the relationship of multiple responses. For example, laboratory studies directed at defining a relationship between mixed-function oxygenase induction in marine animals and long-term consequences in regards to reproduction and disease would greatly strengthen the use of mixed-function oxygenase induction as a monitoring tool.

In designing a monitoring program for determining the long-term impact of waste disposal on marine ecosystems, efforts should be directed at the following:

1. Determine the persistence, the degradation rates, the biogeochemical cycling of contaminants within the sediments, and the flux of these contaminants between sediments, interstitial waters, and biota in the vicinity of the disposal sites.
2. Relate contaminant content in sediments, interstitial waters, and biota to changes in benthic biomass, benthic community structure, and recruitment of benthic populations.
3. Use demersal fish and shellfish populations with limited or no migratory behavior as models, and relate contaminant content (both parent compounds and metabolites) of sediments and interstitial waters with contaminant content of tissues, activity of detoxication enzymes, seasonal alterations in energy reserves, recruitment of juvenile stages, reproductive condition, and the incidence of disease and histopathological conditions.

Such information should provide an analysis of potential long-term change both at the organismal and at the population levels before irreversible damage occurs at the community and the ecosystem levels.

REFERENCES

Aller, R. C. 1977. The Influence of Macrobenthos on Chemical Diagenesis of Marine Sediments. Ph.D. Thesis, Yale University, New Haven, Connecticut, 600 pp.

Anderson, J. W. 1977. Responses to sublethal levels of petroleum hydrocarbons: Are they sensitive indicators and do they correlate with tissue contamination? *In:* Fate and Effects of Petroleum Hydrocarbons in Marine Ecosystems and Organisms, D. A. Wolfe (Ed.). Pergamon Press, New York, pp. 95–114.

Anderson, J. W., S. L. Kiesser, and J. W. Blaylock. 1980. The cumulative effect of petroleum hydrocarbons on marine crustaceans during constant exposure. *Rapports et Procès-Verbaux des Réunions, Conseil International pour L'Exploration de La Mer,* **179,** 62–70.

Atema, J., and L. S. Stein. 1974. Effects of crude oil on the feeding behavior of the lobster *Homarus americanus*. *Environmental Pollution,* **6,** 77–86.

Atema, J., E. B. Karnofsky, and S. Oleszko-Szuts. 1979. Lobster behavior and chemoreception: Sublethal effects of No. 2 fuel oil. *In:* Advances in Marine Environmental Research, F. S. Jacoff (Ed.). U.S. Environmental Protection Agency, Narragansett, Rhode Island, pp. 122–134.

Auguenfeld, J. M. 1980. Effects of Prudhoe Bay crude oil contamination on sediment working rates of *Abarenicola pacifica*. *Marine Environmental Research,* **3,** 307–313.

Baden, S., and L. Hagerman. 1981. Ventilatory responses of the shrimp *Palaemon adspersus* to sublethal concentrations of crude oil extract. *Marine Biology,* **63,** 129–133.

Berdugo, V., R. P. Harris, and S. C. O'Hara. 1977. The effect of petroleum hydrocarbons on reproduction of an estuarine planktonic copepod in laboratory cultures. *Marine Pollution Bulletin,* **8,** 138–143.

Berge, J. A., K. I. Johannessen, and L.-O. Reiersen. 1983. Effects of water soluble fraction of North Sea crude oil on the swimming activity of the sand goby *Pomatoschistus minutus* (Pallas). *Journal of Experimental Marine Biology and Ecology,* **68,** 159–167.

Berman, M. S., and D. R. Heinle. 1980. Modification of the feeding behavior of marine copepods by sublethal concentrations of water-accommodated fuel oil. *Marine Biology,* **56,** 59–64.

Beslier, A., J. L. Birrien, L. Cabioch, C. Larsonneur, and J. LeBorgne. 1980. La pollution des baies de Morlaix et de Lannion par les hydrocarbures de l'*Amoco Cadiz:* Répartition sur les fonds et évolution. *Helgoländer Meeresuntersuchungen,* **33,** 209–224.

Blumer, M. 1969. Oil pollution of the ocean. *In:* Oil on the Sea, D. P. Hoult (Ed.). Plenum Press, New York, pp. 5–13.

Bryan, G. W. 1980. Recent trends in research on heavy metal contamination in the sea. *Helgoländer Meeresuntersuchungen,* **33,** 6–25.

Busdosh, M. 1981. Long-term effects of the water soluble fraction of Prudhoe Bay crude oil on survival, movement, and food search success of the arctic amphipod *Boeckosimus (Onisimus) affinis. Marine Environmental Research,* **5,** 167–180.

Cabioch, L., J. C. Dauvin, J. Mora Bermudez, and C. Rodriguez Babio. 1980. Effets de la marée noire de l'*Amoco Cadiz* sur le benthos sublittoral du nord de la Bretagne. *Helgoländer Meeresuntersuchungen,* **33,** 192–208.

Cairns, J. 1983. Are single species toxicity tests alone adequate for estimating environmental hazard? *Hydrobiologia,* **100,** 47–57.

Capuzzo, J. M. 1981. Predicting pollution effects in the marine environment. *Oceanus,* **24**(1), 25–33.

Capuzzo, J. M., B. A. Lancaster, and G. Sasaki. 1984. The effects of petroleum hydrocarbons on lipid metabolism and energetics of larval development and metamorphosis in the American lobster (*Homarus americanus*). *Marine Environmental Research,* **14,** 201–228.

Ceas, M. P. 1974. Effects of 3-4-benzopyrene on sea urchin egg development. *Acta Embryologiae Experimentalis,* **3,** 267–272.

Clarke, A. 1979. Living in cold water: K-strategies in antarctic benthos. *Marine Biology,* **55,** 111–119.

Conan, G. 1982. The long-term effects of the *Amoco Cadiz* oil spill. *Philosophical Transactions of the Royal Society of London,* **297B,** 323–333.

Cowles, T. J. 1983. Effects of exposure to sublethal concentrations of crude oil on the copepod *Centropages hamatus.* II. Activity patterns. *Marine Biology,* **78,** 53–57.

Cucci, T. L., and C. E. Epifanio. 1979. Long-term effects of water-soluble fractions of Kuwait oil on the larval and juvenile development of the mud crab *Eurypanopeus depressus. Marine Biology,* **55,** 215–220.

DiMichele, L., and M. H. Taylor. 1978. Histopathological and physiological responses of *Fundulus heteroclitus* L. to naphthalene exposure. *Journal of the Fisheries Research Board of Canada,* **35,** 1060–1066.

Dunbar, M. J. 1968. Ecological Development of Polar Regions. Prentice-Hall, Englewood Cliffs, New Jersey, 199 pp.

Eldridge, M. B., T. Echeverra, and J. A. Whipple. 1977. Energetics of Pacific Herring (*Clupea harengus pallasi*) embryos and larvae exposed to low concentrations of benzene, a monoaromatic component of crude oil. *Transactions of the American Fishery Society,* **106,** 452–461.

Elmgren, R., and J. B. Frithsen. 1982. The use of experimental ecosystems for evaluating the environmental impact of pollutants: A comparison of an oil spill in the Baltic Sea and two long-term, low-level oil addition experiments in mesocosms. *In:* Marine Mesocosms: Biological and Chemical Research in Experimental Ecosystems, G. D. Grice and M. R. Reeve (Eds.). Spinger-Verlag, New York, pp. 153–165.

Elmgren, R., S. Hansson, U. Larsson, B. Sundelin, and P. D. Boehm. 1983. The *Tsesis* oil spill: Acute and long-term impact on the benthos. *Marine Biology,* **73,** 51–65.

Englehardt, F. R., M. P. Wong, and M. E. Duey. 1981. Hydromineral balance and gill morphology in rainbow trout *Salmo gairdneri,* acclimated to fresh and sea water as affected by petroleum exposure. *Aquatic Toxicology,* **1,** 175–186.

Farrington, J. W., E. D. Goldberg, R. W. Risebrough, J. H. Martin, and V. T. Bowen. 1983. U.S. "Mussel Watch" 1976–1978: An overview of the trace-metal, DDE, PCB, hydrocarbon, and artificial radionuclide data. *Environmental Science and Technology,* **17,** 490–496.

Fletcher, G. L., M. J. King, J. W. Kiceniuk, and R. F. Addison. 1982. Liver hypertrophy in winter flounder following exposure to experimentally oiled sediments. *Comparative Biochemistry and Physiology,* **73C,** 457–462.

Forns, J. M. 1977. The effects of crude oil on larvae of lobster *Homarus americanus. In:* Proceedings of 1977 Oil Spill Conference (Prevention, Behavior, Control, Cleanup). American Petroleum Institute, Washington, D.C., pp. 569–573.

Fries, C. R., and M. R. Tripp. 1977. Cytological damage in *Mercenaria mercenaria* exposed to phenol. *In:* Fate and Effects of Petroleum Hydrocarbons in Marine Ecosystems and Organisms, D. A. Wolfe (Ed.). Pergamon Press, New York, pp. 174–181.

Gardner, G. R. 1975. Chemically induced lesions in estuarine or marine teleosts. *In:* The Pathology of Fishes, W. E. Ribelin and G. Migaki (Eds.). University of Wisconsin Press, Madison, Wisconsin, pp. 657–693.

Gearing, J. N., and P. J. Gearing. 1983. Suspended load and solubility affect sedimentation of petroleum hydrocarbons in controlled estuarine ecosystems. *Canadian Journal of Fisheries and Aquatic Sciences,* **40,** (Supplement No. 2), 54–62.

Gearing, P. J., J. N. Gearing, R. J. Pruell, T. L. Wade, and J. G. Quinn. 1980. Partitioning of No. 2 fuel oil in controlled estuarine ecosystems: Sediments and suspended particulate matter. *Environmental Science and Technology,* **14,** 1129–1136.

Giere, O., and D. Hauschildt. 1979. Experimental studies on the life cycle and production of the littoral oligochaete *Lumbricillus lineatus* and its response to oil pollution. *In:* Cyclic Phenomena in Marine Plants and Animals, E. Naylor and R. G. Hartnoll (Eds.). Pergamon Press, Oxford, England, pp. 113–122.

Gilfillan, E. S. 1980. The use of scope-for-growth measurements in monitoring petroleum pollution. *Rapports et Procès-Verbaux des Réunions, Conseil International pour L'Exploration de La Mer,* **179,** 71–75.

Gilfillan, E. S., D. Mayo, S. Hanson, D. Donovan, and L. C. Jiang. 1976. Reduction in carbon flux in *Mya arenaria* caused by a spill of No. 6 fuel oil. *Marine Biology,* **37,** 115–123.

Gilfillan, E. S., D. W. Mayo, D. S. Page, D. Donovan, and S. Hanson. 1977. Effects of varying concentrations of petroleum hydrocarbons in sediments on carbon flux in *Mya arenaria. In:* Physiological Responses of Marine Biota to Pollutants, F. J. Vernberg, A. Calabrese, F. P. Thurberg, and W. B. Vernberg (Eds.). Academic Press, New York, pp. 299–314.

Gillfillan, E. S., and J. H. Vandermeulen. 1978. Alterations in growth and physiology of soft shell clams, *Mya arenaria,* chronically oiled with Bunker C from Chedabucto Bay, Nova Scotia, 1970–76. *Journal of the Fishery Research Board of Canada,* **35,** 630–636.

Glémarec, M., and E. Hussenot. 1981. Definition d'une succession écologique en milieu meuble anormalement enrichie en matières organiques à la suite de la catastrophe de l'*Amoco Cadiz. In:* Proceedings of the International Symposium on the *Amoco Cadiz:* Fate and Effects of the Oil Spill. Centre Nationale pour l'Exploitation des Océans, Paris, pp. 499–512.

Gordon, D. C., J. Dade, and P. D. Keizer. 1978. Importance of sediment working by the deposit feeding polychaete, *Arenicola marina,* on the weathering rate of sediment-bound oil. *Journal of the Fisheries Research Board of Canada,* **35,** 591–603.

Grassle, J. F., R. Elmgren, and J. P. Grassle. 1981. Response of benthic communities in MERL experimental ecosystems to low level, chronic additions of No. 2 fuel oil. *Marine Environmental Research,* **4,** 279–297.

Haensly, W. E., J. M. Neff, J. R. Sharp, A. C. Morris, M. F. Bedgood, and P. D. Boehm. 1982. Histopathology of *Pleuronectes platessa* L. from Aber Wrac'h and Aber Benoît, Brittany, France: Long-term effects of the *Amoco Cadiz* crude oil spill. *Journal of Fish Diseases*, **5**, 365–391.

Hauschildt-Lillge, D. 1982. Long-term effects of petroleum hydrocarbons on the life cycle and productivity of the littoral oligochaete *Lumbricillus lineatus*. *Netherlands Journal of Sea Research*, **16**, 502–510.

Hawkes, J. M., and C. M. Stehr. 1982. Cytopathology of the brain and retina of embryonic surf smelt (*Hypomesus pretiosus*) exposed to crude oil. *Environmental Research*, **27**, 164–178.

Hellstrom, T., and K. B. Doving. 1983. Perception of diesel oil by cod (*Gadus morhua*). *Aquatic Toxicology*, **4**, 303–315.

Hooftman, R. N., and G. J. Vink. 1981. Cytogenetic effects on the eastern mud-minnow, *Umbra pygmaea*, exposed to ethyl methane-sulphonate, benzo[*a*]pyrene and river water. *Ecotoxicology and Environmental Safety*, **5**, 261–269.

Jacobson, S. M., and D. B. Boylan. 1973. Effect of seawater soluble fraction of kerosene on chemotaxis in a marine snail, *Nassarius obsoletus*. *Nature*, **241**, 213–215.

Kalke, R. D., T. A. Duke, and R. W. Flint. 1982. Weathered IXTOC I oil effects on estuarine benthos. *Estuarine and Coastal Shelf Science*, **15**, 75–84.

Kittredge, J. S., F. T. Takahashi, and F. O. Sarinana. 1974. Bioassays indicative of some sublethal effects of oil pollution. *In:* Proceedings of Marine Bioassays Workshop. Marine Technology Society, Washington, D.C., pp. 891–897.

Krebs, C. T., and K. A. Burns. 1977. Long term effects of an oil spill on populations of the salt-marsh crab *Uca pugnax*. *Science*, **197**, 484–487.

Kurelec, B., Z. Matijasevic, M. Rijavic, M. Alacevic, S. Britvic, W. E. G. Muller, and R. K. John. 1979. Induction of benzo[*a*]pyrene monoxygenase in fish and the *Salmonella* test as a tool for detecting mutagenic/carcinogenic xenobiotics in the aquatic environment. *Bulletin of Environmental Contamination and Toxicology*, **21**, 799–807.

Laughlin, R. B., Jr., L. G. L. Young, and J. M. Neff. 1978. A long-term study of the effect of water-soluble fractions of No. 2 fuel oil on the survival, development rate, and growth of the mud crab *Rhithropanopeus harrisii*. *Marine Biology*, **47**, 87–95.

Lee, R. F. 1981. Mixed function oxidases (MFO) in marine invertebrates. *Marine Biology Letters*, **2**, 87–105.

Linden, O. 1976. The influence of crude oil and mixtures of crude oil/dispersants on the ontogenic development of the Baltic herring, *Clupea harengus membras* L. *Ambio*, **5**, 136–140.

Linden, O., R. Laughlin, Jr., J. R. Sharp, and J. M. Neff. 1980. The combined effect of salinity, temperature and oil on the growth pattern of embryos of the killifish *Fundulus heteroclitus* Walbaum. *Marine Environmental Research*, **3**, 129–144.

Longwell, A. C. 1977. A genetic look at fish eggs and oil. *Oceanus*, **20**(4), 45–58.

Lowe, D. M., M. N. Moore, and K. R. Clarke. 1981. Effects of oil on digestive cells in mussels: Quantitative alterations in cellular and lysosomal structure. *Aquatic Toxicology*, **1**, 175–186.

Malins, D. C. 1982. Alterations in the cellular and subcellular structure of marine teleosts and invertebrates exposed to petroleum in the laboratory and field: a critical review. *Canadian Journal of Fisheries and Aquatic Sciences*, **39**, 877–889.

Malins, D. C., and H. O. Hodgins. 1981. Petroleum and marine fishes: A review of uptake, disposition, and effects. *Environmental Science and Technology*, **15**, 1272–1280.

Malins, D. C., M. S. Myers, and W. T. Roubal. 1983. Organic free radicals associated with idiopathic liver lesions of English sole (*Parophrys vetulus*) from polluted marine environments. *Environmental Science and Technology*, **17**, 679–685.

Maynard, D. J., and D. D. Weber. 1981. Avoidance reactions of juvenile coho salmon (*Oncorhynchus kisutch*) to monocyclic aromatics. *Canadian Journal of Fisheries and Aquatic Sciences*, **38**, 772–778.

McCain, B. B., and D. C. Malins. 1982. Effects of petroleum hydrocarbons on selected demersal fish and crustaceans. *In:* Ecological Stress and New York Bight: Science and Management, G. F. Mayer (Ed.). Estuarine Research Federation, Columbia, South Carolina, pp. 315–326.

McIntyre, A. D., and J. B. Pearce (Eds.). 1980. Biological Effects of Marine Pollution and the Problems of Monitoring. *Rapports et Procès-Verbaux des Réunions, Conseil International pour L'Exploration de La mer*, **179**, 1–346.

Means, J. C., J. J. Hassett, S. G. Wood, and W. L. Banwart. 1979. Sorption properties of energy-related pollutants and sediments. *In:* Polynuclear Aromatic Hydrocarbons, P. W. Jones and P. Leber (Eds.). Ann Arbor Science, Ann Arbor, Michigan, pp. 327–340.

Minchew, C. D., and J. D. Yarbrough. 1977. The occurrence of fin rot in mullet (*Mugil cephalus*) associated with crude oil contamination of an estuarine-pond ecosystem. *Journal of Fish Biology*, **10**, 319–323.

Moore, M. N., and R. Clarke. 1982. Use of microstereology and quantitative cytochemistry to determine the effects of crude oil-derived aromatic hydrocarbons on lysosomal structure and function in a marine bivalve mollusc, *Mytilus edulis*. *Histochemistry Journal*, **14**, 713–718.

Morel, F. M. M., and S. L. Schiff. 1983. Geochemistry of municipal waste in coastal waters. *In:* Ocean Discharge of Municipal Waste: Impact on Coastal Waters, E. P. Myers (Ed.). Massachusetts Institute of Technology Sea Grant Program, Cambridge, Massachusetts, pp. 249–421.

Neff, J. M. 1979. Polycyclic Aromatic Hydrocarbons in the Aquatic Environment. Applied Sciences Publishers, Ltd., London, 266 pp.

Neff, J. M. 1985. The use of biochemical measurements to detect pollutant-mediated damage to fish. *In:* Aquatic Toxicology and Hazard Assessment, Seventh Symposium, R. D. Cardwell, R. Purdy, and R. C. Bahner (Eds.). Special Technical Publication No. 854, American Society for Testing and Materials, Philadelphia, pp. 155–183.

Olla, B. L., A. J. Bejda, and W. H. Pearson. 1983. Effects of oiled sediment on the burrowing behavior of the hard clam *Mercenaria mercenaria*. *Marine Environmental Research*, **9**, 183–193.

Olsen, C. R., N. H. Cutshall, and J. L. Larsen. 1982. Pollutant–particle associations and dynamics in coastal marine environments: a review. *Marine Chemistry*, **11**, 501–533.

Oviatt, C., J. Frithsen, J. Gearing, and P. Gearing. 1982. Low chronic additions of No. 2 fuel oil: Chemical behavior, biological impact and recovery in a simulated estuarine environment. *Marine Ecology Progress Series*, **9**, 121–136.

Paris, D. F., W. C. Steen, and G. L. Baughman. 1978. Role of physicochemical properties of Aroclors 1016 and 1242 in determining their fate and transport in aquatic environments. *Chemosphere*, **4**, 319–325.

Payne, J. F. 1976. Field evaluation of benzopyrene hydroxylase induction as a monitor for marine petroleum pollution. *Science*, **191**, 945–946.

Pearson, W. H., S. E. Miller, J. W. Blaylock, and B. L. Olla. 1981. Detection of the water soluble fraction of crude oil by the blue crab *Callinectes sapidus*. *Marine Environmental Research*, **5**, 3–11.

Prouse, N. J., and D. C. Gordon, Jr. 1976. Interactions between the deposit feeding polychaete *Arenicola marina* and oiled sediment. *In:* Sources, Effects and Sinks of Hydrocarbons in the Aquatic Environment. American Institute of Biological Sciences, Washington, D.C., pp. 407–422.

Rabalais, S. C., C. R. Arnold, and N. S. Wohlschlag. 1981. The effects of IXTOC I oil on the eggs and larvae of red drum (*Sciaenops ocellata*). *Texas Journal of Science*, **33**, 33–38.

Rice, S. D. 1973. Toxicity and avoidance tests with Prudhoe Bay oil and pink salmon fry. *In*: Proceedings of 1973 Oil Spill Conference. American Petroleum Institute, Washington, D.C., pp. 667–670.

Sanders, H. L., J. F. Grassle, G. R. Hampson, L. S. Morse, S. Garner-Price, and C. C. Jones. 1980. Anatomy of an oil spill: long-term effects from the grounding of the barge *Florida* off West Falmouth, Massachusetts. *Journal of Marine Research,* **38,** 265–380.

Smith, R. L., and J. A. Cameron. 1979. Effect of water soluble fraction of Prudhoe Bay crude oil on embryonic development of Pacific herring. *Transactions of American Fishery Society,* **108,** 70–75.

Solangi, M. A., and R. M. Overstreet. 1982. Histopathological changes in two estuarine fishes, *Menidia beryllina* (Cope) and *Trinectes maculatus* (Bloch and Schneider), exposed to crude oil and its water soluble fractions. *Journal of Fish Diseases,* **5,** 13–35.

Spaulding, M. L., S. B. Saila, E. Lorda, H. Walker, E. Anderson, and J. C. Swanson. 1983. Oil-spill fishery impact assessment model: application to selected Georges Bank fish species. *Estuarine and Coastal Shelf Science,* **16,** 511–541.

Spies, R. B., J. S. Felton, and L. Dillard. 1982. Hepatic mixed-function oxidases in California flatfishes are increased in contaminated environments and by oil and PCB ingestion. *Marine Biology,* **70,** 117–127.

Stegeman, J. J. 1980. Mixed function oxygenase studies in monitoring for effects of organic pollution. *Rapports et Procès-Verbaux des Réunions, Conseil International pour L'Exploration de La Mer,* **179,** 33–38.

Stegeman, J. J. 1981. Polynuclear aromatic hydrocarbons and their metabolism in the marine environment. *In*: Polycyclic Hydrocarbons and Cancer, Vol. 3, H. V. Gelboin and P. O. P. Ts'O (Eds.). Academic Press, New York, pp. 1–59.

Stott, G. G., W. E. Haensly, J. M. Neff, and J. R. Sharp. 1983. Histopathologic survey of ovaries of plaice *Pleuronectes platessa* L. from Aber Wrac'h and Aber Benoît, Brittany, France: Long-term effects of the *Amoco Cadiz* crude oil spill. *Journal of Fish Diseases,* **6,** 429–437.

Stumm, W., and J. J. Morgan. 1981. Aquatic Chemistry, 2nd edition, Wiley-Interscience, New York, 780 pp.

Thomas, M. L. H. 1978. Comparison of oiled and unoiled intertidal communities in Chedabucto Bay, Nova Scotia. *Journal of the Fishery Research Board of Canada,* **35,** 707–716.

U.S. National Research Council. 1985. Petroleum in the Marine Environment. National Academy Press, Washington, D.C., 601 pp.

U.S. National Research Council. 1984. Ocean Disposal Systems for Sewage Sludge and Effluent. National Academy Press, Washington, D.C., 126 pp.

Vandermeulen, J. H., and J. M. Capuzzo. 1983. Understanding sublethal pollutant effects in the marine environment. Paper #9, *In:* Ocean Waste Management: Policy and Strategies, M. A. Champ and M. Trainor (Eds.). Background Papers of Symposium, May 2–6 1983, University of Rhode Island, Kingston, Rhode Island, 11 pp.

Varanasi, U. and D. J. Gmur. 1980. Metabolic activation and covalent binding of benzo[*a*]pyrene to deoxyribonucleic acid catalyzed by liver enzymes of marine fish. *Biochemical Pharmacology,* **29,** 753–761.

Varanasi, U., and D. J. Gmur. 1981a. Hydrocarbons and metabolites in English sole (*Parophrys vetulus*) exposed simultaneously to [3H]benzo[*a*]pyrene and [14C]naphthalene in oil-contaminated sediment. *Aquatic Toxicology,* **1,** 49–68.

Varanasi, U., and D. J. Gmur. 1981b. *In vivo* metabolism of naphthalene and benzo[*a*]pyrene by flatfish. *In:* Chemical Analysis and Biological Fate: Polynuclear Aromatic Hydrocarbons, M. Cooke and A. J. Dennis (Eds.). Battelle Press, Columbus, Ohio, pp. 367–376.

Varanasi, U., D. J. Gmur, and M. M. Krahn. 1980. Metabolism and subsequent binding of benzo[*a*]pyrene to DNA in pleuronectid and salmonid fish. *In:* Polynuclear Aromatic Hydrocarbons: Biological Effects, A. Bjorseth and J. Dennis (Eds.). Battelle Press, Columbus, Ohio, pp. 455–470.

Varanasi, U., J. E. Stein, and T. Hom. 1981. Covalent binding of benzo[*a*]pyrene to DNA in fish liver. *Biochemistry and Biophysics Research Communications,* **103,** 780–787.

Veith, D. G., D. L. Defoe, and B. V. Bergstedt. 1979. Measuring and estimating the bioconcentration factor of chemicals in fish. *Journal of the Fishery Research Board of Canada,* **36,** 1040.

Weber, D. D., D. J. Maynard, W. D. Gronlund, and V. Konchin. 1981. Avoidance reactions of migrating adult salmon to petroleum hydrocarbons. *Canadian Journal of Fisheries and Aquatic Sciences,* **38,** 779–781.

Wells, P. G., and J. B. Sprague. 1976. Effects of crude oil on American lobster (*Homarus americanus*) larvae in the laboratory. *Journal of the Fishery Research Board of Canada,* **33,** 1604–1614.

Widdows, J. 1982. Field measurement of the biological impacts of pollutants. *In:* Assimilative Capacity of the Oceans for Man's Wastes. SCOPE/ICSU, Academia Sinica, Taipei, Republic of China, pp. 111–129.

Widdows, J., T. Bakke, B. L. Bayne, P. Donkin, D. R. Livingstone, D. M. Lowe, M. N. Moore, S. V. Evans, and S. L. Moore. 1982. Responses of *Mytilus edulis* on exposure to the water accommodated fraction of North Sea oil. *Marine Biology,* **67,** 15–31.

PART II: TRACE METALS IN THE MARINE ENVIRONMENT

Chapter 2

Neritic–Oceanic Trends in Trace-Metal Toxicity to Phytoplankton Communities

William G. Sunda

National Marine Fisheries Service
Beaufort, North Carolina

ABSTRACT

The effects of trace-metal additions on carbon fixation by phytoplankton communities was investigated in four neritic and oceanic seawater samples from the Gulf of Mexico. Both short-term (6–12 h) and long-term (2–4 d) effects were examined. Clean techniques were used to minimize inadvertent sample contamination. Copper additions ≥4 nmol liter^{-1} inhibited carbon fixation in oceanic seawater after both short-term and long-term exposures. The toxicity of low (4–12 nmol liter^{-1}) Zn concentrations was apparent in the oceanic sample only after long-term (2–4 d) exposure, and inhibition of carbon fixation increased with time in all samples tested. The toxicity of added Cd, tested in only one neritic sample, was similar to that of Zn. Both Cu and Zn were considerably more toxic in low-productivity oceanic samples than in high-productivity neritic samples. This higher toxicity of Cu and Zn in low-productivity seawater was attributed to a number of factors: (1) lower organic complexation; (2) lower ambient concentrations of Cu and Zn; and (3) competitive interferences with the uptake and metabolism of nutrient metals such as Mn and Fe, present at low concentrations in oceanic waters.

2.1. INTRODUCTION

The industrial and technological revolutions have resulted in the contamination of marine waters by various heavy metals such as Cu, Zn, Cd, Pb, Cr, and Hg. In the case of Pb, the contamination has been global, resulting in order-of-magnitude increases in its oceanic surface concentrations (Schaule and Patterson, 1981). Contamination by most other metals generally has been localized in coastal environments receiving point or multiple-point discharges from municipal and industrial sources, including ocean dumping.

There is controversy over the relative economic and environmental costs of nearshore and offshore dumping of toxic wastes (Swanson and Devine, 1982). Decisions on where to discharge such wastes, however, must be based on sound scientific information concerning the effects of increased pollutant concentrations on biological communities. There is a need to determine not only the levels at which adverse effects occur but also to determine the factors that control these effects. Because of large regional differences in seawater chemistry (Gordon et al., 1982; Bruland and Franks, 1983; Sunda and Ferguson, 1983) and in the composition of biological communities, there is reason to hypothesize that there will be regional differences in adverse biological effects of increased toxic-metal concentrations.

2.2. MATERIALS AND METHODS

Seawater was collected at four stations in the Gulf of Mexico during February 1981 (Fig. 2.1). The seawater sampled

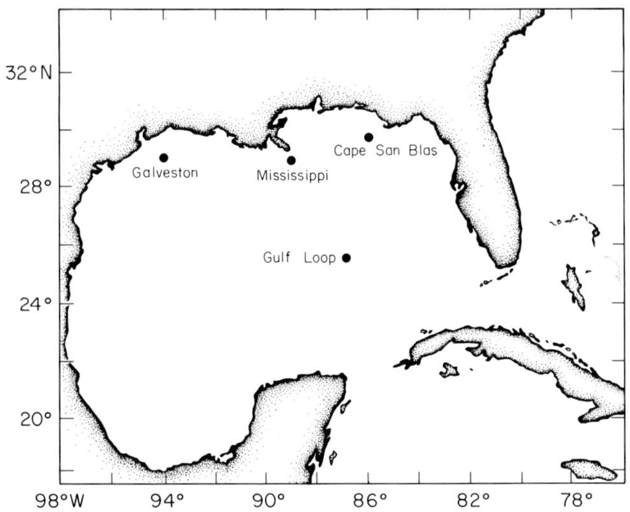

Figure 2.1. Stations sampled in the Gulf of Mexico during February 1981.

ranged from high-productivity, low-salinity water from the Mississippi discharge plume to low-productivity, high-salinity water from within the loop of the Gulf Loop Current (Table 2.1). The station that had the next to highest productivity, off Galveston, Texas, was shallow (15 m) and well-mixed to the bottom. The fourth station, 120 km south of Cape San Blas, Florida, had a high salinity (36.3‰) and next to lowest productivity.

and Company, Wilmington, Delaware) bottles and polyethylene carboys were washed with detergent, rinsed with distilled water, and then filled with 1.2*N* HCl. The acid was left in the containers for at least one week. They were then rinsed three times with distilled water and given a final rinse with distilled water from a quartz sub-boiling still. The containers were air-dried in a laminar-flow hood with their caps slightly ajar and taken to sea in large polyethylene bags. All bottles and carboys received a rinse with sample water before final filling.

Seawater was collected in 30-liter Go-Flo® bottles (General Oceanics, Miami, Florida) suspended on a Kevlar® (E.I. DuPont de Nemours and Company, Wilmington, Delaware) plastic hydrowire, that was attached to a portable winch with a stainless steel pulley. Seawater was transferred via Teflon® tubing to 20-liter polyethylene carboys in a laminar flow hood located in a trace-metal clean van. Plastic spigots on the carboys were used to dispense 250-ml subsamples into 250-ml Teflon® bottles. Equivalent 1-liter subsamples in 1-liter bottles were used in the Cape San Blas experiment to test for possible effects of bottle size. At the lowest-productivity station, Gulf Loop, the water was amended with 4 μmol liter^{-1} $NaNO_3$, 0.4 μmol liter^{-1} NH_4Cl, 0.4 μmol liter^{-1} Na_2HPO_4, and 4 μmol liter^{-1} Na_2SiO_3 (before dispensing) in order to have sufficient nutrients for phytoplankton growth. At the other low-productivity station, Cape San Blas, seawater from a depth of 10 m was mixed with seawater

Table 2.1. Environmental Conditions at Stations in the Gulf of Mexico

Station[a]	Sample Depth (m)	Salinity (‰)	Ambient Temperature (°C)	Experimental Temperature (°C)	Productivity (μg C liter^{-1} h^{-1}) 4 h[c]	6–12 h[b]
Gulf Loop	25	36.0	25.3	22 ± 2	0.21	0.27
Cape San Blas	10	36.3	20.9	20 ± 1	—[d]	0.41
	70	36.3	18.7			
Galveston	4	35.4	13.4	23 ± 2	3.2	3.6
Mississippi River Plume	4	33.6	14.6	17 ± 1	—[d]	7.9

[a] For station locations see Fig. 2.1.
[b] Computed for initial light period in present experiments.
[c] Measured in samples collected from the depth of maximum productivity (P. Ortner, personal communication). Samples were incubated at saturating light [400 μE (einsteins) m^{-2} s^{-1}] for 4 h at ambient seawater temperature.
[d] No data available.

To avoid inadvertent sample contamination, trace-metal clean procedures were used throughout to collect and process seawater samples (Boyle and Edmond, 1975; Bruland et al., 1978; Fitzwater et al., 1982; Sunda and Ferguson, 1983). Stringent procedures were also adopted to ensure cleanliness of sample containers. All Teflon® (E.I. Du Pont de Nemours

from 70 m in a ratio of 1:4. This procedure, which simulated what might occur during upwelling, was also used to ensure adequate phytoplankton growth during the experiment. Nutrient addition and mixing procedures were not used at the two high-productivity stations.

Trace-metal stock solutions were prepared from analytical grade salts and quartz redistilled water adjusted to pH 4 with HCl. Stock solutions were prepared 1 week before the cruise and stored in acid-cleaned, 60-ml Teflon® bottles. Usually, each stock solution was made so that a 50-μl addition (automatic pipet) into a 250-ml sample would yield the desired added concentration. In the case of Fe additions, however, a concentrated stock solution of 0.1 mol liter^{-1} $FeCl_3$ and 0.1 mol liter^{-1} HCl was diluted 1:1000 just before addition to the experimental seawater in order to avoid formation of colloidal iron hydroxide in the dilute stock solution. The ^{14}C stock solutions (1.85×10^{13} mBq liter^{-1}) were made up according to clean procedures outlined by Fitzwater et al. (1982).

Seawater subsamples were amended with the various trace-metal additions and allowed to incubate in light for 0.5 h. There were two replicates per treatment. A ^{14}C bicarbonate tracer was then added (Fitzwater et al., 1982) and the bottles were placed in an incubator for the 3- to 4-day duration of the experiment. The bottles were illuminated (270 μE m^{-2} s^{-1}) on a 12-h light, 12-h dark cycle with a bank of fluorescent lights. The duration of the initial light period was 12 h at Gulf Loop, 8 h at Galveston and Mississippi, and 6 h at Cape San Blas. This period was dependent on the time of sample collection (0400–1000 h).

The experimental bottles were sampled daily for carbon fixation 1 h after the end of each light period. Sample bottles were shaken vigorously and subsampled into a graduated cylinder. Light intensity was kept to a bare minimum during the 15–30 s required for subsampling each bottle. The subsamples were filtered through glass fiber filters (Gelman A/E, Ann Arbor, Michigan); the filters were then fumed with concentrated HCl and counted for ^{14}C by liquid scintillation counting. The ^{14}C counts were corrected for quench by the external standards method and converted to fixed carbon values in units of mg C liter^{-1}.

Seawater subsamples were dispensed and trace metals were added within the anteroom of a class-1000 clean van. The addition of ^{14}C and subsampling for fixed carbon was carried out under a laminar flow hood in the ship's oceanographic laboratory.

2.3. RESULTS

The addition of Cu, Zn, and Cd reduced carbon fixation at all stations tested (Figs. 2.2–2.7). The effects of Cu and Zn were examined at all four stations, whereas the effect of Cd was investigated only at the Galveston station. In general, Cu was more toxic than was Zn or Cd, and the latter two metals had similar toxicity (Fig. 2.3). An apparent stimulation in carbon fixation during the initial 8-h exposure to 12 nmol liter^{-1} Cu at the Galveston station or 12-h exposure to 1.2 nmol liter^{-1} Cu at the Gulf Loop station (Fig. 2.8) may have been due to hormesis (overcompensation for toxic stress) rather than an alleviation of Cu deficiency. In each case, the stimulation did not persist past the first sampling period (Figs. 2.2a and 2.3a).

There were distinct trends over time with respect to the inhibitory effects of Cu and Zn. Copper toxicity was generally apparent in the initial set of subsamples (after 6–12 h of incubation) and persisted with time (Figs. 2.2a, 2.3a, and 2.4a). However, Zn had little or no effect during the initial light period, but became progressively inhibitory with time (Figs. 2.2b, 2.3b, and 2.4b). A similar delayed toxicity also occurred with Cd at the Galveston station (Fig. 2.3c). For both Cu and Zn additions, toxicity was inversely related to the primary productivity at the station (see Table 2.1 for productivity values); that is, both metals were most toxic in the lowest-productivity oceanic Gulf Loop samples and least toxic in the two coastal highest-productivity samples (Figs. 2.8–2.10).

Long-term toxicity differences between stations were larger for Zn than for Cu. After 3 d, 50% inhibition of carbon fixation occurred at ~7 nmol liter^{-1} of added Cu in Gulf Loop seawater and at a 14-fold higher concentration (~100 nmol liter^{-1}) in the Mississippi River plume (Fig. 2.9). In contrast, the additions of Zn required for 50% inhibition on day 3 at these same two stations differed by at least a factor of 170 (Fig. 2.10).

Unlike Cu, Zn, and Cd, additions of Mn alone had little or no effect on carbon fixation (Figs. 2.2d, 2.3d, and 2.7a). A similar lack of effect was observed with additions of Fe to the Galveston and Cape San Blas samples (Figs. 2.3d, 2.5b, and 2.7c) and with addition of 10^{-6} mol liter^{-1} EDTA (ethylenediaminetetraacetic acid), a strong chelating agent, to Galveston (Fig. 2.3d) and Mississippi (data not shown) samples. Both of these additions affected carbon fixation in the Gulf Loop sample (Fig. 2.2c), but the effects occurred only after the second day of the experiment and were opposite in direction (that is, Fe was stimulatory, whereas EDTA was inhibitory). The stimulation by Fe was apparently due to an alleviation of iron deficiency. The delayed inhibition by EDTA probably resulted from an induced trace-metal deficiency.

Previous results with both monoclonal cultures and natural communities have shown that Cu toxicity can be alleviated

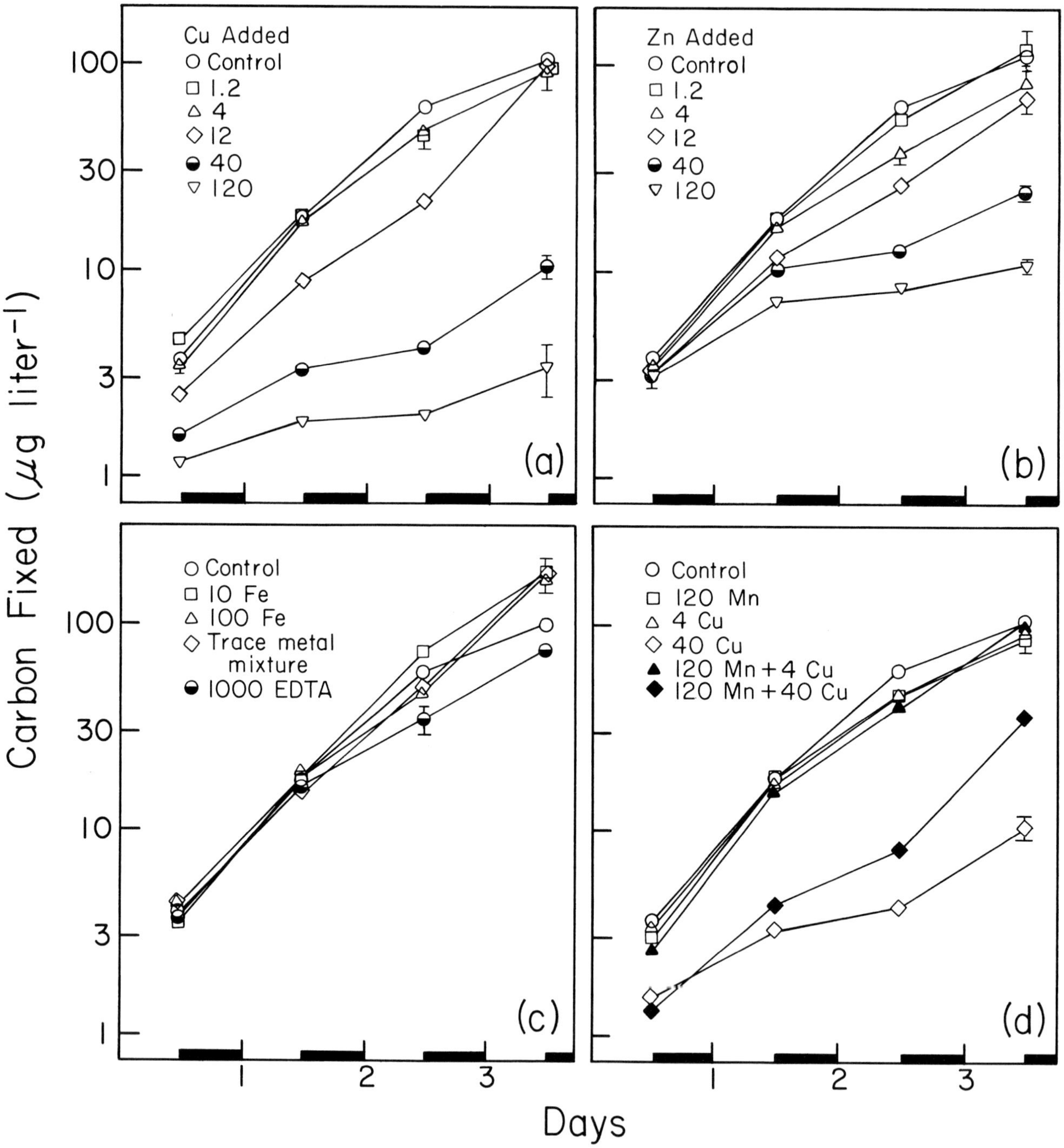

Figure 2.2. Effects of trace-metal and EDTA additions (nmol liter^{-1}) on carbon fixation in a 25-m Gulf Loop seawater sample. The sample was also amended with 4μmol liter^{-1} $NaNO_3$, 4μmol liter^{-1} Na_2SiO_3, 0.4 μmol liter^{-1} NH_4Cl, and 0.4 μmol liter^{-1} Na_2HPO_4. The mean ± range in fixed carbon is plotted for two replicate treatments. Time in days is measured from the addition of ^{14}C bicarbonate. Light and dark regions on the x axis represent light and dark periods of the 12:12 h light–dark cycle. The trace-metal mix (c) has the following composition: 1.2 nmol liter^{-1} $CuSO_4$, 4 nmol liter^{-1} $ZnCl_2$, 20 nmol liter^{-1} $MnCl_2$, 1.2 nmol liter^{-1} $CoCl_2$, and 100 nmol liter^{-1} $FeCl_3$.

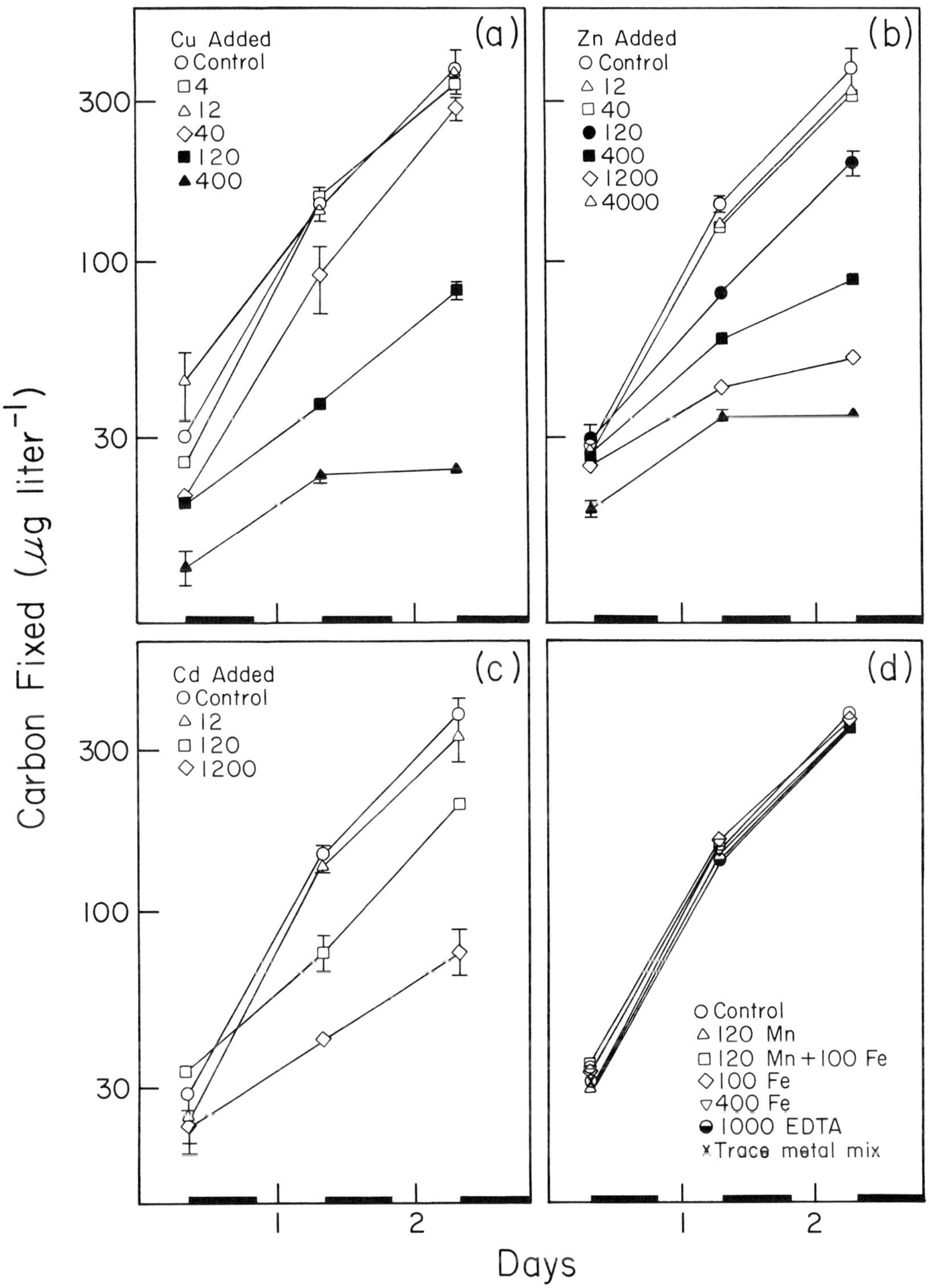

Figure 2.3. Effect of trace-metal and EDTA additions on carbon fixation (nmol liter^{-1}) in the Galveston seawater sample. Composition of trace-metal mix (d) is given in the legend of Fig. 2.2.

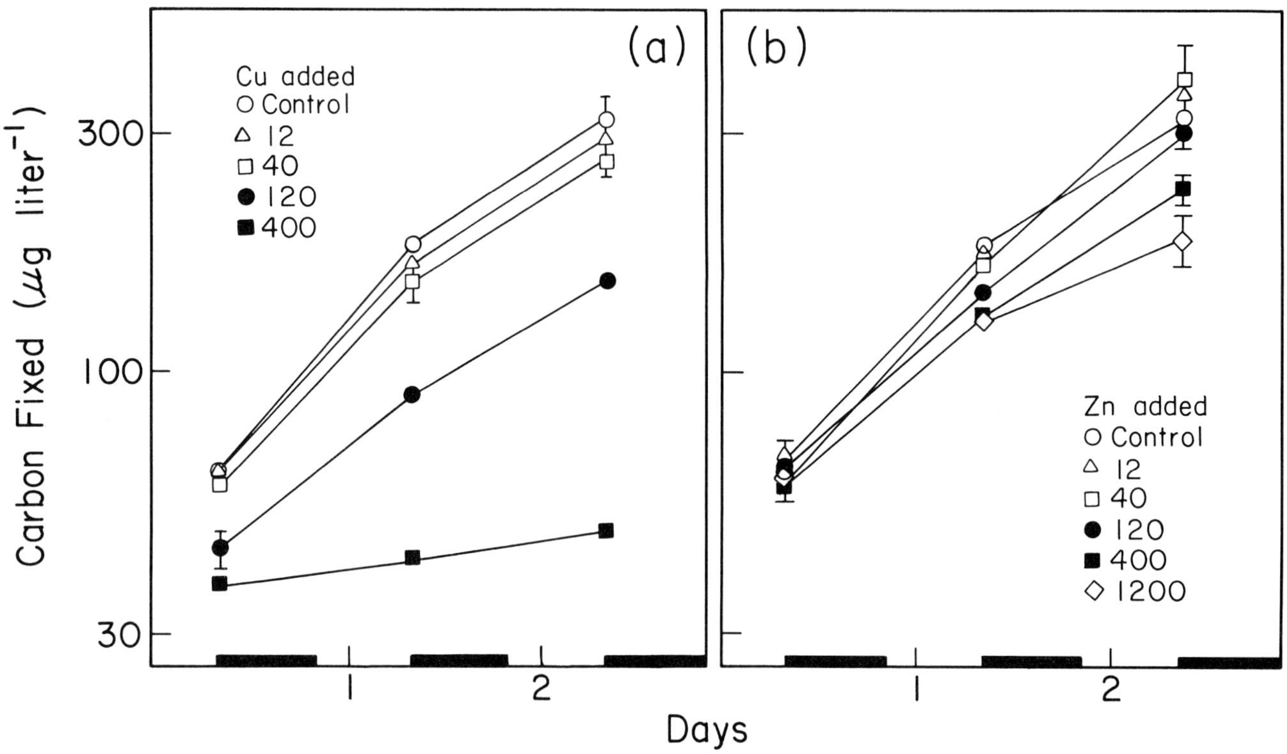

Figure 2.4. Effect of Cu and Zn additions (nmol liter^{-1}) on carbon fixation in water from the Mississippi River plume (salinity 33.6‰).

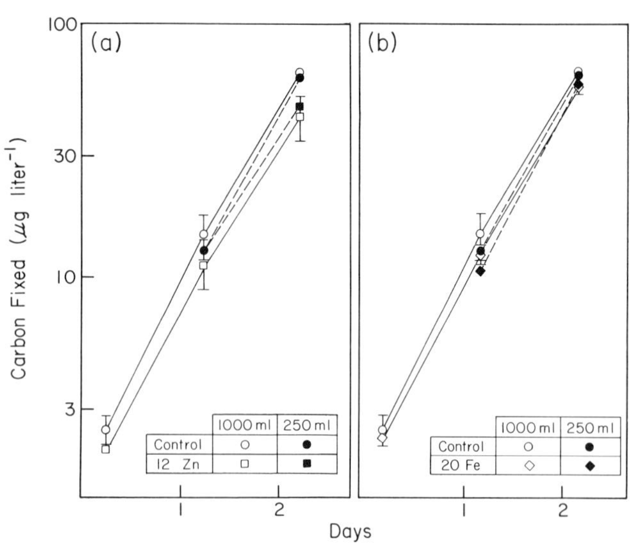

Figure 2.5. Effect of Zn and Fe additions (nmol liter^{-1}) on carbon fixation in water from the Cape San Blas experiment. Results are compared for two different bottle sizes, 250 and 100 ml. Because of insufficient carbon fixation, 6-h subsamples were not collected for the 250-ml bottles. The seawater is a 4:1 mixture of samples collected from 70 and 10 m, respectively.

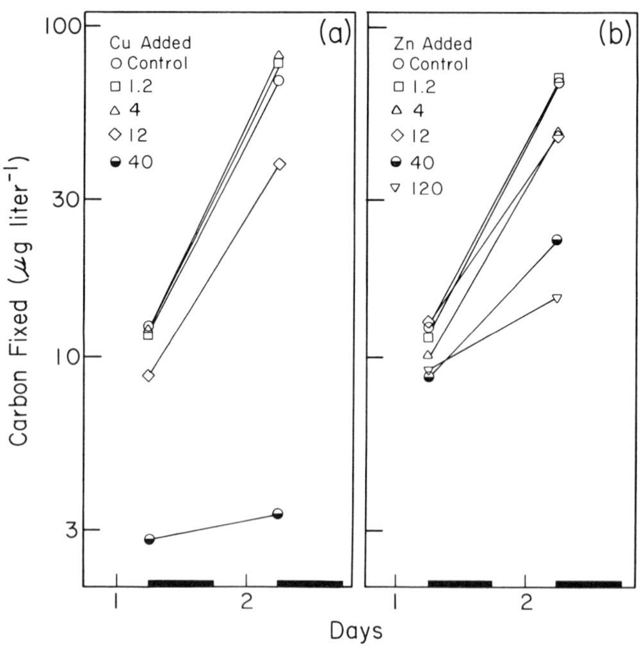

Figure 2.6. Effect of Cu and Zn additions (nmol liter^{-1}) on carbon fixation in water from the Cape San Blas experiment (250 ml bottles).

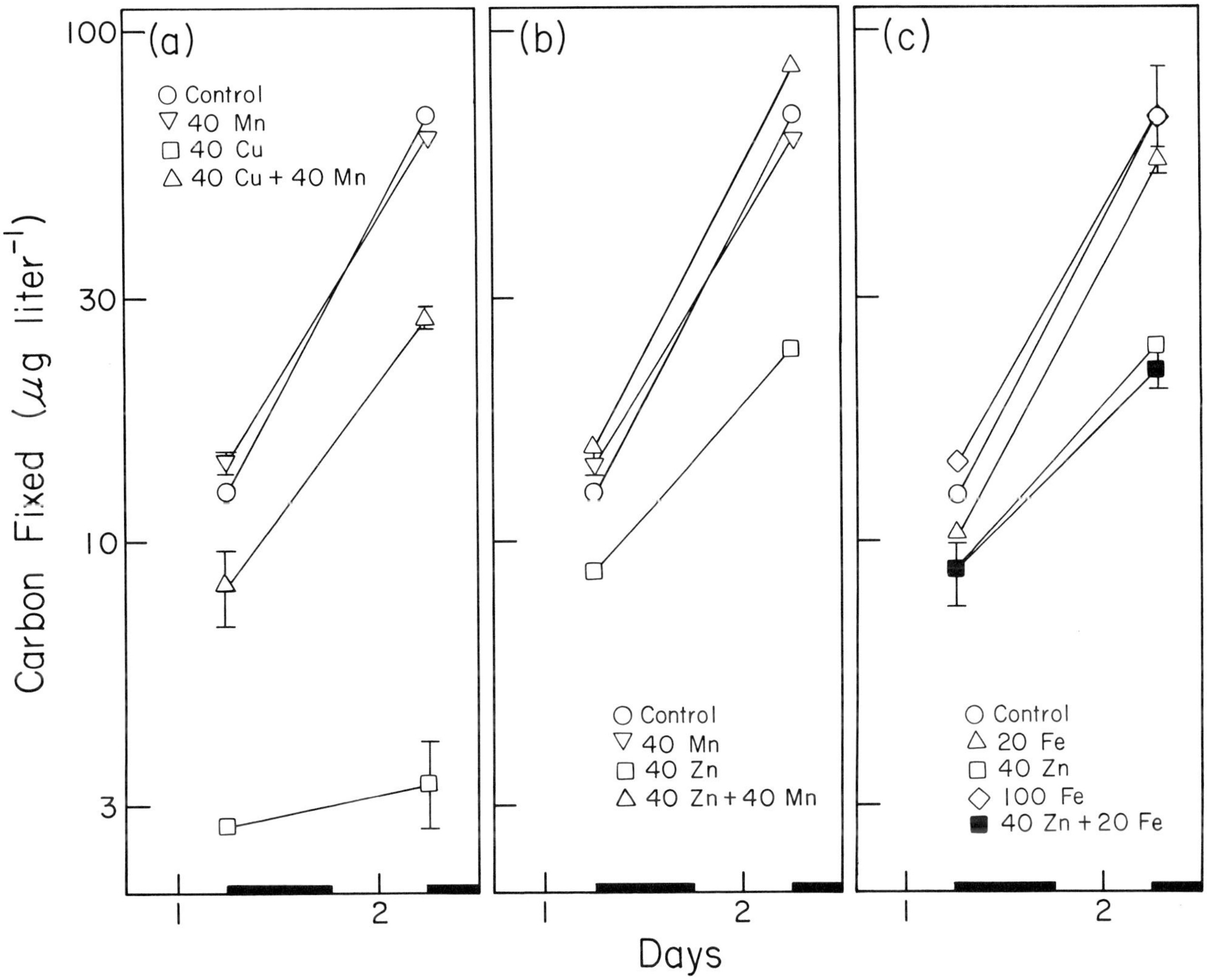

Figure 2.7. Effect of Mn and Fe on the toxicity of Cu and Zn in the Cape San Blas experiment (250 ml bottles). All units are in nmol $liter^{-1}$.

by additions of Mn and Fe (Jackson and Morgan, 1978; Sunda et al., 1981). In the present experiments, the addition of Mn had no appreciable effect by itself but partially reversed the long-term (2–4 d) toxicity of Cu in the Gulf Loop and Cape San Blas experiments (Figs. 2.2d and 2.7a) and fully reversed the toxicity of Zn in the Cape San Blas experiment (Fig. 2.7b). Manganese, however, did not reverse the short-term (12 h) toxicity of Cu (Fig. 2.2d). In the Cape San Blas experiment, the addition of Fe had no effect on the toxicity of Zn (Fig. 2.7c).

2.4. DISCUSSION

Copper and Zn were most toxic to phytoplankton in low-productivity oceanic seawater where nanomolar additions, which approach measured ambient concentrations, inhibited carbon fixation. Using similar clean techniques, Fitzwater et al. (1982) also noted inhibition by nanomolar Cu additions in an oceanic sample, but equivalent sensitivity to Zn has not been observed previously.

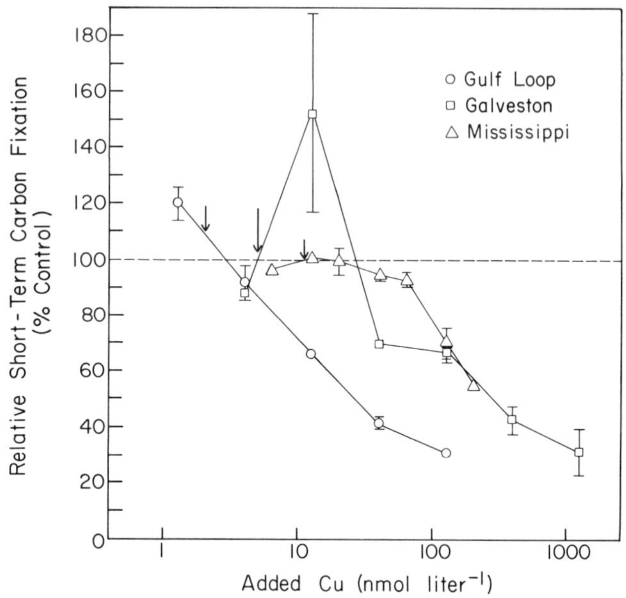

Figure 2.8. Relative short-term (8–12 h) inhibition of carbon fixation by Cu additions. Arrows indicate approximate ambient total Cu concentrations at each station as measured by anodic stripping voltametry in acidified (pH 2) samples collected at similar times and locations as samples used in the experiments.

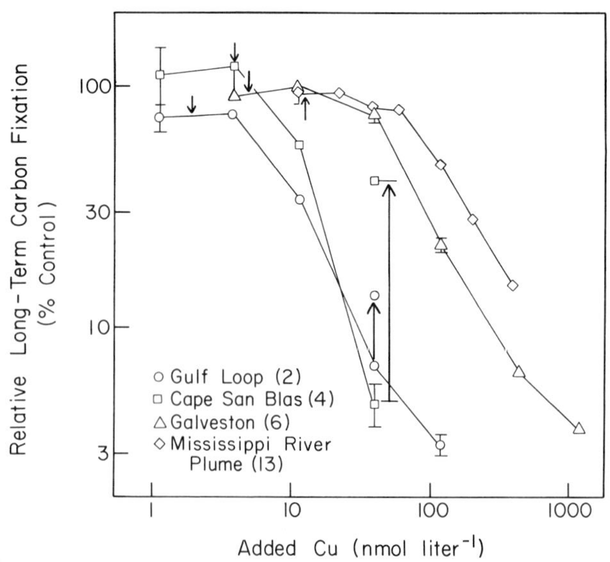

Figure 2.9. Long-term (third sampling period) inhibition of carbon fixation by added Cu at the four stations. Short arrows and numbers in parentheses designate approximate ambient Cu concentrations (nmol liter^{-1}). The ameliorating effect of Mn is shown by the long arrow.

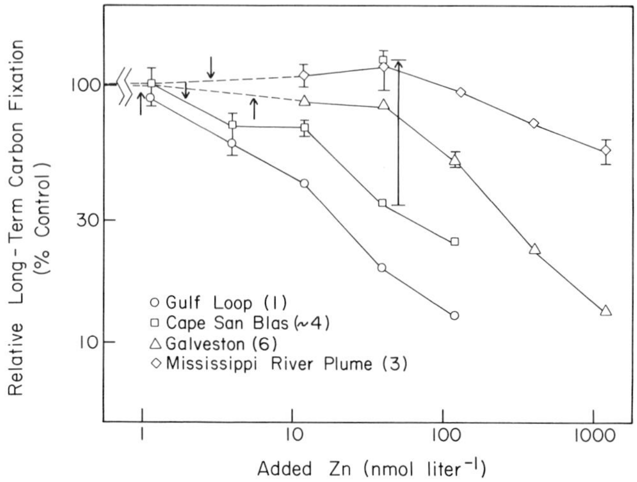

Figure 2.10. Relative long-term (third sampling period) inhibition of carbon fixation by added Zn at the four stations. Short arrows and numbers in parentheses designate approximate ambient Zn concentrations (pH 4.8, nmol liter^{-1}). Reversal of Zn toxicity by 40 nmol liter^{-1} of added Mn in the Cape San Blas experiment is shown by the long arrow.

There were consistent neritic–oceanic trends in Cu and Zn toxicity that appear to have resulted from a number of factors. One of these is differences in organic complexation, which is probably more important for Cu than for Zn because of the higher tendency of Cu toward complex formation. Bacterial bioassays of Cu complexation conducted on a previous cruise to the Gulf of Mexico showed an appreciably higher complexation capacity in high-productivity Mississippi River plume water than in low-productivity water from the Gulf Loop station (Fig. 2.11). Because of such differences in complexation, Cu additions to Gulf Loop seawater should result in higher free cupric ion activities, and therefore higher toxicity than equivalent additions to Mississippi plume water. Higher levels of organic complexation in Mississippi plume water correlated with higher concentrations of organic matter and humic compounds (Sunda and Ferguson, 1983; Harvey et al., 1983), which appear to result largely from high levels of primary productivity. Input of organic matter (such as humic substances) from the Mississippi River may have also contributed to the higher organic complexation.

A second factor contributing to the oceanic–neritic trend in the toxicity of Cu and Zn additions is the ambient concentrations of these metals in seawater. Measurements with anodic stripping voltametry in acidified samples indicate up to six-fold higher concentrations of these metals in near-surface seawater from the Mississippi and Galveston stations than in seawater from the Gulf Loop station (S. Piotrowicz,

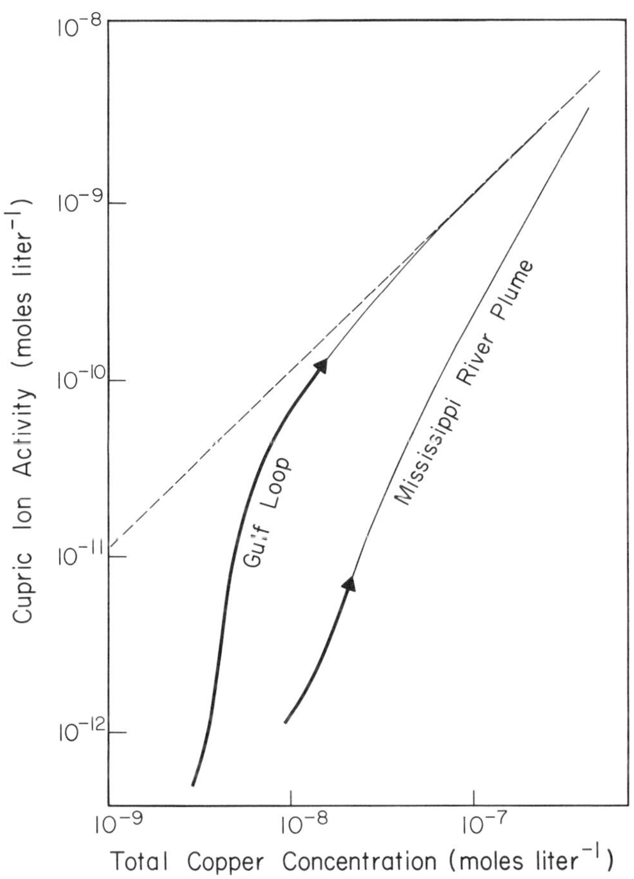

Figure 2.11. Relationship between cupric ion activity and total Cu concentration in Gulf of Mexico samples, as estimated by bacterial bioassay (April 1980) (Sunda and Ferguson, 1983). The Gulf Loop curve shows the relationship between cupric ion activity and Cu concentration for low-productivity, high salinity (36.1‰) seawater (pH 8.2). The Mississippi River Plume illustrates the same relationship for high-productivity, low-salinity seawater (corrected from pH 8.1 to pH 8.2). The dashed line shows the relationship between cupric ion activity and total Cu concentration that would exist in the absence of any organic complexation. Heavy solid arrows give the predicted change in cupric ion activity in the two seawater samples resulting from a 12 nmol liter^{-1} addition of Cu.

personal communication). The higher trace-metal concentrations in neritic waters would result in these samples requiring a higher minimum addition of metals before toxic effects could be observed; that is, it would be impossible to observe biological effects unless the added metal concentration approached or exceeded ambient concentrations.

A third and perhaps most important factor in the neritic–oceanic pattern in Cu and Zn toxicity appears to involve competitive interactions of these metals with nutrient metals, such as Mn, present at low concentrations in oceanic seawater. An addition of 40 nmol liter^{-1} Mn in the Cape San Blas experiment either largely or wholly reversed inhibition caused by an equivalent amount of Cu and Zn. A similar antagonism between Mn and Cu has been observed with cultures of *Chaetoceros socialis* and with natural community inocula grown in nutrient-rich seawater from a depth of 800 m (Sunda, et al., 1981). Subsequent experiments with both coastal and oceanic diatoms (*Thalassiosira pseudonana* and *T. oceanica*) in defined culture media indicate that Cu competitively interferes with cellular Mn accumulation, resulting in Mn deficiency (Sunda and Huntsman, 1983). Zinc and Cd, likewise, have been shown to inhibit competitively the cellular uptake of Mn (Hart et al., 1979; Sunda and Huntsman, unpublished data).

In the above laboratory experiments, the three toxic metals caused little or no inhibition of growth until 1–2 d after exposure. This delayed response may reflect the time required for dilution of cellular Mn pools once accumulation of Mn has been blocked by the toxic metals. A similar delayed pattern of inhibition was observed in the present experiments with Zn, Cd, and EDTA. Inhibition by EDTA is also apparently linked to an induction of trace-metal deficiency through reduction in free-ion activities of one or more nutrient metals (for example, Cu, Zn, Co, and Fe). For Cu the temporal pattern is less clear, and both short-term (6–12 h) and long-term (2–4 d) inhibition was observed. In the Gulf Loop experiment, Mn reversed only the long-term Cu inhibition but had no effect on short-term toxicity. Short-term toxicity of Cu thus does not appear to involve Mn nutrition, but instead may result from interferences with various membrane functions other than Mn transport [such as major ion pumps (Huntsman and Sunda, 1980)].

The above results, in conjunction with neritic–oceanic patterns in Mn distributions, suggest that low concentrations of Mn in Gulf Loop and Cape San Blas seawater are a major factor in the higher long-term toxicity of Zn and Cu at these stations. There are no reliable Mn data for the Gulf of Mexico. For the northwestern Atlantic, where surface Mn concentrations are inversely related to salinity, there is a pattern consistent with large-scale conservative mixing of Mn-rich continental waters with low-Mn oceanic seawater (Bruland and Franks, 1983). Dissolved Mn concentrations varied from 20 nmol liter^{-1} in 30‰ salinity coastal seawater (from off Massachusetts) to 2 nmol liter^{-1} in 36‰ salinity Sargasso seawater. The toxicity of both Cu and Zn additions were lowest in the Mississippi River plume (33.6‰ salinity), which should have contained relatively high Mn concentrations because of high dissolved concentrations of Mn (~ 200 nmol liter^{-1}) in Mississippi River water (Trefry and Presley, 1976).

Neritic–oceanic differences in trace-metal toxicity resulting from competition between toxic metals and nutrient metals may not be restricted to Mn. High levels of Cu can induce Zn deficiency (Price and Quigley, 1966; Rueter and Morel, 1981) and high levels of Cd can cause Fe deficiency (Foster and Morel, 1982; Harrison and Morel, 1983). Also, in higher plants a number of metals (Co, Cu, Zn, Cr, Ni, and Mn) are known to bring about Fe deficiency (Hewitt, 1948). As with Mn, low-salinity coastal seawater contains Fe and Zn concentrations that are one or more orders of magnitude higher than those present in oceanic waters (Betzer and Pilson, 1970; Gordon et al., 1982; Bruland and Franks, 1983). However, unlike Mn, additions of Fe had no effect on Zn toxicity in the Cape San Blas experiment and therefore an Fe antagonism was not a factor in the high sensitivity to Zn in this experiment. Whether or not the low availability of Fe contributed to Cu and Zn sensitivity in the Gulf Loop experiment is not known; however, apparent Fe deficiency in the control culture in this experiment suggests this possibility.

Differences in community species composition are a final factor that may have contributed to the neritic–oceanic pattern in trace-metal sensitivity. An examination of control organisms at the end of the experiments indicated differences in species dominance. Controls from the Mississippi plume experiment were dominated by *Skeletonema costatum*, whereas those from the Gulf Loop and Galveston stations primarily contained species of *Chaetocerus* and pennate diatoms. In the Cape San Blas experiment, the controls were dominated by coccolithophores.

Species often differ in their sensitivities to trace-metal deficiency and toxicity, and a neritic–oceanic pattern in trace-metal micronutrient limitation among species has been reported (Brand et al., 1983). In general, neritic species are more readily limited by low Fe, Zn, or Mn, which coincides with the neritic–oceanic pattern in the concentrations of these metals. Similar experiments examining Cu toxicity have shown oceanic species, on an average, to be slightly more sensitive; however, the variability within neritic and oceanic groups was much larger than the small difference in average sensitivities between these groups (Gavis et al., 1981). The extent to which differences in species composition affected the neritic–oceanic pattern in the present experiments is unknown. The dominance of *Skeletonema costatum* in the Mississippi River plume experiment, however, may have contributed to low toxicity of Cu in the Mississippi River plume water. This species has been shown to be relatively insensitive to Cu toxicity (Morel et al., 1978; Gavis et al., 1981).

2.5. CONCLUSIONS

Copper and Zn additions had a much greater inhibitory effect on carbon fixation by phytoplankton in low-productivity oceanic seawater than in high-productivity coastal seawater. This neritic–oceanic pattern appears to be related to at least three factors: (1) higher levels of organic complexation in neritic seawater, a factor that is particularly important for Cu because of its high affinity toward complex formation; (2) higher concentrations of Cu and Zn in coastal seawater, which effectively increase the minimum threshold of addition at which inhibitory affects can be observed; and (3) low availability of essential nutrient metals such as Mn (and perhaps Fe) in oceanic seawater, whose uptake and metabolism by organisms are subject to competitive interference by Cu and Zn. Differences in species composition among stations also may have been a contributing factor.

Because of such factors, the area where toxic-metal wastes are discharged can be as important as how much is discharged. Higher-productivity coastal seawater appears to have a much higher assimilative capacity, but this fact must be weighed against the relative importance of such regions to overall fishery productivity. Finally, it is now apparent that a major mechanism of trace-metal toxicity is competitive interference with the uptake and metabolism of essential metals. Consequently, the relative balance among metals in wastes can be a critical factor in the adverse impact of these wastes on phytoplankton communities.

ACKNOWLEDGMENTS

This research was conducted on the *R/V Researcher* from 27 January to 26 February 1981 in cooperation with scientists from the Ocean Chemistry and Biological Laboratory of the Atlantic Oceanographic and Meterological Laboratories (AOML), Miami, Florida. I thank Peter Crumley (Beaufort Laboratory) and Brian Wren (AOML) for assistance in conducting the experiments at sea. Thanks also go to Dr. Steven Piotrowicz for the use of clean facilities and for anodic-stripping trace-metal analyses, to Dr. Peter Ortner for primary productivity measurements, and to Dr. Randy Ferguson and Jo Ann Lewis for logistical support in preparing for the cruise. This research was supported by a contract from the Ocean Assessments Division, National Ocean Services, U.S. National Oceanic and Atmospheric Administration.

REFERENCES

Betzer, P. R., and M. E. Q. Pilson. 1970. Concentrations of particulate iron in Atlantic open-ocean water. *Journal of Marine Research*, **28**, 251–267.

Boyle, E., and J. M. Edmond. 1975. Copper in surface waters south of New Zealand. *Nature*, **253**, 107–109.

Brand, L. E., W. G. Sunda, and R. R. L. Guillard, 1983. Limitation of marine phytoplankton reproductive rates by zinc, manganese and iron. *Limnology and Oceanography*, **28**, 1182–1198.

Bruland, K. W., and R. P. Franks. 1983. Mn, Ni, Cu, Zn, and Cd in the western North Atlantic. *In:* Trace Metals in Sea Water, C. S. Wong, E. A. Boyle, K. W. Bruland, J. D. Burton, and E. D. Goldberg (Eds.). Plenum Press, New York, pp. 395–414.

Bruland, K. W., G. A. Knauer, and J. H. Martin. 1978. Zinc in northeast Pacific water. *Nature*, **271**, 741–743.

Fitzwater, S. E., G. A. Knauer, and J. H. Martin. 1982. Metal contamination and its effect on primary production measurements. *Limnology and Oceanography*, **27**, 544–551.

Foster, P. L., and F. M. M. Morel. 1982. Reversal of cadmium toxicity in a diatom: An interaction between cadmium activity and iron. *Limnology and Oceanography*, **27**, 745–752.

Gavis, J., R. R. L. Guillard, and B. L. Woodward. 1981. Cupric ion activity and the growth of phytoplankton clones isolated from different marine environments. *Journal of Marine Research*, **39**, 315–333.

Gordon, R. M., J. H. Martin, and G. A. Knauer. 1982. Iron in north-east Pacific waters. *Nature*, **299**, 611–612.

Harrison, G. I., and F. M. M. Morel. 1983. Antagonism between cadmium and iron in the marine diatom *Thalassiosira weissflogii*. *Journal of Phycology*, **19**, 495–507.

Hart, B. A., P. E. Bertram, and B. D. Scaife. 1979. Cadmium transport by *Chlorella pyrenoidosa*. *Environmental Research*, **18**, 327–335.

Harvey, G. R., D. A. Boran, L. A. Chesal, and J. M. Tokar. 1983. The structure of marine fulvic and humic acids. *Marine Chemistry*, **12**, 119–132.

Hewitt, E. J. 1948. Relation of manganese and some other metals to the iron status of plants. *Nature*, **161**, 489–490.

Huntsman, S.A., and W. G. Sunda. 1980. The role of trace metals in regulating phytoplankton growth. *In:* The Physiological Ecology of Phytoplankton, I. Morris (Ed.). Blackwell Scientific Publishers, London, pp. 285–328.

Jackson, G. A., and J. J. Morgan. 1978. Trace metal–chelator interactions and phytoplankton growth in seawater media: theoretical analysis and comparison with reported observation. *Limnology and Oceanography*, **23**, 268–282.

Morel, N. M. L., J. G. Rueter, and F. M. M. Morel. 1978. Copper toxicity to *Skeletonema costatum* (Bacillariophyceae). *Journal of Phycology*, **14**, 43–48.

Price, C. A., and J. W. Quigley. 1966. A method for determining quantitative zinc requirements for growth. *Soil Science*, **101**, 11–16.

Rueter, J. G., Jr., and F. M. M. Morel. 1981. The interaction between zinc deficiency and copper toxicity as it affects the silicic acid uptake mechanism in *Thalassiosira pseudonana*. *Limnology and Oceanography*, **26**, 67–73.

Schaule, B. K., and C. C. Patterson, 1981. Lead concentrations in the northeast Pacific: evidence for global anthropogenic perturbations. *Earth and Planetary Science Letters*, **54**, 97–116.

Sunda, W. G., and R. L. Ferguson. 1983. Sensitivity of natural bacterial communities to additions of copper and to cupric ion activity: a bioassay of copper complexation in seawater. *In:* Trace Metals in Sea Waer, C. S. Wong, E. A. Boyle, K. W. Bruland, J. D. Burton, and E. D. Goldberg (Eds.). Plenum Press, New York, pp. 871–891.

Sunda, W. G., and S. A. Huntsman. 1983. The effect of competitive interactions between manganese and copper on cellular manganese and growth in estuarine and oceanic species of the diatom *Thalassiosira*. *Limnology and Oceanography*, **28**, 924–934.

Sunda, W. G., R. T. Barber, and S. A. Huntsman. 1981. Phytoplankton growth in nutrient rich seawater: importance of copper–manganese cellular interactions. *Journal of Marine Research*, **39**, 567–586.

Swanson, R. L., and M. Devine. 1982. Ocean Dumping Policy. *Environment*, **24**, 14–20.

Trefry, J. H., and B. J. Presley. 1976. Heavy metal transport from the Mississippi River to the Gulf of Mexico. *In:* Marine Pollutant Transfer, H. L. Windom, and R. A. Duce (Eds.). D. C. Heath and Company, Lexington, Massachusetts, pp. 39–76.

Chapter 3

Mercury Tolerance in Killifish: a Stage-Specific Phenomenon

Judith S. Weis, Mark Renna, and Swati Vaidya

Department of Zoology and Physiology
Rutgers University
Newark, New Jersey

Peddrick Weis

Department of Anatomy
University of Medicine and Dentistry of New Jersey
Newark, New Jersey

ABSTRACT

Previous work has shown that embryos of killifish (*Fundulus heteroclitus*) from a polluted environment (Piles Creek, a tributary of the Arthur Kill in Linden, New Jersey) exhibit enhanced tolerance to methylmercury compared to populations from cleaner habitats. This differs from other studies on fish populations from metal-contaminated environments in which stress, rather than increased metal tolerance, was demonstrated. The present study compares methylmercury tolerance by larvae and adult killifish from Piles Creek with those from Southampton, New York, a more pristine area. Larval tolerance, as determined by deaths in 0.05 $\mu g\ g^{-1}$ methylmercury, was not significantly different between the two populations when measured by LT_{50} (time to death of 50% of the organisms) or by analysis of covariance. Thus, the enhanced tolerance of embryos observed in previous studies did not persist into the larval stages. Tolerance by adults was measured by fin regeneration rate and mortality in 0.05 $\mu g\ g^{-1}$ methylmercury. There was no evidence of increased resistance to methylmercury on the part of Piles Creek fish. The Piles Creek controls regenerated more slowly than Southampton controls, and the regeneration of methylmercury-treated groups was comparably depressed. In addition, the mortality rate of the Piles Creek fish was much higher. It appears that, rather than being tolerant to methylmercury, the Piles Creek adults were under stress and weaker than fish from Southampton.

3.1. INTRODUCTION

Sewage sludge from industrialized areas contains heavy metals, and organisms from dumpsite areas have been found to have elevated levels of heavy metals in their tissues. Some populations living in metal-polluted environments have become resistant. For example, Bryan (1976) studied the polychaete *Nereis diversicolor* and found populations from sediments rich in a particular metal to be resistant to that metal; resistance to Cu and Zn, and possibly to Ag and As, was demonstrated. Differences in tolerance were believed to be genetically based since the responses to high metal concentrations were not reversible (Bryan and Hummerstone, 1971).

Fraser et al. (1978) demonstrated resistance to Pb in the isopod *Asellus aquaticus* from sites with high Pb concentrations. Brown (1976) found that *Asellus meridianus* strains tolerant to Pb and Cu could survive and accumulate concentrations that proved lethal to nontolerant animals. Tolerance persisted into subsequent generations bred in the laboratory, indicating a genetic basis for the resistance. Moraitou-Apostolopoulou et al. (1979) determined that a population of the copepod *Acartia clausi* from a chronically Cd-polluted area was more resistant to Cd than was a population from a nonpolluted area, but did not determine whether genetic adaptation was involved.

Levinton (1980) discussed pollution as a selective agent that causes the more resistant genotypes to be prevalent in polluted areas. Luoma (1977) discussed the potential use of toxin resistance as a tool in assessing contaminant effects in natural ecosystems and hypothesized that greater resistance to a toxicant in a population from one area as compared to another area is direct evidence that the toxicant is exerting selective pressure at the first site. Luoma also stated that the level of resistance of the organisms should reflect the degree of contamination of the environment.

Although these ideas are conceptually appealing, one does not always find more resistant organisms in contaminated areas. Most studies of fish have not shown increased metal tolerance in populations from metal-contaminated areas. Rahel (1981) found that shiners (*Notropus cornutus*) from a Zn-contaminated stream were not more Zn-tolerant in terms of survival and swimming stamina than were those from uncontaminated areas. These results and a literature search led Rahel to conclude that, while fish may have the genetic potential to evolve metal tolerance, populations under chronic metal stress do not exhibit such tolerance but instead suffer from chronic stress, which weakens their ability to withstand further stress.

In our previous work with *Fundulus heteroclitus* (killifish), we demonstrated that embryos from a clean environment showed considerable variation in tolerance to methylmercury. Some females produced eggs that were very tolerant, while others produced susceptible eggs (Weis et al., 1982a). The abnormalities caused by methylmercury exposure were craniofacial, cardiovascular, and skeletal. We devised indexes to rate the relative severity of these defects, thus allowing quantification of teratogenic responses within and between populations.

F. heteroclitus are also found in Piles Creek in Linden, New Jersey, an area polluted with metals and oil. This creek is a tributary of the Arthur Kill, a highly polluted portion of the Newark Bay Estuary (Koepp et al., 1980). The marsh surrounding the creek is ringed by industrial sites, including a generating station, an oil refinery, a sewage plant, and several chemical plants, and it is near a major highway. The creek begins at the marsh edge and does not receive diluting freshwater from outside the industrialized area (Luther et al., 1982).

In this population of *Fundulus heteroclitus,* very few females produced susceptible eggs, and most were tolerant with respect to the incidence of malformations (Weis et al., 1981). Mitton and Koehn (1975) believe that the polygynous mating system of this species allows for rapid evolutionary response to a variable environment, which means this species has a high potential for adapatation to stress. These data are in apparent contradiction to Rahel's (1981) results with *Notropus*. In this chapter we report on comparative methylmercury susceptiblity of *F. heteroclitus* larvae and adults from Piles Creek as compared with those from Southampton, New York, a less polluted area.

3.2. MATERIALS AND METHODS

3.2.1. Larval Studies

Mortality was used as the parameter of larval tolerance. Adult *F. heteroclitus* from Piles Creek and from Bullhead Bay, Southampton, were collected in seines and traps. Gravid females were stripped into fingerbowls of seawater and fertilized with milt stripped from males. Eggs from each female were maintained separately because of the variablity in tolerance between individuals (Weis et al., 1982a) in 50 ml of seawater at 30‰ salinity and 24°C. The water was changed daily for the first 4 d of development. To stimulate hatching, water was changed daily starting on day 12 of development. When at least 20 had hatched, larvae were transferred in groups of ten (from the same clutch) to 100 ml of seawater (salinity 30‰; 24°C). Experimental groups were dosed with 0.05 μg g^{-1} methylmercury in 0.2% $NaHCO_3$. Controls were not treated. Solutions were changed daily, and dead larvae were removed and recorded for up to 8 d (larvae were not fed during the experiment, as yolk sac absorption provided nutrition until this time). For each group an LT_{50} (time to death of 50% of the organisms) and the mean LT_{50} for both the Piles Creek and Southampton fish were calculated. In addition, analysis of covariance was used to compare slopes and Y intercepts of regressions generated by the number dead versus day for each population.

3.2.2. Adult Studies

Mortality and fin regeneration were used as parameters in studying methylmercury tolerance by adults. Adult fish were collected from the two sites and acclimated in all-glass aquaria for several days. Ten to 14 fish of comparable size were placed in each aquarium in the experiment. All tanks were aerated but not filtered. On day zero, the bottom half of the caudal fin was amputated just distal to the caudal peduncle. Experimental fish were placed in artificial seawater (RilaMix®, Rila Co., Teaneck, New Jersey) of 20‰ salinity and dosed with 0.05 $\mu g\ g^{-1}$ methylmercury. Control fish were kept under the same conditions, but without the toxicant. Water was changed and redosed two times per week for two weeks. Before each change of water, fish were fed commercial fish food pellets or flakes. Regeneration of the amputated fin was measured at 1 week, 1.5 weeks, and 2 weeks by using a calibrated ocular micrometer in a stereomicroscope; all deaths were recorded.

One experiment used fish 45–55 mm in length, and a second used fish 50–60 mm in length. In the second experiment water was conditioned by having other killifish live in it for a minimum of 4 d while the water was filtered through undergravel filters.

For comparison of regenerate length, the means of two groups on any one day were compared by Student's t test. For the comparison of regenerative growth curves, analysis of covariance was used. Since fin regeneration growth curves are sigmoidal (Zimmerer, 1980), data used were taken from the linear portion of the sigmoid curve. Analysis of covariance can test for differences in elevation (Y intercept) and differences in slope. The Y intercept reflects the time needed for wound healing and blastema formation, while the slope represents the rate of fin elongation (Zimmerer, 1980).

3.3. RESULTS

3.3.1. Larval Studies

The LT_{50} among 34 batches of Southampton larvae exposed to 0.05 $\mu g\ g^{-1}$ methylmercury ranged from 1.2–6.0 d, and averaged 3.79 ± 0.21 d. The LT_{50} among 29 batches of Piles Creek larvae exposed to 0.05 $\mu g\ g^{-1}$ methylmercury ranged from 1.4–6.7 d and averaged 3.87 ± 0.23 d. The distributions are given in Fig. 3.1. There was no significant difference in the values for the two populations.

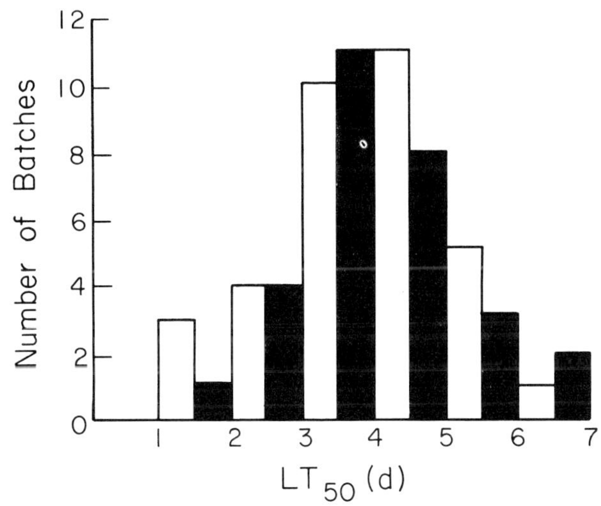

Figure 3.1. LT_{50} values for batches of *F. heteroclitus* larvae from Southampton (unshaded) and Piles Creek (shaded) after exposure to 0.05 $\mu g\ g^{-1}$ methylmercury.

Analysis of covariance of the data also reveals that no significant difference exists between Y intercepts ($F = 2.465$, $p > 0.05$), indicating no significant difference in time at which larvae began to die. There was also no significant difference in the slopes ($F = 0.238$), indicating that death rates were comparable. However, when the data were arcsine transformed, the analysis of covariance showed a significant difference in the Y intercepts ($F = 26.248$), although the slopes were still not significantly different ($F = 3.929$).

3.3.2. Adult Studies

Regenerative growth rates are summarized in Table 3.1 and Fig. 3.2. In the first experiment (unconditioned water) methylmercury-exposed fish in both populations regenerated more slowly than the controls, although not always significantly. Experimental fish from the two populations demonstrated comparable rates of fin regeneration; however, high mortality in the Piles Creek fish (85% versus 15% for Southampton) could make these results inconclusive. In Fig. 3.2 it appears as if growth of the Piles Creek fish in methylmercury approached that of controls by day 14, but it should be noted

Table 3.1. Effects of Methylmercury on Fin Regeneration in Two Populations of *Fundulus heteroclitus*

			Length of Fin Regeneration[a] (mm)			
Experiment	Population	Day	Exposed Organisms[b]	Control Organisms	*t* Value	*F* Test[c]
1[d]	Piles Creek	7	0.20 ± 0.19 (10)	0.73 ± 0.13 (11)	2.30, $p<0.05$	$F_1 = 0.103, p>0.05$
		11	0.27 ± 0.24 (6)	0.91 ± 0.21 (11)	2.01, $p>0.05$	$F_2 = 3.96, p>0.05$
		14	0.53 ± 0.44 (3)	0.94 ± 0.22 (10)	0.83, $p>0.05$	
	Southampton	7	0.91 ± 0.14 (17)	1.16 ± 0.09 (14)	1.50, $p>0.05$	$F_1 = 4.40, p<0.05$
		11	1.44 ± 0.19 (16)	2.15 ± 0.17 (14)	2.78, $p<0.01$	$F_2 = 17.7, p<0.001$
		14	1.50 ± 0.23 (16)	2.52 ± 0.20 (11)	3.35, $p<0.005$	
2[e]	Piles Creek	7	1.10 ± 0.17 (4)	0.93 ± 0.08 (10)	0.90, $p>0.05$	$F_1 = 1.42, p>0.05$
		11	2.20 ± 0.17 (4)	2.01 ± 0.13 (10)	0.89, $p>0.05$	$F_2 = 0.098, p>0.05$
		14	2.38 ± 0.16 (4)	2.61 ± 0.13 (10)	1.12, $p>0.05$	
	Southampton	7	0.92 ± 0.12 (10)	1.16 ± 0.14 (10)	1.30, $p>0.05$	$F_1 = 0.027, p>0.05$
		11	2.17 ± 0.19 (10)	2.34 ± 0.10 (10)	0.81, $p>0.05$	$F_2 = 2.69, p>0.05$
		14	2.90 ± 0.20 (9)	3.10 ± 0.10 (10)	0.89, $p>0.05$	

[a] Mean ± standard error; number of experimental organisms in parentheses.
[b] Organisms exposed to 0.05 μg g^{-1} methylmercury.
[c] Analysis of covariance; F_1 = comparison between slopes, F_2 = comparison between *Y* intercepts.
[d] Unconditioned water.
[e] Water was conditioned by having other killifish live in it for a minimum of 4 d.

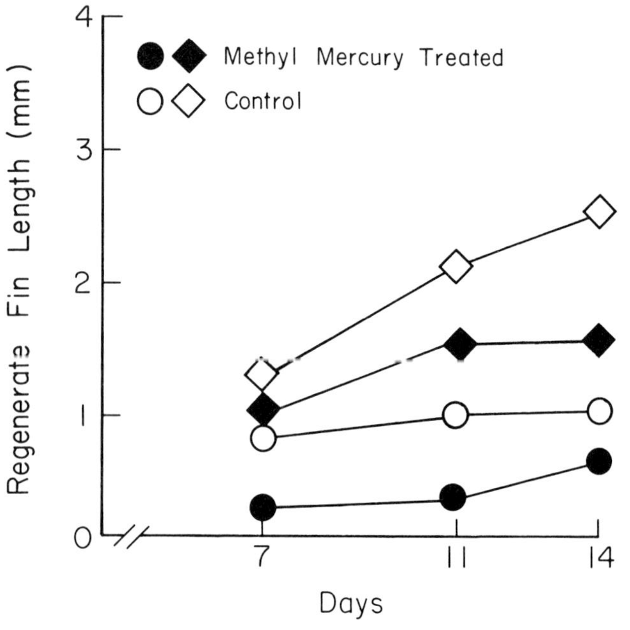

Figure 3.2. Caudal fin regeneration in *F. heteroclitus* from Piles Creek (○, ●) and Southampton (◇, ◆) under control conditions and after addition of 0.05 μg g^{-1} methylmercury.

that the mean value of the experimental organisms was generated by only three surviving fish, presumably the most tolerant individuals. In clean water, Southampton fish regenerated significantly faster than did Piles Creek (Table 3.2).

In the second experiment (conditioned water), the methylmercury-exposed fish in each population generally did not show a decrease in fin regeneration rate (Table 3.1); however, experimental Piles Creek fish demonstrated 67% mortality over 2 weeks, compared to 22% in Southampton fish. Again, regeneration was significantly faster for Southampton controls than for Piles Creek controls (Table 3.2).

3.4. DISCUSSION

Mortality of Piles Creek versus Southampton larvae in 0.05 μg g^{-1} methylmercury was not significantly different in terms of LT_{50} or analysis of covariance, although when the data were arcsine transformed, a difference in *Y* intercept was noted. Thus, the difference in tolerance of larvae was minor compared to the striking tolerance differences between embryos of two populations studied by Weis et al. (1981).

Table 3.2. Comparison of Southampton and Piles Creek *Fundulus heteroclitus* Fin Regeneration[a]

		F_1	F_2
Experiment 1	SH versus PC control organisms	10.6, $p < 0.003$	45.7, $p < 0.0005$
	SH versus PC experimental organisms	0.106, $p > 0.05$	3.62, $p > 0.05$
Experiment 2	SH versus PC control organisms	1.19, $p > 0.05$	13.1, $p < 0.001$
	SH versus PC experimental organisms	2.42, $p > 0.05$	0.308, $p > 0.05$

[a] Analysis of covariance; F_1 = comparison between slopes, F_2 = comparison between Y intercepts; Southampton (SH) and Piles Creek (PC).

Much of the embryonic tolerance can be attributed to a reduced chorionic permeability and uptake of methylmercury into the more tolerant embryos.

In adults, the rate of fin regeneration of both populations of fish exposed to 0.05 $\mu g\ g^{-1}$ methylmercury in unconditioned water was comparably depressed. Fin regeneration was not significantly depressed by this dosage in conditioned water in either population. The Piles Creek fish demonstrated a higher mortality in both experiments, indicating a lower tolerance to methylmercury. Southampton controls showed a significantly faster rate of fin regeneration than did Piles Creek controls in both experiments; this may be considered evidence that the Piles Creek fish are under chronic stress. Stress caused by pollutants in Piles Creek may have reduced the energy available for regenerative growth. Instead of exhibiting increased tolerance to methylmercury, the Piles Creek adults appeared to be less tolerant than the ones from Southampton. This is similar to the findings of Rahel (1981) for a Zn-contaminated population of *Notropus cornutus*. Similarly, McFarlane and Franzin (1978) found that white suckers (*Catostomus commersoni*) from metal-contaminated lakes exhibited signs of stress and weakness compared to those inhabiting a cleaner lake. Comparisons of regeneration and mortality of Southampton and Piles Creek populations of *F. heteroclitus* similarly indicate that the latter was a population under chronic stress and was not more tolerant to methylmercury.

One possible mechanism involved in metal tolerance is the synthesis of metal-binding proteins, or metallothioneins, which bind the metal, thus reducing the toxicity. These low-molecular-weight proteins are rich in cysteine, in which free thiol groups readily bind the metal, making it metabolically unavailable (Cherian and Goyer, 1978). Synthesis of these proteins is believed to be responsible for the increase in tolerance developed in animals previously exposed to low levels of metals (Pascoe and Beattie, 1979; Dixon and Sprague, 1981). These proteins do not appear to be involved in the increased methylmercury tolerance of Piles Creek embryos since only barely detectable amounts were found in eggs from both populations (Weis et al., 1982b).

Calabrese et al. (1982) suggest that adult marine animals may be more Hg-tolerant than are juvenile stages, possibly because of the acquisition of a sequestering mechanism. This does not appear to be the case for the Piles Creek fish because the adults were less tolerant than the population from the uncontaminated site. There is a limit to the ability of metallothioneins to bind metals, beyond which the toxicants may react with nonspecific proteins and cause adverse effects. This may be the situation for the Piles Creek fish.

Moraitou-Apostolopoulou et al. (1982) found that a chronically polluted population of the shrimp *Palaemon elegans* exhibited significantly increased Cd tolerance (as measured by LC_{50}, the concentration needed to kill half of a group of organisms), but only in the smaller size class. In the larger individuals no significant difference in Cd tolerance between the two populations was seen.

3.5. CONCLUSIONS

Although embryos of *F. heteroclitus* from a contaminated area (Piles Creek) have greatly increased methylmercury tolerance (Weis et al., 1981), the larvae have only slightly increased tolerance. The adults from Piles Creek seem to have been stressed by their environment so that they are weakened and less tolerant of further metal stress. Although fish from dumpsite areas have not been studied specifically in this regard, the findings of weakened fish populations from metal-contaminated waters would probably apply to these populations also. The elevated incidence of disease among fish populations living near dumpsites (Ziskowski and Murchelano, 1975; Ziskowski et al., 1980) attests to their weakened condition.

ACKNOWLEDGMENTS

This study was part of project RF-4 supported by grant AA-D-0065 from the Office of Sea Grant, U.S. National Oceanic and Atmospheric Administration. This chapter is publication NJSG No. 83-99. We thank T. Haresign, H. Reisman, and A. Siegel of the Department of Natural Sciences of Southampton College for their hospitality.

REFERENCES

Brown, B. E. 1976. Observations on the tolerance of the isopod *Asellus meridianus* Rac. to copper and lead. *Water Research,* **10,** 555–559.

Bryan, G. W. 1976. Some aspects of heavy metal tolerance in aquatic organisms. *In:* Effects of Pollutants on Aquatic Organisms, A. Lockwood (Ed.). Cambridge University Press, Cambridge, Massachusetts, pp. 7–34.

Bryan, G. W., and L. G. Hummerstone. 1971. Adaptation of the polychaete *Nereis diversicolor* to estuarine sediments containing high concentrations of heavy metals. I. General observations and adaptation to copper. *Journal of the Marine Biological Association of the United Kingdom,* **51,** 845–863.

Calabrese, A., E. Gould, and F. P. Thurberg. 1982. Effects of toxic metals in marine animals of the New York Bight: some laboratory observations. *In:* Ecological Stress and New York Bight: Science and Management, G. F. Mayer (Ed.). Estuarine Research Federation, Columbia, South Carolina, pp. 281–297.

Cherian, M., and R. Goyer. 1978. Metallothioneins and their role in the metabolism and toxicity of metals. *Life Sciences,* **23,** 1–10.

Dixon, D. G., and J. B. Sprague. 1981. Copper bioaccumulation and hepatoprotein synthesis during acclimation to copper by juvenile rainbow trout. *Aquatic Toxicology,* **1,** 69–81.

Fraser, J., T. Parker, and E. Verspoor. 1978. Tolerance to lead in the freshwater isopod *Asellus aquaticus. Water Research,* **12,** 845–863.

Koepp, S. J., J. M. McCormick, and S. L. Cheng. 1980. Distribution, flux, and biological effects of heavy metal pollutants in the Newark Bay Estuary. *In:* Water Conference, L. L. Ciaccio, and A. C. Cristini (Eds.). Ramapo College, Mahwah, New Jersey, pp. 186–201.

Levinton, J. S. 1980. Genetic divergence in estuaries. *In:* Estuarine Perspectives, V. Kennedy (Ed.). Academic Press, New York, pp. 509–520.

Luoma, S. N. 1977. Detection of trace contamination effects in aquatic ecosystems. *Journal of the Fisheries Research Board of Canada,* **34,** 436–439.

Luther, G. W., III, A. L. Meyerson, K. Rogers, and F. Hall. 1982. Tidal and seasonal variations of sulfate ion in a New Jersey marsh system. *Estuaries,* **5,** 189–196.

McFarlane, G. A., and W. G. Franzin. 1978. Elevated heavy metals: a stress on a population of white suckers *Catostomus commersoni* in Hammell Lake, Saskatchewan. *Journal of the Fisheries Research Board of Canada,* **35,** 963–970.

Mitton, J. B., and R. Koehn. 1975. Genetic organization and adaptive responses of allozymes to ecological variables in *Fundulus heteroclitus. Genetics,* **79,** 97–111.

Moraitou-Apostolopoulou, M., G. Verriopoulos, and P. Lentzou. 1979. Effects of sublethal concentrations of cadmium as possible indicators of cadmium pollution for two populations of *Acartia clausi* (Copepoda) living at two differentially polluted areas. *Bulletin of Environmental Contamination and Toxicology,* **23,** 642–649.

Moraitou-Apostolopoulou, M., G. Verriopoulos, and I. Rogdakis. 1982. Evaluation of the stress exerted by a polluted environment to a marine organism by comparative toxicity tests. *Bulletin of Environmental Contamination and Toxicology,* **28,** 416–423.

Pascoe, D., and J. H. Beattie. 1979. Resistance to cadmium by pretreated rainbow trout alevins. *Journal of Fish Biology,* **14,** 303–308.

Rahel, F. J. 1981. Selection for zinc tolerance in fish: results from laboratory and wild populations. *Transactions of the American Fisheries Society,* **110,** 19–28.

Weis, J. S., P. Weis, M. Heber, and S. Vaidya. 1981. Methylmercury tolerance of killifish (*Fundulus heteroclitus*) embryos from a polluted vs. non-polluted environment. *Marine Biology,* **65,** 283–287.

Weis, J. S., P. Weis, and M. Heber. 1982a. Variation in response to methylmercury by killifish (*Fundulus heteroclitus*) embryos. *In:* Aquatic Toxicology and Hazard Assessment: Fifth Conference, J. G. Pearson, R. Foster, and W. Bishop (Eds.). ASTM STP 766, American Society for Testing and Materials, Philadelphia, pp. 109–119.

Weis, J. S., P. Weis, M. Heber, and S. Vaidya. 1982b. Investigations into mechanisms of methylmercury in killifish (*Fundulus heteroclitus*) embryos. *In:* Physiological Mechanisms of Marine Pollutant Toxicity, W. Vernberg, A. Calabrese, F. Thurberg, and J. Vernberg (Eds.). Academic Press, New York, pp. 311–330.

Zimmerer, E. J. 1980. Environmental Effects on Growth and Fin Regeneration in Poeciliid Fishes. Ph.D. Dissertation, Rutgers University, Newark, New Jersey, 132 pp.

Ziskowski, J. J., and R. A. Murchelano. 1975. Fin erosion in winter flounder. *Marine Pollution Bulletin,* **6,** 26–29.

Ziskowski, J. J., V. T. Anderson, Jr., and R. A. Murchelano. 1980. A bent fin ray condition in winter flounder, *Pseudopleuronectes americanus,* from Sandy Hook and Raritan Bays, New Jersey, and Lower Bay, New York. *Copeia,* **1980,** 895–899.

Chapter 4

Polychaetes as Indicators of Heavy-Metal Availability in Marine Deposits

Geoffrey W. Bryan and Peter E. Gibbs

Marine Biological Association of the United Kingdom
The Laboratory, Plymouth, United Kingdom

ABSTRACT

Polychaetes are common in both uncontaminated and contaminated sediments. Previous studies have related heavy-metal concentrations in sediments with those in polychaete body tissues, but few species have been investigated to date. Not all polychaetes are useful indicators of contamination. Some polychaetes incorporate heavy metals into particular body structures; for example, the Cu level in *Melinna palmata* is high, usually about 1000 μg g^{-1}, chiefly because this metal is accumulated in the epidermal cells of the gills. Some species exhibit wide variation in the levels of certain metals; for example, at one site, the Cu concentrations in *Owenia fusiformis* ranged from 75 to 1340 μg g^{-1} dry wt and the Zn concentrations ranged from 144 to 769 μg g^{-1}. Even closely related species have different capacities to accumulate metals or metalloids, a notable feature found in the genus *Tharyx*, at least two species of which exhibit arsenic levels differing by one order of magnitude. Most polychaetes show some ability to regulate certain metals, especially Zn; for example, in *Glycera convoluta*, Zn levels may increase by only a factor of two in response to a change of over two orders of magnitude in the sediment. Of the species studied to date, the most promising indicators appear to be the members of the family Nephtyidae, a group found worldwide in estuaries and coastal waters. They are easy to clean prior to analysis and accumulate most heavy metals. The present study has demonstrated that significant relationships exist between concentrations in the body tissues of *Nephtys hombergii* and those of surface-sediment extracts of Ag, Co, Cu, Fe, Pb, and Zn, indicating that this species responds to environmental differences. Many factors have to be considered in the selection of polychaetes as indicators of heavy-metal contamination. In some species, high concentrations reflect the detoxication of bioavailable forms of the metal in the environment, and in others, they reflect the essential utilization of the metal by polychaete.

4.1. INTRODUCTION

Polychaetes are one of the most common macroinvertebrates in uncontaminated and contaminated sediments. Previous studies have related concentrations of heavy metals in burrowing polychaetes with those of the sediments. For the most part, these investigations have concerned estuarine forms, especially nereid species such as *Nereis diversicolor*. It is now known that *N. diversicolor* accumulates metals including Ag, Cd, Cu, and Pb in amounts that appear to reflect the availabilities of these metals in the environment. Thus, this species provides a useful basis for comparative studies of polychaete species (Luoma and Bryan, 1982; Bryan, 1985).

Although nereid polychaetes have been used as indicators of heavy-metal bioavailability in estuaries, they are relatively scarce in offshore communities; therefore, when investigating offshore disposal sites, it is necessary to consider species from other families. In order to assess heavy-metal concentrations in species from offshore disposal sites in terms of the availability of metals and their potential effects on the organism and its predators, samples must be obtained from localities covering a wide range of ambient metal concentrations. Such studies can be time-consuming and expensive; however, the distribution of many polychaete species extends from offshore sites (including disposal areas) to the lower reaches of estuaries, some of which are contaminated with heavy metals. By analyzing these species, the influence of large changes in sediment–metal levels can be studied and metal concentrations in worms from dumpsites can be put into perspective. It should then be possible to determine (1) which polychaetes are most useful as indicators of metal contamination, (2) whether metals at disposal sites are likely to affect the polychaete fauna, and (3) whether predators feeding on worms are likely to become affected.

Clearly, not all polychaetes have potential as useful indicators. Many are too small, fragile, or uncommon to be readily analyzed. The ideal indicator species is a sedentary deposit-feeder, preferably one that constructs a durable tube that will withstand shipboard washing and sieving. The worm should be fairly hardy, relatively short, not subject to frequent regeneration, and it should have a long life span, perhaps >3 y, and should be common in a wide range of sediments.

Few polychaetes meet all of these requirements, but members of the families Ampharetidae and Oweniidae are obvious candidates. Two species, *Melinna palmata* and *Owenia fusiformis,* from sites around southwestern England have been most examined. Other deposit-feeding species have been analyzed, but most have not been discovered in a sufficiently wide range of metal concentrations to be relevant to the present investigations; however, the data on two cirratulid species, *Tharyx marioni* and *Cirriformia tentaculata,* can be included. For comparative purposes, carnivorous species have been investigated, notably glycerids and nephtyids. *Nephtys hombergii*, which is both widespread and common, had proved to be particularly useful.

4.2. MATERIALS AND METHODS

Worms were collected offshore by using an anchor dredge (Forster, 1953). The main site sampled was the Rame Ground, an area used for sewage-sludge and dredged-material disposal; here, nine stations along a 3.5-km transect were sampled in depths from 50 to 54 m (Fig. 4.1). Further dredge samples were taken from Cawsand Bay (8-m depth) and in the Fal Estuary (25-m depth). Other worm samples were obtained by digging intertidal deposits at uncontaminated (Yealm Estuary) and contaminated (Restronguet Creek) sites.

Prior to analysis, the worms were cleaned by exposing them to acid-washed sand covered with seawater for up to 6 d (depending on the species) and to seawater for an additional day. The samples were dried at 80°C and digested with nitric acid. Following evaporation, they were redissolved in 1 *N* hydrochloric acid. Oxidized surface sediments associated with the worms were sieved through 100-μm plastic mesh and then extracted for 2 h with 1 *N* hydrochloric acid (sediment dilution, 30:1)(Luoma and Bryan, 1981). Analyses of solutions were carried out mainly by flame atomic absorption spectrophotometry (AAS), but Ag and Cd were sometimes measured by carbon furnace AAS. Organic matter in the sediments was determined from the loss in weight produced by heating the samples at 400°C for 6 h. A correction factor was applied to cover weight losses due to salt in the sediment.

4.3. OBSERVATIONS AND RESULTS

4.3.1. Ampharetidae

The deposit-feeding ampharetid *Melinna palmata,* which has been reported from sewage-sludge disposal areas (Eagle et al., 1978) and *Owenia fusiformis* were collected on the Rame Ground in the region of the dredged-material dumpsite. Remarkably, concentrations of Cu in pooled samples of *M. palmata* from this area are almost two orders of magnitude higher than in *O. fusiformis,* but *M. palmata* contains less Cu than *O. fusiformis* in samples from Place Cove (Tables 4.1 and 4.2). Relationships between concentrations of metals in *M. palmata* and sediment extracts are illustrated in Fig. 4.2a. Levels of Cu in the worms were generally high and not related to those of the sediments. Gibbs et al. (1981) have shown that 30 to 40% of the body Cu in *M. palmata* occurs in the branchiae, which have concentrations between 4 and 12 mg g^{-1}; this Cu reduces the palatability of the tissues and is accumulated as a chemical defense against predation. In the present study, levels of Mn in *M. palmata* were unrelated to sediment concentrations, however, concentrations of Zn in the worms increased slightly with increasing sediment levels, whereas levels of Fe rose steeply. Values for the concentration of Ag in the worms are unrelated to those of the sediments, but are directly related ($p < 0.001$) to the concentration of Cu in the tissue.

The results for *M. palmata* reveal at least three properties regarded as undesirable in an indicator of metal bioavailability: (1) the worms contain a very high, somewhat variable concentration of Cu, which has an essential role; (2) concentrations of Ag in the body were related to the high level of Cu and not to Cu concentrations in the environment; (3) the level of Zn, and possibly Mn, did not appear to respond to metal variations in the environment.

Melinna cristata also contains a high concentration of Cu and probably behaves similarly to *Melinna palmata*, however, *Ampharete acutifrons* (Gibbs et al., 1981) and *Amphicteis gunneri* both contain low levels of Cu. The latter species, from the vicinity of the dredged-material dumpsite off Rame Head, contained 56 times less Cu than *M. palmata* from the

Table 4.1. Heavy-Metal Concentrations (μg g^{-1}, dry wt) in *Melinna palmata, Amphicteis gunneri, Cirriformia tentaculata,* and Sediment

Sampling Site[a]	Organism	Number of Animals	Mean Dry Wt (mg)	Metal Concentration[b] Ag	Cd	Cu	Fe	Mn	Zn	Sediment Organic Matter Content (%)
Rame Ground (depth ~50 m)	*Melinna palmata*									
RM1		10	20.0	2.82		947	268	23.7	114	
				(0.08)		(14.3)	(3640)	(121)	(47)	2.0
RM3		10	18.5	3.16		752	279	14.3	124	
				(0.08)		(17.3)	(3700)	(112)	(53)	2.5
RM5		10	17.8	3.43		894	241	19.7	127	
				(0.08)		(16.3)	(3690)	(119)	(51)	2.5
RM7		10	16.6	4.49		725	245	18.4	97	
				(0.07)		(14.7)	(3310)	(122)	(51)	2.3
RM9		10	19.5	2.51		942	277	16.5	109	
				(0.09)		(21.6)	(3770)	(121)	(60)	3.0
Cawsand Bay		17	9.1	5.49		1190	260	12.9	105	
(depth ~8 m)				(0.02)		(13.7)	(3600)	(159)	(55)	1.7
Place Cove		7	20.3	0.72		314	410	13.6	121	
		3	26.8	1.42		400	486	12.1	121	
				(0.03)		(10.6)	(1430)	(80)	(42)	1.6
Carrick Roads		21	8.0			1280	630	15.8	121	
(depth ~25 m)						(210)	(8210)	(132)	(300)	~6
Rame Ground (depth ~50 m)	*Amphicteis gunneri*									
RM7		2	17.2	0.33		12.9	155	12.8	312	
				(0.007)		(14.7)	(3310)	(122)	(51)	2.3
Yealm Estuary	*Cirriformia tentaculata*	5 ♂	71	0.13	0.22	22.0	954	6.7	63	
		5 ♀	59	0.24	0.27	22.4	1190	13.6	86	
				(0.08)	(0.22)	(17)	(3770)	(171)	(50)	3.4
Stonehouse Pool		22 ♂	42	0.51		37.3	841	13.4	82	
(Tamar Estuary)		22 ♀	29	0.53		40.5	936	21.3	90	
				(0.21)		(92)	(8450)	(238)	(253)	4.0
Restronguet Creek		11 ♂	137	0.50	0.42	194	752	6.8	176	
		8 ♀	200	0.51	0.41	243	628	16.5	172	
				(0.88)	(1.19)	(1460)	(20900)	(189)	(2390)	5.0

[a] For sampling site locations see Fig. 4.1.

[b] Surface sediment concentrations in parentheses.

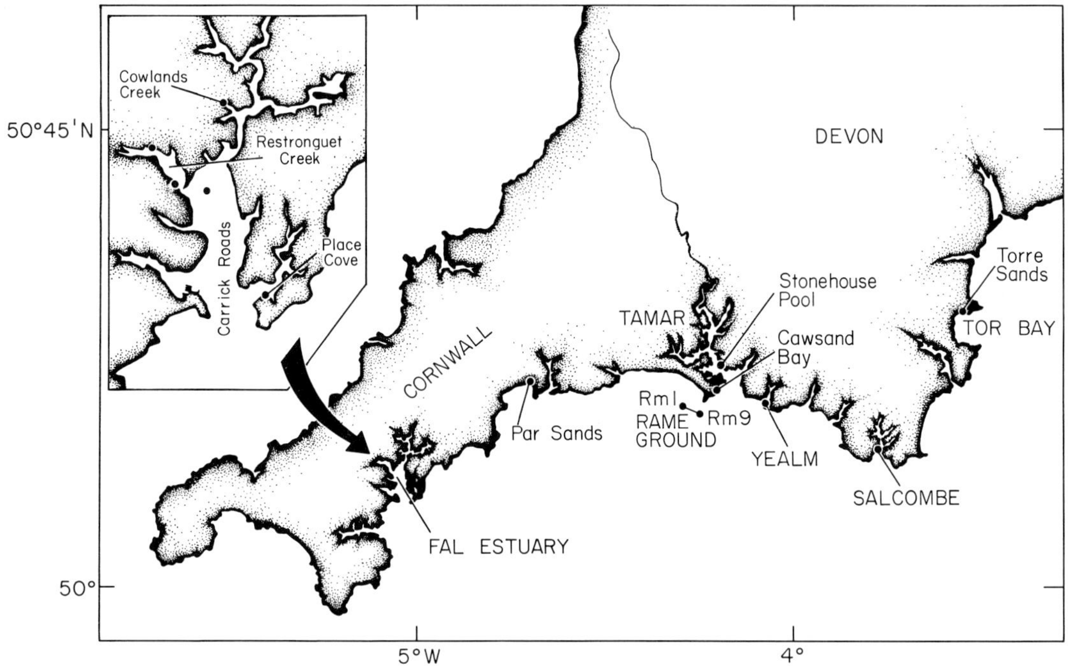

Figure 4.1. Polychaete and sediment sampling sites.

Table 4.2. Heavy-Metal Concentrations (μm g⁻¹, dry wt) in *Owenia fusiformis* (whole) and Surface Sediment

Sampling Site[a]	Number of Animals	Mean Dry Wt (mg)	Metal Concentration[b] Ag	Cu	Fe	Mn	Zn	Sediment Organic Matter Content (%)
Rame Ground (depth ~50 m)								
RM1	6	41	0.24	11.9	341	8.5	72	
			(0.08)	(14.3)	(3640)	(121)	(47)	2.0
RM3	9	46	0.13	12.6	659	10.9	76	
			(0.08)	(17.3)	(3700)	(112)	(53)	2.5
RM5	6	29	0.30	10.8	272	5.2	74	
			(0.08)	(16.3)	(3700)	(119)	(51)	2.5
RM7	11	36		12.0	155	7.4	86	
			(0.07)	(14.7)	(3310)	(122)	(51)	2.3
RM9	6	40		9.3	618	11.8	79	
			(0.09)	(21.6)	(3770)	(121)	(60)	3.0
Cawsand Bay	3	40	0.51	32.6	352	4.3	247	
(depth ~8 m)	3	8.5	1.71	34.7	368	7.1	306	
			(0.02)	(13.7)	(3600)	(159)	(55)	1.7
Par Sands	6	17.3	0.87	206	370	5.8	192	
	5	17.3	2.47	641	354	6.3	213	
			(0.03)	(9.2)	(1180)	(102)	(44)	0.8
Place Cove	16	8.2 ± 4.9		526 ± 449	373 ± 132		430 ± 194	
		[1.2–15]		[74.8–1340]	[254–726]		[144–769]	
				(55.6)	(5870)		(149)	3.8

[a] For sampling site locations see Fig. 4.1.

[b] Surface sediment concentrations are in parentheses and are for 1*N* HCl extracts of the <100-μm fraction; for Place Cove samples the range of values is in brackets.

same site (Table 4.1), but at present there are insufficient data to draw conclusions about the suitability of these ampharetids as indicators.

4.3.2. Oweniidae

Owenia fusiformis is common in the sewage-sludge and dredged-material dumpsites off Plymouth (Eagle et al., 1979) and at other disposal sites around Britain (Eagle et al., 1978; Talbot et al., 1982). The concentrations of five metals found in pooled samples of *O. fusiformis* from the Rame Ground proved remarkably constant (Table 4.2). Worms from Cawsand Bay in Plymouth Sound contained much higher concentrations of Cu and Zn, although the concentrations in the sediment were similar to those on the Rame Ground. Plymouth Sound receives metals from the River Tamar, the catchment of which includes the old mining area of East Cornwall and West Devon; thus, higher concentrations of dissolved Cu and Zn might be expected in Cawsand Bay than farther offshore. Still higher concentrations of Cu occured in worms from Par Sands, which receives some industrial contamination, although, again, this was not reflected by the Cu levels in the sediment. One Cu value for pooled *O. fusiformis* from Par was three times higher than the other, suggesting considerable individual variation. This variability was examined further in worms from Place Cove at the mouth of the Fal Estuary system. Although the Place Cove sediments are not highly enriched with metals, the Fal Estuary system receives high inputs of Cu and Zn (Bryan and Gibbs, 1983) and, logically, of the samples analyzed, Place Cove should have yielded the most contaminated worms. When the mean concentrations of Cu and Zn are considered, this is certainly true (Table 4.2), but individual variation at Place Cove is enormous; values for Cu ranged from 75 to 1340 $\mu g\ g^{-1}$, and values for Zn ranged from 144 to 769 $\mu g\ g^{-1}$. This variability is not due to contamination by uncleared sediment particles in the gut (compare values for sediment and worms in Table 4.2) and does not appear to relate to the size of organism, suggesting that the metabolism of Cu and Zn in *O. fusiformis* is cyclic and includes periods of storage followed by rapid depuration. Thus, although *O. fusiformis* may possibly reflect the availability of dissolved metals, the variations between individuals from the more contaminated areas seems to preclude its use as an indicator, unless large numbers of specimens are analyzed.

4.3.3. Cirratulidae

Relationships between element concentrations in the deposit feeder *Tharyx marioni* and those of the sediments are shown in Fig. 4.2b. Although the results cover a wide variety of sites and sediment concentrations, levels of As, Fe, and Zn in the worms do not vary significantly. Concentrations of Cu in the animals increase with rising sediment concentrations, but the slope (0.201) is small. Concentrations of Mn in *T. marioni* show a significant inverse relationship to the sediment values. *Tharyx marioni* is so effective at regulating metal concentrations in its tissues that it is totally unsuitable as an indicator. The most curious feature of this species is its high, relatively constant concentration of As, a large proportion of which is found in the extensible feeding palps (Gibbs et al., 1983). The role of As in this species, however, remains unknown. High values were not found in a second species, *Tharyx sp. A* (a new, as yet undescribed species) although both were collected at the same site.

Other species belonging to the family Cirratulidae may be more useful as indicators. In *Cirriformia tentaculata*, concentrations of Cu were appreciably higher in worms from contaminated areas (Table 4.1). Concentrations of Fe in the worms appeared unrelated to those of the sediments, but levels of Mn, for which appreciably higher values were found in females, increased with those of the sediments. Concentrations of Ag, Cd, and Zn were higher in worms from the more contaminated sites, but did not increase in direct proportion to the sediment levels.

4.3.4. Glyceridae

Various species of *Glycera*, including *G. rouxi* (Eagle et al., 1978) and *G. lapidum* (Eagle et al., 1979), have been found in offshore disposal areas. Single samples of these species have been analyzed by Gibbs and Bryan (1980), who compared them with two other species, *G. gigantea* and *G. convoluta*. The latter species appears to reflect changes in the availability of Cu in sediments, but concentrations of Zn and Mn show evidence of regulation.

Studies of Cu in glycerid worms have revealed yet another factor that must be recognized when using some species as indicators: each of the four jaws on the proboscis of *Glycera* species contains ~1.5% of Cu, and together the four jaws contain over 50% of the Cu in worms from uncontaminated sites. The Cu level in the jaws appears to be unaffected by environmental levels and, since the tip of the jaw in *G. gigantea* contains ~13% of the total Cu in the jaws, it is thought that the metal is utilized by the worm to harden its jaws (Gibbs and Bryan, 1980). Because levels of Cu in the soft tissues appear to reflect differences in the bioavailability of the metal, it is recommended that the jaws be removed

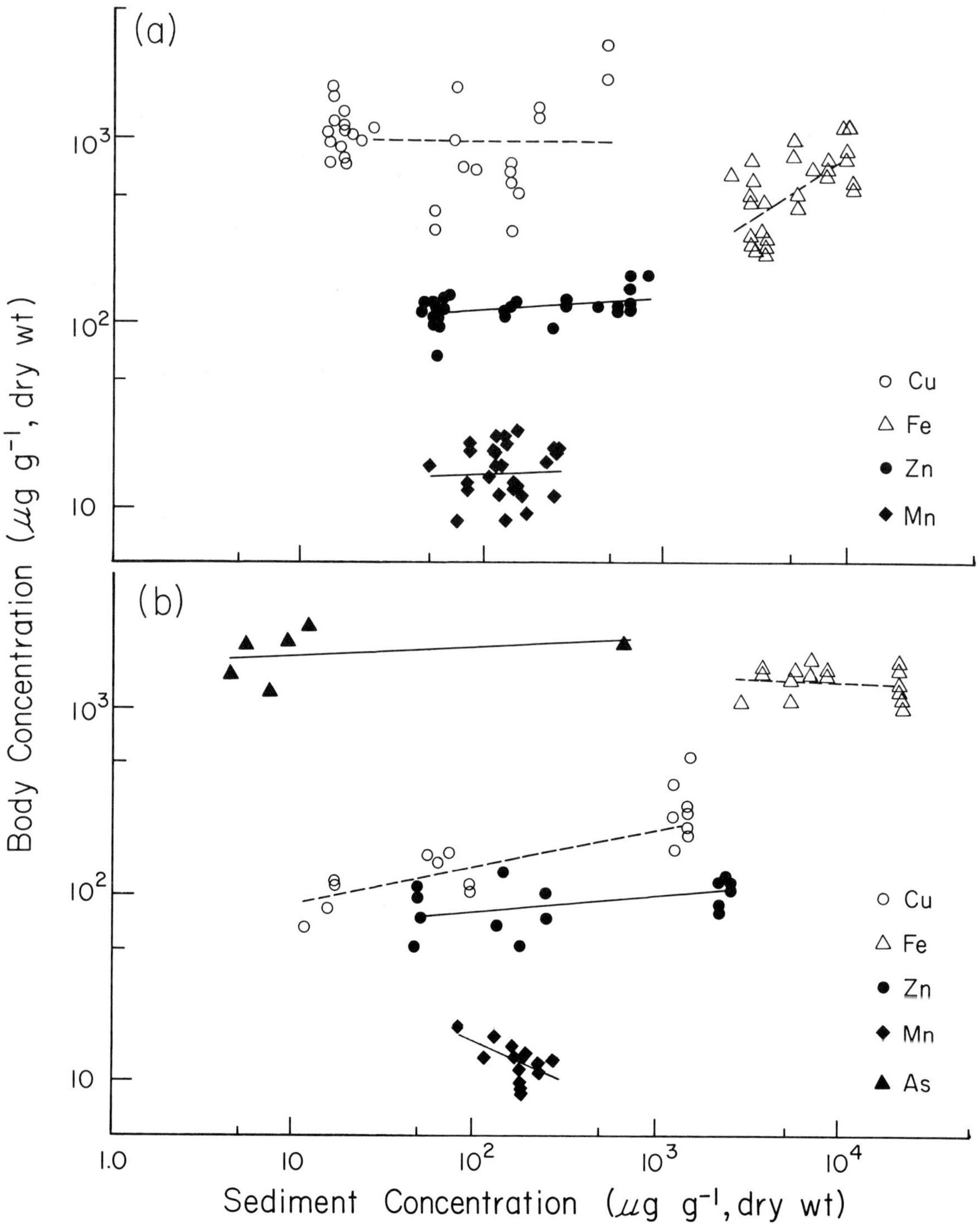

Figure 4.2. Relationships between concentrations of metals in tissues and in sediments (HCl extract) in (a) *Melinna palmata* and (b) *Tharyx marioni*. Equations of significant relationships are: $\log_{10}$Fe *M. palmata* = 0.592 $\log_{10}$Fe sediment + 0.490 (r = 0.620, $p < 0.001$); $\log_{10}$Zn *M. palmata* = 0.071 $\log_{10}$Zn sediment + 1.918 (r = 0.358, $p < 0.05$); $\log_{10}$Cu *T. marioni* = 0.201 $\log_{10}$Cu sediment + 1.745 (r = 0.881, $p < 0.001$); $\log_{10}$Mn *T. marioni* = −0.416 $\log_{10}$Mn sediment + 2.039 (r = −0.541, $p < 0.05$). Slopes of other relationships are not significantly different from zero.

before using glycerid worms as indicators. A similar situation exists in nereid polychaetes, in which the jaws appear to be hardened with Zn. The same problem does not arise with *Nephtys* species, since heavy metals in the jaws account for only a negligible portion of the total body burden.

4.3.5. Nephtyidae

Nephtys hombergii is one of the most common species in the vicinity of sewage-sludge dumpsites around the United Kingdom, including that off Plymouth (Eagle et al., 1978, 1979; Norton et al., 1981). It is also common in estuaries, including those draining the metalliferous mining regions of Cornwall and Devon. Examples of some of the more extreme metal concentrations observed in *N. hombergii* and in sediment extracts are presented in Table 4.3. Concentrations of some metals in the worms, notably Ag, Cd, Cr, Fe and Pb, vary by about one order of magnitude, whereas Cu levels range over more than two orders of magnitude. Relationships between concentrations in worms and sediments are shown in Fig. 4.3. One of the factors likely to be responsible for some of the variability in these relationships is the size of the worms analyzed. Results from two localities (Fig. 4.4), however, suggest that concentrations found in *N. hombergii* are unrelated to size differences. Similarly, although *N. hombergii* is a long-lived species with a life span of ~5 y (Olive, 1977), age is unlikely to be an important factor.

Of the less abundant metals, significant worm–sediment relationships are found for Ag and Co (Fig. 4.3b), but not for Cd. In this respect, *N. hombergii* resembles *N. diversicolor* (Luoma and Bryan, 1982; Bryan, 1985). It is thought that Cd levels in both species depend largely on those of the overlying water; thus the high concentration of *N. hombergii* from the Bristol Channel (Table 4.3) reflects the high Cd levels in the water observed by Morris and Bale (1975). Copper concentrations in *N. hombergii* are related very significantly to those of the sediments, where Zn, although significantly related, shows signs of being regulated (Fig. 4.3a). In *N. hombergii* from the most Cu-contaminated site (Restronguet Creek), the anterior end was blackened, and x-ray microanalysis indicated the presence of Cu sulfide particles mainly in the cuticle. Analysis by AAS of the high-Cu worms showed that Cu concentrations in the anterior region were more than two times those of the posterior region (Table 4.3). Much of this difference may be due to the presence of copper sulfide at the anterior end, although it seems unlikely from the body color that much of the Cu in the tail occurs as sulfide. It is possible that the presence of sulfide represents a detoxication mechanism, but more likely

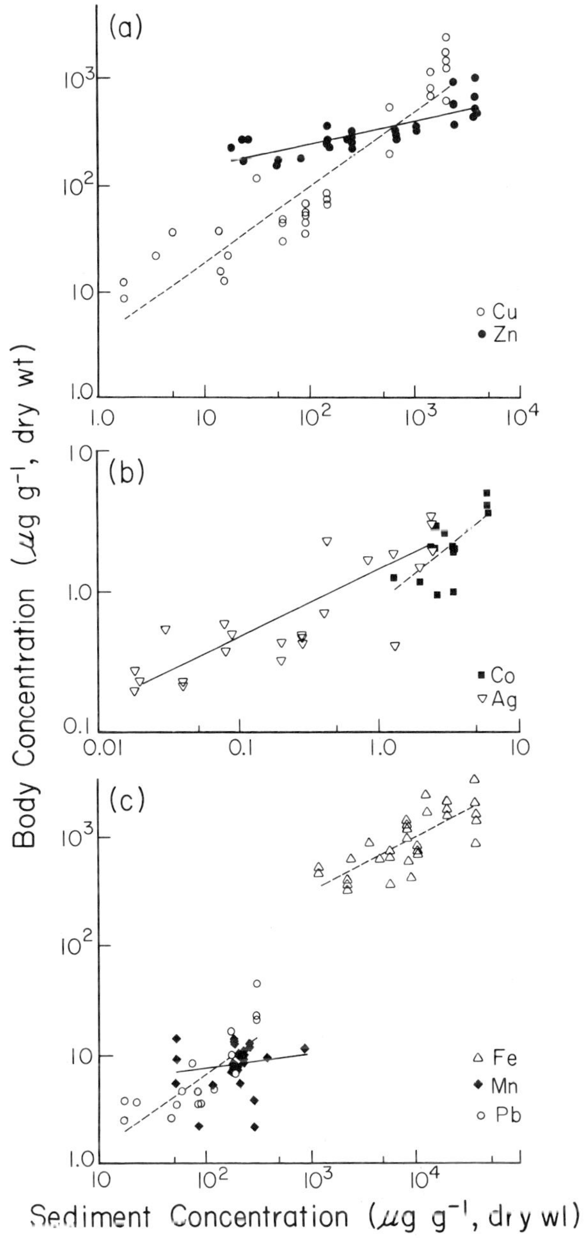

Figure 4.3 *Nephtys hombergii*: relationships between concentrations of metals in tissues and in sediment (HCl extract). Equations of significant relationships are: $\log_{10}$Cu *Nephtys hombergii* = 0.693 $\log_{10}$Cu sediment + 0.592 (r = 0.921, $p<0.001$); $\log_{10}$Zn *N. hombergii* = 0.209 $\log_{10}$Zn sediment + 1.955 (r = 0.812, $p<0.001$); $\log_{10}$Fe *N. hombergii* = 0.494 $\log_{10}$Fe sediment + 1.011 (r = 0.748, $p<0.001$); $\log_{10}$Pb *N. hombergii* = 0.690 $\log_{10}$Pb sediment − 0.546 (r = 0.798, $p<0.001$) $\log_{10}$Ag *N. hombergii* = 0.407; $\log_{10}$Ag sediment + 0.127 (r = 0.782, $p<0.001$); $\log_{10}$Co *N. hombergii* = 0.785 $\log_{10}$Co sediment − 0.088 (r = 0.685, $p<0.01$). Slope for Mn is not significantly different from zero.

Table 4.3. Heavy-Metal Concentrations (μg g^{-1}, dry wt) in *Nephtys* and Surface Sediment

Sampling Site[a]	Organism	Number of Samples	Mean Dry Wt (mg)	Metal Concentration[b] Ag	Cd	Co	Cr	Cu	Fe	Mn	Ni	Pb	Zn	Sediment Organic Matter Content (%)
Torre Sands	*Nephtys hombergii*	5	214	0.02	0.98	1.49	0.21	12.7	400	3.8	2.5	3.5	260	
				(0.018)	(0.093)	(2.06)	(2.3)	(1.75)	(2320)	(289)	(3.1)	(17)	(25)	0.9
Cawsand Bay (depth ~8 m)		5	262	0.23	0.22			38.5	616	8.6		4.6	178	
				(0.02)	(0.09)			(14)	(2490)	(183)		(59)	(52)	1.8
Bristol Channel (Weston)		7	56	2.28	3.61			117	410	11.5		4.6	259	
				(0.44)	(0.50)			(33)	(9130)	(891)		(122)	(232)	4.7
Restronguet Creek		4	84	3.27	0.57	3.98	2.43	2230	3260	7.5	4.7	44.7	518	
				(2.5)	(0.30)	(6.1)	(6.8)	(2010)	(37900)	(209)	(14.2)	(309)	(3790)	~6
Restronguet Creek	*Nephtys hombergii* (body sections)[c]													
	head		34.8					1470	2180	5.7			389	
	middle		41.4					1120	1640	5.3			269	
	tail		23.8					634	1160	4.2			262	
Salcombe	*N. hombergii*	10	154	0.54	1.89	1.26	0.59	36.5	364	5.4	2.96	3.7	268	
	N. cirrosa	4	178	0.24	0.93	1.04	0.55	16.3	468	4.2	3.04	3.8	259	
	N. caeca ciliata	1	915	0.24	0.93	1.10	0.54	17.4	551	5.5	2.01	3.9	191	
		4	103	0.33	0.78	0.81	1.28	28.5	341	3.4	4.97	9.0	328	
				(0.03)	(0.096)	(1.3)	(2.5)	(5.2)	(2310)	(214)	(4.7)	(22)	(28)	1.4
Place Cove	*N. hombergii*	2	84	0.69	0.54			30.0	353	2.2		8.4	355	
	N. cirrosa	30	25	0.20	0.42			29.9	249	4.5		4.9	307	
				(0.04)	(0.36)			(55.6)	(5870)	(86)	(1.3)	(75)	(149)	3.8
Place Cove[d]	*N. hombergii*	3	450	0.31	0.29			22.9	459	5.6		3.4	223	
	N. cirrosa	6	27	0.19	0.25			23.0	512	8.3		5.5	235	
				(0.004)	(0.023)			(3.6)	(1250)	(51)		(8.6)	(19)	1.0

[a] For sampling site locations see Fig. 4.1.
[b] Surface sediment values are in parentheses, and are for 1*N* HCl extracts of the <100-μm fraction.
[c] Body section weights given as percentage of total dry weight.
[d] Unsieved sediment.

it results from the exposure of the head end to hydrogen sulfide when the animal burrows into reducing sediments.

As in *N. diversicolor* (Bryan and Hummerstone, 1973a; Bryan 1974), concentrations of Pb in *N. hombergii* relate significantly to those of the sediments, but levels of Mn in the body tended to be low and not significantly related to sediment levels (Fig. 4.3c). Concentrations of Fe in *N. hombergii* showed a greater tendency to increase with those of the sediments than they did in *N. diversicolor* (Bryan and Gibbs, 1980). In *N. hombergii*, the polychaete with the highest concentrations of Fe (Table 4.3), this was visible at the anterior end as a surface layer of oxide and may be produced by oxidation following contact with high levels of dissolved Fe in the pore water of reducing sediments.

Concentrations of Ag, Cd, Co, Cu, and Pb in *N. hombergii* reflect the bioavailabilities of these metals in the environment. To a lesser degree, this applies also to Fe and Zn, but not to Mn. Because *N. hombergii* is a carnivore rather than a deposit feeder, it is perhaps surprising that the concentrations in the tissues related to those of the sediments. Such relationships may result if metals are absorbed from solution in pore water or through direct scavenging of metals from sediment particles in contact with the body surface. If concentrations in the prey are related directly to those of the sediments, this might produce the same result in *N. hombergii*. There is no guarantee of this, however, because concentrations in some deposit-feeding prey species are poorly related to those of the sediments.

Several other *Nephtys* species were analyzed, including *Nephtys caeca* and *N. cirrosa*. Concentrations of heavy metals in both of these species are comparable with those of *N. hombergii* from the same site (Table 4.3). It seems reason-

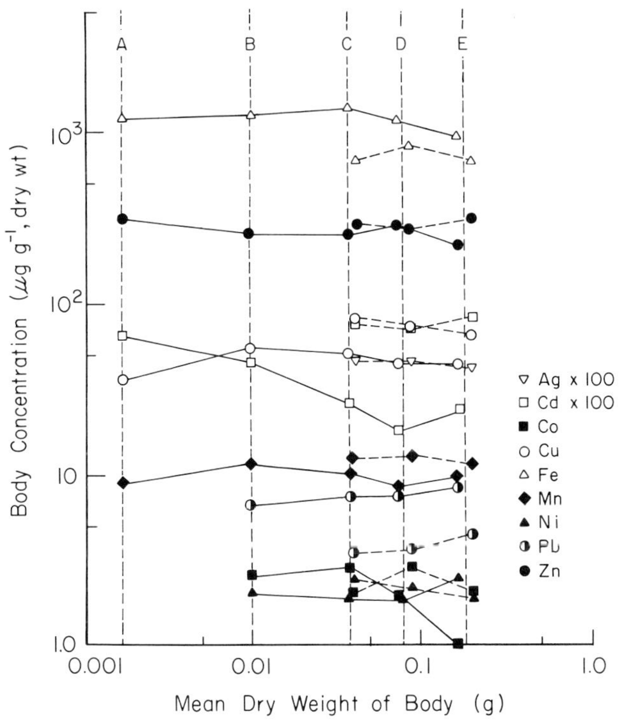

Figure 4.4. The influence of body weight on metal concentrations in tissues of *Nephtys hombergii* from Stonehouse Pool (solid line) and Cowlands (dashed line). The vertical lines A–E represent the number of individuals sampled. For Stonehouse Pool, A, B, C, D, and E are samples of 23, 26, 30, 10 and 1, respectively. For Cowlands Creek C, D, and E are samples of 12, 8, and 8, respectively.

able to suppose that many nephtyids may reflect environmental differences in the same way as *N. hombergii*, but this remains to be confirmed.

4.4. DISCUSSION

Many species of polychaetes are unsuitable as indicators of heavy-metal contamination in the environment. Some are too uncommon, too fragile, or too difficult to clean; others, although often abundant and widespread over areas used for waste disposal, have proved unsuitable for a variety of reasons. First, some groups of species contain certain heavy metals at very high concentrations, and these have an essential role in particular body structures. For example, Cu is accumulated by the ampharetid *Melinna palmata*, notably in the branchiae, to the extent that the whole-body concentration of Cu is raised to about 1 mg g^{-1}, irrespective of the environmental concentrations. Second, some species exhibit remarkably wide variations in metal concentrations between individuals; such is the case for *Owenia fusiformis*, in which at one site the range in body Cu was 75–1340 μg g^{-1} and that of Zn was 144–769 μg g^{-1}. Third, even species within the same genus may have markedly different capacities to accumulate certain metals or metalloids. A notable example is the cirratulid genus *Tharyx*, in which *Tharyx marioni* accumulated high concentrations of As (~2 mg g^{-1}), but another species did not. Finally, some metals appear to be regulated by polychaetes against environmental fluctuations.

Much of what is known regarding the use of polychaetes as indicators of the bioavailability of heavy metals is summarized in Table 4.4. The disposition of species toward different metals has been rated as poor, moderate, or good. Where rated as poor, the worms usually regulate the element, and thus the slope of the relationship between concentrations in the worms and those of the environment is at best rather small. For example, *Nephtys hombergii* is regarded as a poor indicator of Zn because the slope of the worm–sediment relationship (log transformed) is only 0.209 (Fig. 4.3a), but the slope is 0.69 for Cu and exceeds 0.4 for Ag (Figs. 4.3a, b), placing *N. hombergii* in the good and moderate indicator categories for these respective metals. Although the slope of the relationship for Pb is also 0.69 (Fig. 4.3c), *N. hombergii* is not regarded as being more than a moderately good indicator, since the Pb concentration in the tissues is generally much lower than that of the surrounding sediments.

Based on the data in Table 4.4, it appears that there is no suitable indicator species for Mn. Since the pore water of uncontaminated sediments often contains concentrations of Mn on the order of micrograms per milliliter, polychaetes are almost certainly adapted to deal with very high concentrations. In *Nereis diversicolor* there are two pools of Mn with very different exchange rates (Bryan and Hummerstone, 1973a). In the rapidly exchanging body pool, the concentration of Mn depends not only on the concentration in the sediment pore water but also on its salinity, since low salinity enhances the accumulation of Mn. Much of this exchangeable Mn is lost when *N. diversicolor* is cleaned for one week prior to analysis, leaving a slowly exchanging pool of Mn that shows signs of being regulated and includes a significant amount of Mn incorporated into the jaws (Bryan and Gibbs, 1980).

With the exception of *Capitella capitata, Owenia fusiformis,* and possibly, *Arenicola marina,* most of the species listed in Table 4.4 appear to regulate Zn. For example, in *Glycera convoluta* the concentration in the tissues increases by a factor of about two in response to a change of over two

Table 4.4. Polychaetes as Potential Heavy-Metal Indicators [a]

Organism	Rating as Indicator of Metal Availability: Poor	Moderate	Good	Feeding Habit	Remarks	Reference
Phyllodocidae						
Phyllodoce maculata			Cu	Carnivore	Uptake from solution	McLusky and Phillips (1975)
Nereidae						
Nereis diversicolor	Fe, Mn, Zn	As, Co, Hg, Pb; and possibly Cr, Ni, Sn	Ag, Cd, Cu	Omnivore, filter feeder	Cd relates to dissolved metal but others to sediment levels; high Zn in jaws; estuarine	Bryan and Hummerstone (1971, 1973a, 1973b); Bryan et al. (1980); Langston (1980); Luoma and Bryan (1982); Bryan (1985)
Nereis virens	Zn	Cd, Pb		Omnivore, herbivore	Compared to sediment levels; high Zn in jaws	Ray et al. (1981)
Perinereis cultrifera	Fe, Mn, Zn		Cu	Herbivore	Compared to sediment levels; high Zn in jaws	Bryan and Gibbs (1983)
Nephtyidae						
Nephtys hombergii	Mn, Zn	Ag, Co, Pb; possibly Fe	Cu; possibly Cd	Carnivore	Cd relates to dissolved metal but others to sediment levels	The present study
Glyceridae						
Glycera convoluta	Fe, Mn, Zn	Cu		Carnivore	Compared to sediment levels; high Cu in jaws	Gibbs and Bryan (1980)
Glycera americana				Carnivore	Compared to sediment levels	Cross et al. (1970)
Onuphidae						
Diopatra cuprea	Fe, Mn, Zn			Omnivore	Compared to sediment levels	Cross et al. (1970)
Arabellidae						
Arabella iricolor	Fe, Mn, Zn			Carnivore	Compared to sediment levels	Cross et al. (1970)
Dorvilleidae						
Ophryotrocha diadema		Cd		Omnivore	Uptake from solution and diet	Klöckner (1979)
Cirratulidae						
Cirriformia tentaculata	Fe, Mn, Zn	Ag, Cu		Deposit-feeder, surface	Compared to sediment levels; females higher in Mn	The present study
Cirriformia spirabrancha		Cu		Deposit-feeder, surface	Uptake from solution	Milanovich et al. (1976)
Tharyx marioni	Ag, Cu, Fe, Mn, Zn			Deposit-feeder, surface	Compared to sediment levels; high As in palps	The present study
Capitellidae						
Capitella capitata	Ni	Pb, Zn	Cd, Co, Hg	Deposit-feeder	Uptake from diet	Windom et al. (1982)
Oweniidae						
Owenia fusiformis	possibly Fe	possibly Cu, Zn		Deposit-feeder, surface	Highly variable; may reflect water levels	The present study
Arenicolidae						
Arenicola marina	Mn	Cd; possibly Cu, Pb, Zn		Deposit-feeder	Compared to sediment levels	Packer et al. (1980)
Pectinariidae						
Pectinaria californiensis	Cd	Pb	Hg	Deposit-feeder	Uptake from solution	Anderlini et al. (1975)
Ampharetidae						
Melinna palmata	Ag, Cu, Mn, Zn	Fe		Deposit-feeder, surface	Compared to sediment levels; high Cu in gills	The present study

[a] Feeding habits are based on Fauchald and Jumars (1979).

orders of magnitude in the sediment (Gibbs and Bryan, 1980). Iron also appears to be regulated by most species, although in *Melinna palmata* a slope of 0.59 was observed for the worm–sediment relationship (Fig. 4.2a) and in *Nephtys hombergii* the value was 0.49 (Fig. 4.3c). In the latter species, the indicator rating is partly dependent on the deposition of iron oxide over the anterior body surface rather than on its incorporation into the tissues.

Most polychaete species appear to have some potential as indicators of Cu contamination, but *Tharyx marioni* and *Melinna palmata* are notable exceptions. *Melinna palmata* is also unsuitable as an indicator of Ag. Reference has been made to the high concentration of As distinguishing *Tharyx marioni;* however, because As, Cu, Fe, Mn, and Zn are regulated, the species has no value as an indicator. In contrast, concentrations of As in *Nereis diversicolor,* although much lower, do appear to reflect environmental variations (Langston, 1980).

With the single exception of Ni in *Capitella capitata,* the available evidence suggests that concentrations of Cd, Co, Cr, Hg, Ni, Pb, and Sn in polychaetes tend to reflect their bioavailabilities in the environment. There are a number of problems, however, in using polychaetes as indicators for these metals. First, none of the metals are strongly accumulated by polychaetes, so concentrations in the tissues tend to be rather low, except at very high environmental levels. Thus, concentrations of Cr, Pb, and Sn, in particular, in the worms are usually far exceeded by those of the sediments, making it extremely important that sediment particles are removed from body surfaces and gut prior to analysis. Even after careful sample preparation and analysis, no clear polychaete–environment relationship may emerge. Although polychaete–sediment relations are stressed in this chapter, it may sometimes be more valid to relate concentrations within the worm tissues to other variables. Thus, although concentrations of Cd in *Nereis diversicolor* from estuaries in southwest England are related to those of the sediments (Bryan and Hummerstone, 1973b), concentrations in the same species from the Severn Estuary are best related to those of the overlying water (Bryan, 1985). Furthermore, Cd concentrations in *Capitella capitata* have been related to those of its detritus diet (Windom et al., 1982). Finally, particularly for the less abundant elements, it may often be difficult to discern polychaete–environment relations by direct comparisons. One reason for this difficulty is that the bioavailability of minor heavy metals may be affected by competition with more abundant metals for uptake sites. For example, the rate of Cd accumulation from solution by *Nereis diversicolor* is reduced by increasing the ambient concentration of Zn (Bryan and Hummerstone, 1973b). There is also evidence that the bioavailability of sediment-bound Pb to individuals of the same species is decreased in response to increasing levels of iron oxide or organic matter, which presumably bind the metal more effectively (Luoma and Bryan, 1982; Bryan, 1985).

This chapter has compared concentrations of heavy metals in polychaetes and surface sediments at offshore sites and in uncontaminated and contaminated estuaries. It is clear from the presence of species such as *Nephtys hombergii, Glycera convoluta, Melinna palmata,* and *Tharyx marioni* in estuarine sediments containing high concentrations of elements such as Cu, Zn, Fe, Pb and As that they are unlikely to be vulnerable to heavy metals at offshore disposal sites. The success of several species in colonizing highly metallic sediments is attributable to the development of metal tolerance. This was first observed in *Nereis diversicolor* exposed to high levels of Cu and Zn in the Fal Estuary system, and may reflect the existence of a metal-tolerant strain (Bryan and Hummerstone, 1971; 1973b; Bryan, 1976). Tolerance to Cu has been observed in *Nephtys hombergii* from the same area, and may have been acquired by individual worms through exposure to increasing metal concentrations during colonization of the most contaminated sites (Bryan and Gibbs, 1983). Under experimental conditions, the induction of Cu-tolerance through exposure to sublethal concentrations has been observed in the polychaetes *Neanthes arenaceodentata* (Pesch and Hoffman, 1982) and *Eudistylia vancouveri* (Young and Roesijadi, 1983) but was not found in *Ophryotrocha labronica* (Saliba and Ahsanullah, 1973).

Of the polychaete groups studied to date, the family Nephtyidae seems to contain the most promising indicator species. Compared to the well-studied estuarine species *Nereis diversicolor, Nephtys hombergii,* being a carnivore, has proved much easier to clean prior to analysis and is a better accumulator of a number of heavy metals. Because nephtyids are mobile and generally carnivorous, there is some doubt as to the extent of temporal change in the conditions to which they are exposed at any particular locality. Members of the family are found worldwide, however, and may prove to be the most useful and most convenient of species available for use as indicators.

ACKNOWLEDGMENTS

We are grateful to P. L. Pascoe and G. R. Burt for their assistance with field and laboratory work, and to F. E. Hutchings and the crew of the *R. L. Sepia* for their help with the dredging operations. Part of this work was supported by the U.K. Department of the Environment under Contract DGR 480/51.

REFERENCES

Anderlini, V.C., J. W. Chapman, A. S. Newton, and R. W. Risebrough. 1975. Dredge disposal study, San Francisco Bay and Estuary, Appendix I. *In:* Pollutant Availability Study. U.S. Army Corps of Engineers, San Francisco, 305 pp.

Bryan, G. W. 1974. Adaptation of an estuarine polychaete to sediments containing high concentrations of heavy metals. *In:* Pollution and Physiology of Marine Organisms, F. W. Vernberg and W. B. Vernberg (Eds.). Academic Press, London, pp. 123–135.

Bryan, G.W. 1976. Some aspects of heavy metal tolerance in aquatic organisms. *In:* Effects of Pollutants on Aquatic Organisms, A. P. M. Lockwood (Ed.). Cambridge University Press, Cambridge, pp. 7–34.

Bryan, G.W. 1985. Bioavailability and effects of heavy metals in marine deposits. *In:* Wastes in the Ocean, Vol. 6: Nearshore Waste Disposal, B. H. Ketchum, J. M. Capuzzo, W. V. Burt, I. W. Duedall, P. K. Park, and D. R. Kester (Eds.). Wiley-Interscience, New York, pp. 41–79.

Bryan, G. W. and P. E. Gibbs. 1980. Metals in nereid polychaetes: The contribution of metals in the jaws to the total body burden. *Journal of the Marine Biological Association of the United Kingdom,* **60,** 641–654.

Bryan, G. W., and P. E. Gibbs. 1983. Heavy Metals in the Fal Estuary, Cornwall: A Study of Long-Term Contamination by Mining Waste and Its Effects on Estuarine Organisms, Occasional Publication No. 2. Marine Biological Association of the United Kingdom, Plymouth, 112 pp.

Bryan, G. W., and L. G. Hummerstone. 1971. Adaptation of the polychaete *Nereis diversicolor* to estuarine sediments containing high concentrations of heavy metals. I: General observations and adaptation to copper. *Journal of the Marine Biological Association of the United Kingdom,* **51,** 845–863.

Bryan, G. W., and L. G. Hummerstone. 1973a. Adaptation of the polychaete *Nereis diversicolor* to manganese in estuarine sediments. *Journal of the Marine Biological Association of the United Kingdom,* **53,** 859–872.

Bryan, G. W., and L. G. Hummerstone. 1973b. Adaptation of the polychaete *Nereis diversicolor* to estuarine sediments containing high concentrations of zinc and cadmium. *Journal of the Marine Biological Association of the United Kingdom,* **53,** 839–857.

Bryan, G. W., W. J. Langston, and L. G. Hummerstone. 1980. The Use of Biological Indicators of Heavy Metal Contamination in Estuaries with Special Reference to an Assessment of the Biological Availability of Metals in Estuarine Sediments from South-West Britain. Occasional Publication No. 1, Marine Biological Association of the United Kingdom, Plymouth, 73 pp.

Cross, F. A., T. W. Duke, and J. N. Willis. 1970. Biogeochemistry of trace elements in a coastal plain estuary: distribution of manganese, iron, and zinc in sediments, water, and polychaete worms. *Chesapeake Science,* **11,** 221–234.

Eagle, R. A., P. A. Hardiman, M. G. Norton, and R. S. Nunny. 1978. The Field Assessment of Effects of Dumping Wastes at Sea. 3. A Survey of the Sewage Sludge Disposal Area in Lyme Bay. Fisheries Research Technical Report No. 49, Directorate of Fisheries, Ministry of Agriculture, Food and Fisheries, Lowestoft, U.K., 22 pp.

Eagle, R. A., P. A. Hardiman, M. G. Norton, and R. S. Nunny. 1979. The Field Assessment of Effects of Dumping Wastes at Sea. 4. A Survey of the Sewage Sludge Disposal Area off Plymouth. Fisheries Research Technical Report No. 50, Directorate of Fisheries, Ministry of Agriculture, Food and Fisheries, Lowestoft, U.K., 24 pp.

Fauchald, K., and P. A. Jumars. 1979. The diet of worms: a study of polychaete feeding guilds. *Oceanography and Marine Biology Annual Review,* **17,** 193–284.

Forster, G. R. 1953. A new dredge for collecting burrowing animals. *Journal of the Marine Biological Association of the United Kingdom,* **32,** 193–198.

Gibbs, P. E., and G. W. Bryan. 1980. Copper – the major metal component of glycerid polychaete jaws. *Journal of the Marine Biological Association of the United Kingdom,* **60,** 205–214.

Gibbs, P. E., G. W. Bryan, and K. P. Ryan. 1981. Copper accumulation by the polychaete *Melinna palmata:* an antipredation mechanism? *Journal of the Marine Biological Association of the United Kingdom,* **60,** 707–722.

Gibbs, P. E., W. J. Langston, G. R. Burt, and P. L. Pascoe. 1983. *Tharyx marioni* (Polychaeta): a remarkable accumulator of arsenic. *Journal of the Marine Biological Association of the United Kingdom,* **63,** 313–325.

Klöckner, K. 1979. Uptake and accumulation of cadmium by *Ophryotrocha diadema* (Polychaeta). *Marine Ecology Progress Series,* **1,** 71–76.

Langston, W. J. 1980. Arsenic in U.K. estuarine sediments and its availability to benthic organisms. *Journal of the Marine Biological Association of the United Kingdom,* **60,** 869–881.

Luoma, S. N., and G. W. Bryan. 1981. A statistical assessment of the form of trace metals in oxidized estuarine sediments employing chemical extractants. *The Science of the Total Environment,* **17,** 165–196.

Luoma, S. N., and G. W. Bryan. 1982. A statistical study of environmental factors controlling concentrations of heavy metals in the burrowing bivalve *Scrobicularia plana* and the polychaete *Nereis diversicolor. Estuarine, Coastal and Shelf Science,* **15,** 95–108.

McLusky, D. S., and C. N. K. Phillips. 1975. Some effects of copper on the polychaete *Phyllodoce maculata. Estuarine and Coastal Marine Science,* **3,** 103–108.

Milanovich, F. P., R. Spies, M. S. Guram, and E. E. Sykes. 1976. Uptake of copper by the polychaete *Cirriformia spirabrancha* in the presence of dissolved yellow organic matter of natural origin. *Estuarine and Coastal Marine Science,* **4,** 585–588.

Morris, A. W., and A. J. Bale. 1975. The accumulation of cadmium, copper, manganese and zinc by *Fucus vesiculosus* in the Bristol Channel. *Estuarine and Coastal Marine Science,* **3,** 153–163.

Norton, M. G., R. A. Eagle, R. S. Nunny, M. S. Rolfe, P. A. Hardiman, and B. L. Hampson. 1981. The Field Assessment of Effects of Dumping Wastes at Sea. 8: Sewage Sludge Dumping in the Outer Thames Estuary. Fisheries Research Technical Report No. 62, Directorate of Fisheries, Ministry of Agriculture Food and Fisheries, Lowestoft, U.K., 62 pp.

Olive, P. J. W. 1977. The life-history and population structure of the polychaetes *Nephtys caeca* and *Nephtys hombergi* with special reference to the growth rings in the teeth. *Journal of the Marine Biological Association of the United Kingdom,* **57,** 133–150.

Packer, D. M., M. P. Ireland, and R. J. Wootton. 1980. Cadmium, copper, lead, zinc and manganese in the polychaete *Arenicola marina* from sediments around the coast of Wales. *Environmental Pollution,* **22A,** 309–321.

Pesch, C. E., and G. L. Hoffman. 1982. Adaptation of the polychaete *Neanthes arenaceodentata* to copper. *Marine Environmental Research,* **6,** 307–317.

Ray, S., D. W. McLeese, and M. R. Peterson. 1981. Accumulation of copper, zinc, cadmium and lead from two contaminated sediments by three marine invertebrates – a laboratory study. *Bulletin of Environmental Contamination and Toxicology*, **26**, 315–322.

Saliba, L. J. and M. Ahsanullah. 1973. Acclimation and tolerance of *Artemia salina* and *Ophryotrocha labronica* to copper sulphate. *Marine Biology*, **23**, 297–302.

Talbot, J. W., B. R. Harvey, R. A. Eagle, and M. S. Rolfe. 1982. The Field Assessment of Effects of Dumping Wastes at Sea. 9: Dispersal and Effects on Benthos of Sewage Sludge Dumped in the Thames Estuary. Fisheries Research Technical Report No. 63, Directorate of Fisheries, Ministry of Agriculture Food and Fisheries, Lowestoft, U.K., 52 pp.

Windom, H. L., K. T. Tenore, and D. L. Rice. 1982. Metal accumulation by the polychaete *Capitella capitata:* influences of metal content and nutritional quality of detritus. *Canadian Journal of Fisheries and Aquatic Sciences*, **39**, 191–196.

Young, J. S., and G. Roesijadi. 1983. Reparatory adaptation to copper-induced injury and occurrence of a copper-binding protein in the polychaete, *Eudistylia vancouveri*. *Marine Pollution Bulletin*, **14**, 30–32.

Chapter 5

Bioaccumulation of Hg, Cd, and Pb in *Mytilus edulis* Transplanted to a Dredged-Material Dumpsite

Stephen J. Koepp and Edward D. Santoro

Biology Department
Montclair State College
Upper Montclair, New Jersey

Robin Zimmer

New Jersey Marine Sciences Consortium
Sandy Hook, New Jersey

Joseph Nadeau

Geology Department
Rider College
Lawrenceville, New Jersey

ABSTRACT

This study assesses changes in Hg, Cd, and Pb concentrations in *Mytilus edulis* transplanted to eight New York Bight locations in and around the dredged-material dumpsite. Incubation bags containing 100 mussels were suspended 1 m above the seafloor on platforms designed to allow periodic mussel retrieval by divers. Six platforms were deployed in August 1980, and two platforms were deployed in January 1981 following completion of sand-capping at the dredged-material dumpsite. Four stations were located in and around the dumpsite, three were located in Lower New York Harbor off Jones Beach, New York, and one was located offsh[illegible] Barnegat Light, New Jersey. Stations were sam[illegible] for the first month (August 1980), after [illegible] collections were conducted through July [illegible] soft-tissue samples were analyzed for tota[illegible] Hg, Cd, and Pb. Mercury, Pb, and Cd [illegible] consistently detected in mussels from all stations. Variability in metal concentrations was observed for the mussels deployed in August during the first five weeks of incubation; this was followed by a rise to maximum concentrations in early spring. Mercury concentrations were significantly greater for mussels incubated in the dumpsite (station E); however, at no time were Hg concentrations observed in excess of the 1.0 $\mu g\ g^{-1}$ (wet wt) U.S. Food and Drug Administration guideline. Increases in Cd and Pb appear to reflect long-term urban contamination of the New York Bight and not necessarily dredged-material dumping. The result of sand-capping could not be assessed due to delayed deployment of the sampling platforms and minimal sampling opportunities prior to the termination of the study.

5.1. INTRODUCTION

The New York Bight apex, which incorporates the 25 km coastal zone adjacent to New York Harbor, is one of the most heavily affected inshore areas along the Atlantic coast of the United States. Over the past century, increasing population density and industrialization, particularly involving the Hudson River, Newark, and Raritan Bay estuarine systems, have severely affected the water quality and the physical features of these waters (Pearce, 1980). Only recently have comprehensive research efforts begun to identify effects of this stress on the marine life of the New York Bight apex (Mahoney et al., 1973; Carriker et al., 1982). Concern has been expressed about the environmental impact of ocean disposal of dredged material generated during routine harbor maintenance. Steimle et al. (1982) reported diminished species diversity and trophic alterations in the benthic communities adjacent to the dredged-material dumpsite in the New York Bight apex. Contaminants such as metals, chlorinated hydrocarbons, and petroleum hydrocarbons have been detected in New York Harbor dredged material as well as at the dredged-material dumpsite (Carmody et al., 1973; Mueller et al., 1976).

The Mussel Watch concept stresses the use of mussels and other bivalves as bioindicators to assess the health of coastal waters and to provide early detection of potentially harmful chemical pollutants [U.S. National Academy of Science (NAS), 1980]. The concept has been applied in a series of national and international monitoring efforts (Butler, 1973; Holden, 1973; Goldberg, 1975; Talbot et al., 1976; Phillips, 1976; Goldberg et al., 1978; Risebrough et al., 1980; Farrington et al., 1983). Rationale for use of the mussel includes its ubiquitous distribution, sedentary nature, commercial value, and relatively high bioconcentration potential for a range of contaminants (Farrington et al., 1983). Transplantation of relatively clean mussels to dredged-material disposal areas has been previously reported (Feng et al., 1979).

The purpose of our investigation was to conduct long-range monitoring of Hg, Cd, and Pb concentrations in the blue mussels *Mytilus edulis* transplanted to the vicinity of the dredged-material dumpsite of the New York Bight apex. Adult mussels were obtained from a relatively clean site off southern New Jersey and transplanted onto six platforms in the New York Bight apex in August 1980. Deployment stations were selected on the basis of proximity to the dumpsite and other regional inputs. Following sand-capping operations within the dredged-material dumpsite, two additional mussel platforms were deployed in January 1981. Mussels were periodically retrieved by divers, and composite mussel samples (comprising whole soft tissue) were subsequently analyzed for total wet-tissue concentrations of Hg, Cd, and Pb.

5.2. METHODS

5.2.1. Experimental Design

5.2.1a. Station locations and platform deployment

Station locations are shown in Fig. 5.1. Station A was located near the southern end of the Sandy Hook–Rockaway Point transect on Romer Shoals, where the net current flow is seaward from the Hudson–Raritan Estuary. Station B was located within Gravesend Bay in lower New York Harbor, which is heavily influenced by the Hudson River. Station C was located near Jones Beach, Long Island. Water depths at stations A, B, and C were 7.6, 10.6, and 9.0 m, respectively. Three platforms were deployed in the vicinity of the dumpsite (stations D, E, and F). Station D was situated 3.2 km northeast of the dumpsite, within the Christiaensen Basin; station E was located in the northwest quadrant of the dumpsite, and station F was positioned 1.6 km due west of the dumpsite. Water depths at stations D, E, and F were 33.3, 22.7, and 25.7 m, respectively. In January 1981, a platform was deployed at station X in the recently capped southeast quadrant of the dumpsite, and a control platform (station Y) was deployed 4.8 km east of Barnegat Light, New Jersey.

A large population of blue mussels was located 2.4 km east of Wildwood, New Jersey, in 13.7 m of water. Metal analyses of representative samples proved this population suitable for transplantation studies. Mussels (7200) ranging in length from 50 to 70 mm, were collected, cleaned of epibiota, and transported in ice chests to the deployment base at Sandy Hook, New Jersey. Within 20 h after initial collection, sets of 100 mussels were placed into Vexar® polypropylene mesh bags [9-inch (~23-cm) Vexar tube netting] (Mid-City Steel Co., North Westport, Massachusetts) and

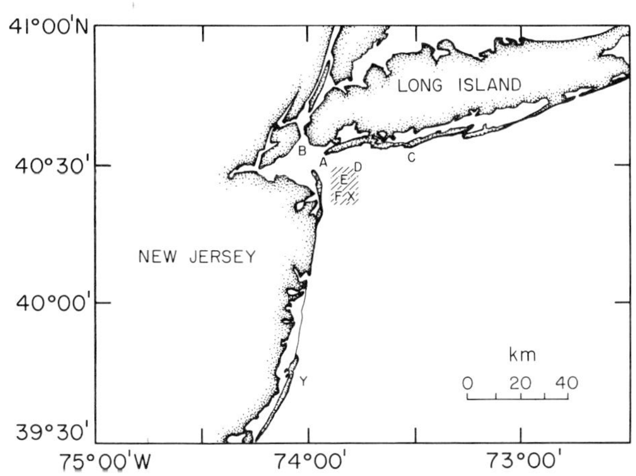

Figure 5.1. Location of mussel biomonitoring stations in the New York Bight. The dredged-material dumpsite is depicted by the shaded area. Transplantation and mussel collection sites are: A, Sandy Hook/Romar Shoals; B, Gravesend Bay; C, Jones Beach; D, ~3 km northeast of the dumpsite; E, center of the dumpsite; F, 1.5 km southwest of the dumpsite; X, sand-capping site; and Y, Barnegat Light.

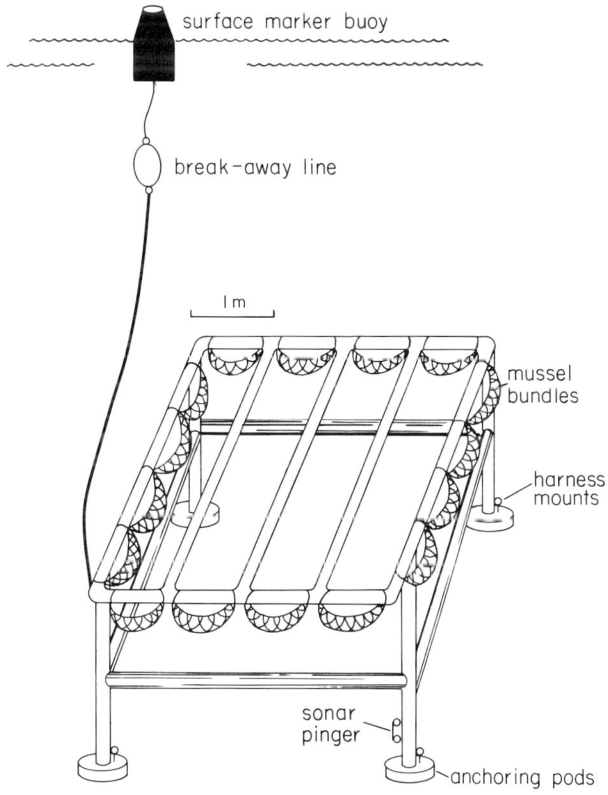

Figure 5.2. Mussel incubation platform.

attached to benthic incubation platforms (Fig. 5.2). Each of the platforms supported 12 bags and a sonar pinger (Model 55, Johnson Scientific Equipment Laboratories, Long Island, New York). The pinger aided in search and recovery operations when we used an acoustic location system (Model M30A5A, Dukane Corp., Chicago). The platforms with mussels were transferred to deployment sites in the New York Bight apex. The mussel incubation bags were suspended 1 m above the sediment.

Following deployment of platforms in late August 1980 and Janaury 1981, individual mussel bags were retrieved weekly for the first month (with the exception of station Y) and bimonthly thereafter until July 1981.

Regular sampling had to be terminated at several stations due to loss of the platforms. The platform at station A was destroyed shortly after deployment, and the station C platform was lost between weeks 4 and 10. Both losses were attributed to high current velocities or severe storm activity. The mussel platform deployed in the northwest quadrant of the dredged-material dumpsite (station E) was lost in March 1981. The platform at station X (sand capping site) was destroyed prior to the final scheduled bimonthly sampling in July 1981. The loss of platforms at stations E and X may have been due to unauthorized dumping activities.

5.2.1b. Sample handling

Bottom-water temperature, salinity, pH, and dissolved oxygen concentrations were determined concomitantly with collection of mussel and sediment samples. During retrieval and transport of the mussels, care was taken to avoid exogenous contamination and maintain viability. Following collection by a diver, each bag was carefully wrapped in aluminum foil and transported within 6 h (at 2–4°C) to the laboratory for pre-analytical measurement and preparation. Specimens were not depurated prior to further processing. The bags were opened, byssal threads were cut, and mussel survival was immediately determined. Live mussels were then randomly split into three subsamples of 25 animals each. This was performed in order to permit statistical comparison of metal residues in and between transplanted mussel populations. Individual mussels were then carefully cleaned of adhering epibiota, measured, weighed, and drained of extraneous water from the mantle cavity. The shells were discarded, and the pooled liquor and soft tissue from each subsample was weighed and placed into a clean glass container. Samples were stored at −20°C prior to chemcial digestion for analysis. Sterile surgical gloves were used by laboratory personnel when handling mussels, and all laboratory equipment was pre-cleaned with methylene chloride, acetone, and distilled water washes.

5.2.2. Laboratory Analyses

The composite sample (liquor and all soft tissues) was homogenized by blending and ultrasonic disruption. The homogenate was placed in glass jars and frozen.

Prior to Hg analysis, each thawed sample was rehomogenized as described above, after which an aliquot was removed and placed into a 300-ml biochemical oxygen demand (BOD) bottle. Twenty-five milliliters of a sulfuric–nitric acid mixture was added, and the bottle was placed in a water bath at 30°C until all solids were dissolved. Upon removal from the water bath, the sample was cooled to 4°C and potassium permanganate was added until the endpoint. The digested samples were allowed to stand overnight at 24°C, after which they were analyzed for total Hg in a Coleman MAS-50 Mercury Analyzer (Perkin-Elmer Corp., Elmwood Park, New Jersey), using the cold vapor technique (Hatch and Ott, 1968).

For analysis of Cd and Pb, 10 g of thawed tissue homogenate was placed into a beaker, acidified with 25 ml of 20% sulfuric acid, and dried in an oven at 110 ± 5°C. The dry samples were then placed into a cold, clean muffle furnace, heated at 270°C for 3 h, and maintained overnight at 450°C to reduce the samples to ash. The samples were then washed with nitric acid, evaporated to dryness, and held at 300°C in a muffle furnace for 30 min. After cooling and drying, the samples were resuspended in a warm aqueous nitric acid solution, quantitatively diluted, and an aliquot removed for analysis. Cadmium and Pb were analyzed using a Perkin-Elmer 303 atomic absorption spectrophotometer (Perkin-Elmer, Elmwood Park, New Jersey) according to the method of Grieg and McGrath (1977).

Detection limits for Hg, Cd, and Pb were 0.01 $\mu g\ g^{-1}$. All metal concentrations were expressed as $\mu g\ g^{-1}$ wet weight. Precision and recovery were determined by adding spikes of known concentrations to duplicate sample aliquots and by utilizing replicate determinations, blanks, and frequent intercalibration exercises with outside laboratories (U.S. National Bureau of Standards and New Jersey Institute of Technology). All recovery measurements were within 0.01 $\mu g\ g^{-1}$.

5.3. RESULTS

With the exception of stations C and F, survival was generally >70% for mussels deployed in August (Table 5.1). Throughout the incubation period mussels retrieved from station F exhibited high mortality. The negligible mortality observed for mussels deployed in January 1981 at stations X and Y may reflect reduced metabolism in response to winter temperatures. No correlation was observed between metal concentration and mussel survival.

Results of testing for Hg, Cd, and Pb in subsamples of source mussels (those organisms used in August and January deployments) are presented in Table 5.2. Concentrations of Hg and Cd were negligible. Lead concentrations averaged slightly above 1.0 $\mu g\ g^{-1}$.

Trend lines based on accumulation of Hg by *Mytilus edulis* over time indicated enhanced biological availability of Hg by mussels incubated in the vicinity of the dumpsite, with the highest concentrations observed among mussels at station E (Fig. 5.3). In comparison, mussels incubated in Gravesend Bay (station B) showed minimal increase in Hg concentration. The patterns for Pb (Fig. 5.4) and Cd (Fig. 5.5) are similar to that observed for Hg. After an initial elevation during the first month of incubation, a decrease in total

Table 5.1. Survival (%) of Transplanted Adult *Mytilus edulis*

Deployment Date	Station[a]	Week 1	2	3	4	10	19	29	36	45
August 1980	A	1[b]	—	—	—	—	—	—	—	—
	B	90	93	91	83	76	87	73	75	74
	C	54	50	90	80	—	—	—	—	—
	D	94	80	73	85	79	89	85	76	86
	E	76	73	84	93	86	72	92	—	—
	F	66	63	62	59	72	71	65	75	60
January 1981	X	100	99	100	98	100	—	—	—	—
	Y	—	—	—	99	100	97	—	—	—

[a] For station locations see Fig. 5.1.
[b] Dash denotes no samples taken.

Table 5.2. Metal Concentrations in Source Mussels prior to Deployment in August 1980 and January 1981[a]

Metal	Source Mussels[b] August 1980	January 1981
Hg	0.02 ± 0.01	0.03 ± 0.01
Pb	0.85 ± 0.10	0.82 ± 0.05
Cd	0.05 ± 0.01	0.05 ± 0.01

[a] Mean values from 25 mussels, three subsamples each, per deployment. Numbers given are in micrograms per gram ± standard deviation.
[b] Stock population obtained from 2 km east of Wildwood, New Jersey.

concentration was followed by a gradual increase (after 5 weeks of incubation). Maximum concentrations both of Pb and Cd in dumpsite mussels were 1.3 and 1.6 times the concentrations of Pb and Cd in source mussels, respectively.

Trend lines (Figs. 5.3, 5.4, and 5.5) indicate no appreciable increase in concentration of the three metals among mussels incubated at stations X and Y. Since the latter mussels were deployed in midwinter, however, these animals would have had reduced metabolism and feeding activity and therefore less opportunity for metal uptake via food.

Duncan's Multiple Range Test for mean concentrations of Hg, Pb, and Cd over the total exposure period are summarized in Tables 5.3, 5.4, and 5.5, respectively. Mercury concentrations in mussels for all stations near the dumpsite (stations D, E, and F) were significantly different ($p < 0.05$) from those for stations outside this area (Table 5.3). Moreover, Hg concentrations for station E mussels were significantly different from those of the two adjacent stations, D and F. The mean Hg concentration was lowest for the Wildwood (Cape May) stock mussel population.

Duncan's Multiple Range Test for Pb (Table 5.4) showed that although stations D, E, and F all had higher mean levels than did stations outside of the dumpsite area, mussel concentrations for station E (dumpsite) were not statistically different from those for stations C and X.

Results for Cd (Table 5.5) were similar to those described for Hg. The dumpsite stations exhibited the highest mean concentration, with station E being significantly greater than all other stations ($p < 0.05$). Interestingly, mussels incubated at station B (Gravesend Bay) had somewhat higher Cd concentrations as compared to other metals.

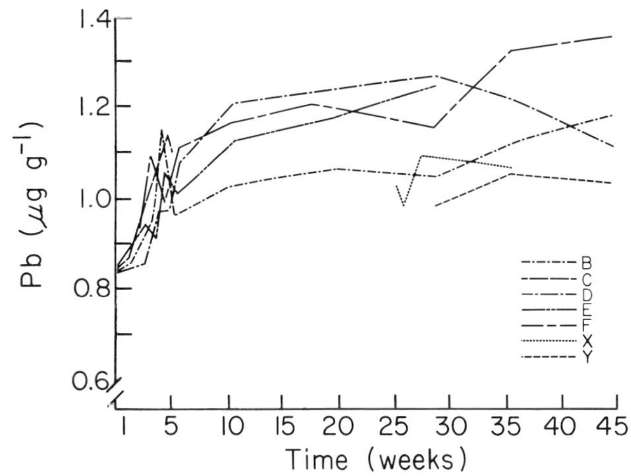

Figure 5.4. Lead accumulation over time for transplanted *Mytilus edulis*. See Fig. 5.3 legend for station location definitions.

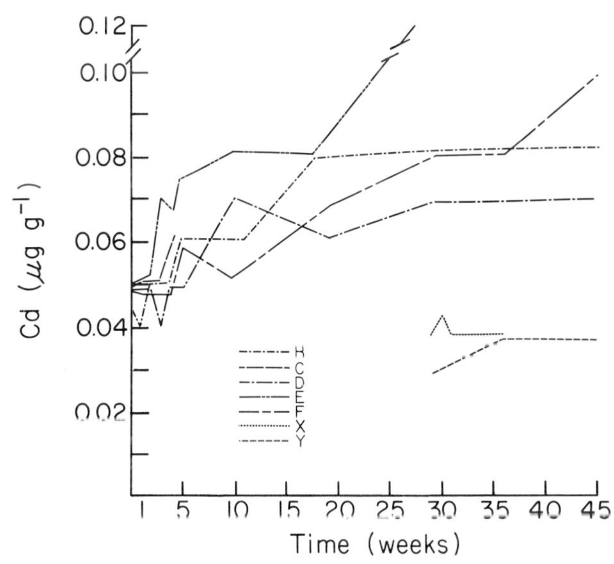

Figure 5.5. Cadmium accumulation over time for transplanted *Mytilus edulis*. See Fig. 5.3 legend for station location definitions.

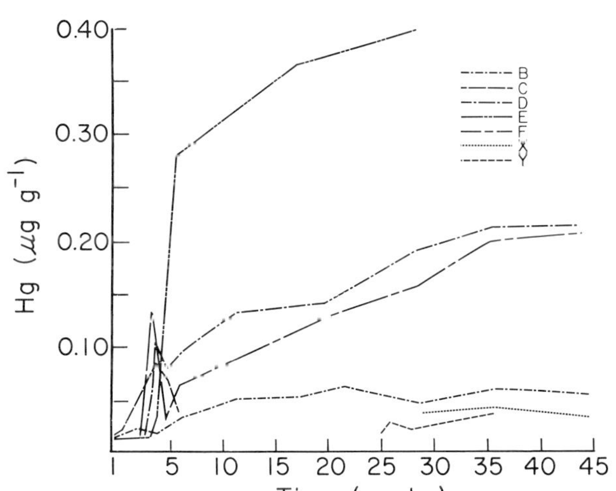

Figure 5.3. Mercury accumulation over time for *Mytilus edulis* transplanted to: B, Gravesend Bay; C, Jones Beach; D, ~3 km northeast of the dumpsite; E, center of the dumpsite; F, 1.5 km southwest of the dumpsite; X, sand-capping site; and Y, Barnegat Light.

5.4. DISCUSSION

In the present study, net increases in concentrations of Hg, and to a lesser degree Pb and Cd, were observed for mussels transplanted at five New York Bight apex stations in August

Table 5.3. Ranking of Hg Concentration in *Mytilus edulis* by Station[a]

Station	E	>	D	>	F	>	B	>	C	>	Y	>	X	>	SM[b]
Mean Concentration ($\mu g\ g^{-1}$)[c]	0.35		0.15		0.12		0.05		0.04		0.04		0.03		0.03
	Duncan's Multiple Range Test[d]														
Station	E		D		F		B		C		Y		X		SM
							———	———	———	———	———				
									———	———	———	———	———	———	———

[a] For station locations see Fig. 5.3.
[b] Stock population (SM) obtained from 2 km east of Wildwood, New Jersey.
[c] Rounded to the nearest significant figure.
[d] Duncan's Multiple Range Tests were performed on data after analysis of variance (ANOVA) detected significant differences at $\alpha < 0.05$ (the mean values used were calculated from the sixth week of exposure to the conclusion of sampling). The mean concentrations of underlined stations are not statistically different at $\alpha = 0.05$.

Table 5.4. Ranking of Pb Concentration in *Mytilus edulis* by Station[a]

Station	F	>	D	>	E	>	B	>	C	>	X	>	Y	>	SM[b]
Mean Concentration ($\mu g\ g^{-1}$)[c]	1.21		1.20		1.16		1.06		1.06		1.06		1.01		0.84
	Duncan's Multiple Range Test[d]														
Station	F		D		E		B		C		X		Y		SM
	———	———	———	———	———		———	———	———	———	———				

[a] For station locations see Fig. 5.3.
[b] Stock population (SM) obtained from 2 km east of Wildwood, New Jersey.
[c] Rounded to the nearest significant figure.
[d] Duncan's Multiple Range Tests were performed on data after ANOVA detected significant differences at $\alpha < 0.05$ (the mean values used were calculated from the sixth week of exposure to the conclusion of sampling). The mean concentrations of underlined stations are not statistically different at $\alpha = 0.05$.

Table 5.5. Ranking of Cd Concentration in *Mytilus edulis* by Station[a]

Station	E	>	B	>	F	>	D	>	C	>	SM[b]	>	X	>	Y
Mean concentration ($\mu g\ g^{-1}$)[c]	0.09		0.07		0.07		0.06		0.06		0.05		0.04		0.04
	Duncan's Multiple Range Test[d]														
Station	E		B		F		D		C		SM		X		Y
			———	———	———	———	———								
							———	———	———						
									———	———	———				
													———	———	———

[a] For station locations see Fig. 5.1.
[b] Stock population (SM) obtained from 2 km east of Wildwood, New Jersey.
[c] Rounded to the nearest signficant figure.
[d] Duncan's Multiple Range Tests were performed on data after ANOVA detected significant differences at $\alpha < 0.05$ (the mean values used were calculated from the sixth week of exposure to the conclusion of sampling). The mean concentrations of underlined stations are not statistically different at $\alpha = 0.05$.

1980. Mean concentrations of Hg and Cd were significantly greater for mussels transplanted in the dredged-material dumpsite. Bioaccumulation is clearly apparent for Hg; the mussels incubated in the dumpsite (station E) had the highest mean Hg concentrations. In no case was the FDA guideline of 1.0 $\mu g\ g^{-1}$ (wet wt) exceeded. Prior investigations have demonstrated that bivalves can concentrate Hg several orders of magnitude in excess of ambient water concentration (O'Connor and Rachlin, 1982).

Previous Mussel Watch investigations have reported generalized low-level Pb and Cd contamination among natural mussel populations of the east coast of the United States (Farrington et al., 1983). The small increase in Pb and Cd concentrations observed in the present study may in part reflect long-term exposure of mussels to the metal contaminated waters of the New York Bight and not solely metal concentrations in dredged material. Lead, in particular, is known to enter the coastal marine environment from a number of sources. Conclusions as to the specific effect of dredged-material disposal on mussel body burden of Pb (as well as other metals) requires knowledge of metal availability and biogeochemistry, both of which are particularly variable in urban coastal waters (Farrington et al., 1983).

The observed fluctuations in metal concentrations during the monitoring period also indicated temporal variation in metal uptake and depuration by the transplanted mussels. The oscillating pattern of metal body burdens during the first five weeks of incubation represents an equilibration phenomenon previously reported in similar bioaccumulation studies involving trace metals (U.S. NAS, 1980). Farrington et al. (1983) noted significant temporal variation in Cd and Pb among natural mussel populations of Narragansett Bay, Massachusetts. Similar seasonality with respect to metal concentration has been reported by Boyden (1977) and Karbe (1977). Phillips (1976) suggested that seasonal changes in the total body burdens of bivalves are minimal and that presumed changes in concentration are actually reciprocal to variations in body weight. This would lead to seasonally maximal concentrations in the period immediately following spawning when body weight is reduced (spring in temperate waters). Moreover, spring coincides with the period of greatest urban runoff, which would increase metal availability (Phillips, 1976). Although mussel weight measured in this study did not statistically correlate with metal body burden, the highest metal concentrations occurred in late winter and early spring. Unfortunately, the loss of station E prevented further evaluation of this pattern.

No significant differences were noted between mussel body burdens and bottom salinity and temperature. Since these parameters were only measured in conjunction with mussel collections, more frequent data are needed to ascertain metal concentrations compared with these parameters. In that regard, Phillips (1977, 1978) has demonstrated that both salinity and temperature affect metal uptake in mussels.

The question of mussel depuration immediately following capture must also be addressed (U.S. NAS, 1980). Contractual obligations relative to the use of these same mussel samples for volatile organic pollutant analyses prevented depuration of the sample. It is possible that a portion of the observed Pb and Cd values for whole soft tissue represent those metals passing through the gut. The absence of concurrent metal analyses for sediments at each station prevents the normalization of mussel metal concentrations with sediment values (U.S. NAS, 1980). Application of the relative gut contribution of 11 and 20%, respectively, for Cd and Pb [estimated previously in mussels by Stephenson et al. (1982)] to our data suggests that the slight elevations may not represent actual bioaccumulation but rather metal flux through the organism. Further research is needed to clearly relate metal dynamics through the bivalve digestive tract to organism uptake.

In December 1980, after several delays, the U.S. Army Corps of Engineers completed sand-capping operations within the southeastern quadrant of the dredged-material dumpsite (Koepp et al., 1982). Due to these delays, previously scheduled deployment of new mussel arrays did not occur until January 1981. As a result of insufficient sampling opportunities and the loss of the platform at station X in May 1981, conclusions as to the efficacy of capping to contain dredged material and thereby inhibit metal uptake by mussels cannot be made. It is of interest that mussel survival was greater for these organisms transplanted in winter than for mussels deployed in summer. Perhaps future biomonitoring efforts using transplanted mussels would benefit from winter deployments to minimize transplantation stress.

5.5. SUMMARY AND CONCLUSIONS

Over the study period, Hg, Cd, and Pb were consistently detected in mussel samples. At no time were Hg concentrations encountered in excess of FDA 1.0 $\mu g\ g^{-1}$ (wet wt) guideline. Although a direct cause-and-effect relationship cannot be rigorously proven, it is probable that the dredged-material dumpsite is a signficant contributor of Hg and, to a lesser degree, Cd and Pb in transplanted mussels. The negligible increase in mussel Cd and Pb concentrations appears to reflect the ubiquitous contamination characteristic of the urban coastal zone. Factors such as temporal variation, temperature, salinity, and mussel condition undoubtedly affect mussel body burdens of these metals.

ACKNOWLEDGMENTS

This research was funded by Contract No. DACW51-80-R-0001 through the U.S. Army Corps of Engineers, New York District. The authors thank the National Marine Fisheries Service for support vessels and divers. We also thank T. Fikslin and D. Fogg for their technical assistance with the manuscript.

REFERENCES

Boyden, C. R. 1977. Effect of size upon metal content of shellfish. *Journal of the Marine Biological Association of the United Kingdom*, **57**, 674–714.

Butler, P.A. 1973. Residues in fish, wildlife and estuaries. *Pesticides Monitoring Journal*, **6**, 283–362.

Carmody, D. J., J. B. Pearce, and W. E. Yasso. 1973. Trace metals in sediments of the New York Bight. *Marine Pollution Bulletin*, **4**, 132–135.

Carriker, M. R., J. W. Anderson, W. P. Davis, D. R. Franz, G. F. Mayer, J. B. Pearce, T. K. Sawyer, J. H. Tietjen, J. F. Timoney, and D. R. Young. 1982. Effects of pollutants on benthos. *In:* Ecological Stress and the New York Bight: Science and Management, G. F. Mayer (Ed.). Estuarine Research Federation, Columbia, South Carolina, pp. 3–21.

Farrington, J. W., E. D. Goldberg, R. W. Risebrough, J. H. Martin, and V. T. Bowen. 1983. U.S. "Mussel Watch" 1976–1978: An overview of the trace-metal, DDE, PCB, hydrocarbon and artificial radionuclide data. *Environmental Science and Technology*, **17**, 490–496.

Feng, S. Y., R. V. Grillo, R. Arimoto, J. K. Watson, E. M. Haddad, and J. R. Ouellette. 1979. Heavy metal concentrations and gonadal development in *Mytilus edulis* and *Modiolus modiolus*. DAMOS Report, Naval Underwater Systems Center, Newport, Rhode Island, 27 pp.

Goldberg, E. D. 1975. The mussel watch—a first step in global marine monitoring. *Marine Pollution Bulletin*, **6**, 111.

Goldberg, E. D., V. T. Bowen, J. W. Farrington, G. R. Harvey, J. H. Martin, P. L. Parker, R. W. Risebrough, W. Robertson, E. Schneider, and E. Gamble. 1978. The Mussel Watch. *Environmental Conservation*, **5**, 101–125.

Greig, R. A., and R. McGrath. 1977. Trace metals in sediments of Raritan Bay. *Marine Pollution Bulletin*, **8**, 188–192.

Hatch, W. R., and W. L. Ott. 1968. Determination of sub-microgram quantities of mercury by atomic absorption spectrophotometry. *Analytical Chemistry*, **40**, 2085–2087.

Holden, A. V. 1973. International cooperative study of organochlorine and mercury residues in wildlife. *Pesticides Monitoring Journal*, **7**, 37–52.

Karbe, L. 1977. Trace elements in mussels (*Mytilus edulis*) from coastal areas of the North Sea and Baltic. Multielement analysis using instrumental analysis. *Journal of Radioanalytical Chemistry*, **37**, 920–927.

Koepp, S. J., E. D. Santoro, R. Zimmer, J. Nadeau, and L. Ciaccio. 1982. Bioaccumulation Monitoring in the New York Bight Using the Blue Mussel (*Mytilus edulis*). Unpublished report, submitted to the U.S. Army Corps of Engineers, New York District. New Jersey Marine Sciences Consortium, Fort Handcock, New Jersey, 42 pp.

Mahoney, J. B., F. H. Midlige, and D. G. Deuel. 1973. A fin rot disease of marine and euryhaline fishes in the New York Bight. *Transactions of the American Fisheries Society*, **102**, 596–605.

Mueller, J. A., J. S. Jeris, A. R. Anderson, and C. Hughes. 1976. Contaminant Inputs to the New York Bight. Technical Memorandum ERL MESA-6, Marine Ecosystems Analysis Program, U.S. National Oceanic and Atmospheric Administration, Boulder, Colorado, 347 pp.

O'Connor, J. M., and J. W. Rachlin. 1982. Perspectives on metals in New York Bight organisms: factors controlling accumulation and body burdens. *In:* Ecological Stress and the New York Bight: Science and Mangement, G. F. Mayer (Ed.). Estuarine Research Federation, Columbia, South Carolina, pp. 655–673.

Pearce, J. 1980. Status of Estuarine Waters between Cape Hatteras and Maine: A Review. International Council for Exploration of the Sea, Charlottenlund, Denmark, C. M. 1980, E-56, 16 pp.

Phillips, D. J. H. 1976. The common mussel *Mytilus edulis* as an indicator of pollution by zinc, cadmium, lead and copper. II. Relationship of metals in the mussel to those discharged by industry. *Marine Biology*, **38**, 71–80.

Phillips, D. J. H. 1977. Effects of salinity on the net uptake of zinc by the common mussel *Mytilus edulis*. *Marine Biology*, **41**, 79–88.

Phillips, D. J. H. 1978. The common mussel *Mytilus edulis* as an indicator of trace metals in Scandinavian waters. II. Lead, iron and manganese. *Marine Biology*, **46**, 147–156.

Risebrough, R. W., B. W. de Lappe, E. F. Letterman, J. L. Lane, M. Firestone-Gills, A. M. Springer, and W. Walder, II. 1980. California State Mussel Watch, Vol. III: Organic Pollutants in Mussels, *Mytilus californianus* and *M. edulis*, along the California Coast. California State Water Resources Control Board, Sacramento, California, 108 pp.

Steimle, F., J. Caracciolo, and J. B. Pearce. 1982. Impacts of dumping on New York Bight apex benthos. *In:* Ecological Stress and the New York Bight: Science and Management, G. F. Mayer (Ed.). Estuarine Research Federation, Columbia, South Carolina, pp. 213–223.

Stephenson, M. D., M. Martin, S. E. Lange, A. R. Flegal, and J. M. Martin. 1982. California Mussel Watch 1980–1981, Vol. II: Trace Metal Concentrations in the California Mussel, *Mytilus californianus*. California State Water Resources Control Board, Sacramento, California, 110 pp.

Talbot, V., R. J. Magoe, and M. Hussain. 1976. Lead in Port Phillip Bay mussels. *Marine Pollution Bulletin*, **7**, 91–94.

U.S. National Academy of Sciences. 1980. The International Mussel Watch. National Academy of Sciences National Research Council, Washington, D. C., 248 pp.

PART III: ORGANIC CONTAMINANTS IN THE MARINE ENVIRONMENT

Chapter 6

Oxygenated Metabolites of DDT and PCBs in Marine Sediments and Organisms

David A. Brown, Richard W. Gossett, and Sophia R. McHugh

Southern California Coastal Water Research Project
Long Beach, California

ABSTRACT

Microorganisms in sediments and many fish and invertebrates can rapidly transform some xenobiotic (synthetic organic) compounds into oxygenated metabolites. These oxygenated metabolites may comprise the major form of organic compound contamination in marine sediments and biota. In marine sediments from the highly contaminated Palos Verdes Shelf in coastal southern California, the oxygenated metabolites [2,2-*bis*(*p*-chlorophenyl) acetic acid (DDA) and 2,2-*bis*(*p*-chlorophenyl) ethanol (DDOH)] comprised 80% of the total of all measured forms of 1,1,1-trichloro-2,2-*bis*(chlorophenyl) ethane (DDT). In addition, these oxygenated metabolites represented 99% of all measured forms of DDT in mussel soft parts, 85% in shrimp muscle, 48% in croaker liver, and 49% in scorpion fish liver. Similar ratios of polychlorinated biphenyl (PCB) parent compounds and their oxygenated metabolites (PCBols) were found in mussel soft parts. Since some of these compounds (for example, DDT) are known to be metabolized very slowly or not at all by mussels and fish, it seems likely that microbial transformations in marine sediments are serving as the main source of these metabolites. There is some evidence to suggest that oxygenated metabolites may not necessarily represent detoxified forms of xenobiotic compounds. Studies of tissue extracts separated into fractions according to molecular weight indicate that metabolites in organisms from relatively uncontaminated control stations are partitioned into a low-molecular-weight glutathione-containing pool. However, in fish from Palos Verdes with fin erosion and other disorders, there is a significant portion of these metabolites bound to both a medium-molecular-weight metallothionein-containing pool and, to a lesser extent, to a high-molecular-weight enzyme-containing pool. Since enzymes are affected and metal binding is decreased in these pools, this may represent toxification from oxygenated metabolites.

6.1. INTRODUCTION

There have been many laboratory studies showing that the rate of conversion of some parent organic contaminants to oxygenated metabolites can be very rapid in marine sediments, invertebrates, and fish (Malins, 1977; Varanasi and Malins, 1977; Lee et al., 1982). Until recently (Brown et al., 1982a), however, there had been no published reports quantifying the portion of organic contaminants present as oxygenated metabolites rather than as parent compounds in samples collected from the marine environment. When this determination was done with croaker (*Genyonemus lineatus*) liver, it was found that the oxygenated metabolites DDA and DDOH accounted for up to 95% of all measured forms of DDT, including the parent compound DDT, its primary derivatives, 1,1,-dichloro-2,2-*bis*(chlorophenyl) ethylene (DDE) and 2,2-*bis*(chlorophenyl)-1,1-dichloroethane (DDD), and their oxygenated metabolites, DDA and DDOH (Brown et al., 1982a). Results from these studies suggest that oxygenated metabolites may represent the major portion of contamination of the marine environment by some organic compounds.

It appears that oxygenated metabolites, rather than parent compounds, are bound to macromolecular sites of toxic action in exposed organisms (Brown et al., 1982a). This higher reactivity of metabolized compounds may be due to the formation of highly reactive epoxide intermediates (Brown et al., 1982a; Jenkins and Brown, 1985). Since oxygenated metabolites may represent both the major portion of contamination of the marine environment by some organic compounds and the form responsible for some of the effects of these compounds, it is important to determine both the source and the degree of contamination.

There is much evidence to support the contention that the conversion of organic contaminants to oxygenated metabolites can occur as a result of both photochemical reactions in the water column and microbial transformations in sediments (Lee et al., 1978, 1982; Hinga et al., 1980; Lee and Ryan, 1983). In experiments using the Marine Ecosystem Research Laboratory (MERL) tanks, Lee et al. (1982) found considerable photo-oxidation in the water column (maximum depth of 5.5 m) of benz(*a*)anthracene, dimethyl-benzanthracene, and pentachlorophenol to more polar hydroxylated metabolites. This conversion was responsible for ~80% of the removal of dimethyl-benzanthracene and pentachlorophenol from the water column. The metabolites and remaining parent compounds [benz(*a*)anthracene and dimethylbenzanthracene] eventually attached to living or nonliving particulate matter in the water column and precipitated to the sediment. Once in contact with the sediment, microbial processes were the major mechanism of further transformation of parent compounds and metabolites and of the compounds' eventual degredation to CO_2. As a consequence of these processes, after 3 d, over 90% of the total dimethylbenzanthracene in sediments was present as metabolites, whereas 20% of the total of benz(*a*)anthracene was present as metabolites in surficial sediments, and over 50% was present as metabolites at a depth of 3 cm in sediment cores. Even with this slower rate of conversion for benz(*a*)anthracene, we estimate that all of the parent compounds would be degraded to CO_2 in 2 y. In a subsequent study, Lee and Ryan (1983) found that the processes of degradation were faster in sediment samples taken from petroleum-contaminated areas and in sediment samples taken during the summer, presumably because of a higher abundance of microbes.

In earlier studies, Lee and Ryan (1978) found that microbial degradation of organochlorine compounds to CO_2 occurred in sediments from a relatively uncontaminated saline (17–24‰) river near Savannah, Georgia, but degradation in water was slow or absent, except for the compound *p*-chlorophenol. In addition, they found that in sediments, hexachlorobiphenyl and hexachlorobenzene were not degraded, but *p*-chlorophenol, trichlorophenol, trichlorophenoxy-acetic acid, chlorobenzene, hexachlorophene, and DDE were degraded, with half-lives of 3 d, 23 d, 35 d, 75 d, 290 d, and 1100 d, respectively.

There is also considerable evidence that at least some organic contaminants can be rapidly converted to oxygenated metabolites in fish. Varanasi et al. (1979) found that 168 h after force-feeding naphthalene to starry flounders (*Platichthys stellatus*), 73, 81, and 98% of the recoverable administered dose was present as metabolites in muscle, liver, and bile tissues, respectively. Roubal et al. (1977) found that the proportions of injected benzene and anthracene present as metabolites were similar to those for naphthalene in these same tissues of coho salmon (*Oncorhynchus kisutch*). Peterson and Guiney (1979) found that up to 82% PCBs present in whole green sunfish (*Lepomis cyanellus*) 7 d after termination of exposure were present as polar metabolites; however, the rate of metabolism of PCBs was inversely related to the degree of chlorination. Similarly, Metcalf et al. (1975) found that trichlorobiphenyl is degraded in fish much more rapidly than is tetrachlorobiphenyl, but that pentachlorobiphenyl and DDE are persistent. These results are in agreement with the review by Franklin et al. (1980), who stated that phenobarbital-like inducers (for example, DDT) do not elevate cytochrome P-450 levels in fish, whereas methylcholanthrene-like inducers (for example, petroleum aromatic hydrocarbons) and PCBs do. Cytochrome P-450 is a component of the mixed-function oxygenase system and catalyzes the conversion of parent compounds to oxygenated metabolites.

Many invertebrates have also demonstrated the ability to convert some parent organic compounds into oxygenated metabolites. Lee (1975) found that copepods, euphausiids, amphipods, and crabs could convert a large portion of benzo(*a*)pyrene, methylcholanthrene, and naphthalene accumulated from surrounding water to oxygenated metabolites within 48 h of exposure. Similar results were found by Sanborn and Malins (1977) in shrimp exposed to naphthalene. Metcalf et al. (1975) reported that in invertebrates, as in fish, there is some degradation of the lesser chlorinated PCBs to metabolites, but much less conversion of the more chlorinated PCBs and DDE.

In this chapter we present results showing the relative amounts of parent compounds and oxygenated metabolites for several organic contaminants in sediments, invertebrates, and fish both from relatively clean and from contaminated sites in coastal southern California. We also present evidence that metabolites can affect macromolecular sites of toxic action in exposed organisms.

6.2. METHODS

6.2.1. Stations

The Palos Verdes station (PV 7–3) was >1 km from the Los Angeles County municipal wastewater discharge zone (Fig. 6.1). Because of its proximity to the discharge zone, PV 7–3 was considered a contaminated area. The Redondo Beach station was considered less contaminated than was PV 7–3 and Catalina Island, off the coast of southern California. Dana Point and Point Dume were considered relatively uncontaminated stations.

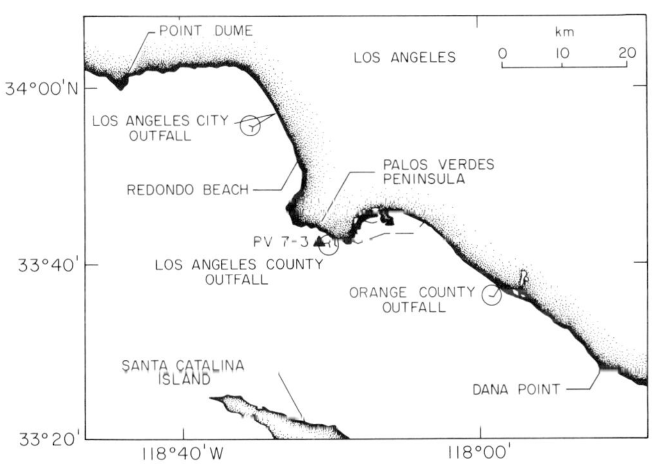

Figure 6.1. Study area showing major municipal wastewater outfalls, sampling station PV 7–3, and landmarks near the other sampling locations.

6.2.2. Sampling and Sample Analyses

Table 6.1 presents sample-collection information. All samples were collected in clean glass jars (kilned at 705°C) with lids lined with aluminum foil. Prior to analysis, tissue samples were stored at −80°C and effluent samples were stored at 4°C (Oshida, 1982).

Cytosol was separated from tissue by homogenizing the tissue in a Sorval Omnimix homogenizer (E. I. Du Pont de Nemours and Co., Inc., Wilmington, Delaware) at 4°C for 30 s at high speed, followed by centrifuging the homogenate at 105,000x*g* for 1 h. Supernatant was applied to a Pharmacia 1.6-cm column (Pharmacia, Uppsala, Sweden) packed with Sephadex G-75 (Pharmacia, Uppsala, Sweden) to separate it into cytosolic pools, including the high-molecular-weight enzyme-containing (ENZ) pool, the medium-molecular-weight metallothionein-containing (MT) pool, and the low-molecular-weight glutathione-containing (GSH) pool. The position of each pool was determined by measuring Cd, Cu, and Zn concentrations in each fraction of column eluate (Brown et al., 1977; Brown and Parsons, 1978; Jenkins et al., 1982). Fractions comprising each pool were then combined for analysis of oxygenated metabolites (Brown et al., 1982a).

For analysis of parent organic compounds and metabolites in whole tissue, 5 g of tissue was homogenized in 20 ml of deionized water and the blade was rinsed two times with 20 ml of deionized water. Sodium hydroxide (1.2 g) was added to the homogenate to produce a 2% NaOH solution. When cytosolic pools were analyzed for oxygenated metabolites, NaOH was added to each mixture to produce a 2% solution. Cytosolic pool solutions and whole tissue solutions were each extracted three times with 50 ml of hexane–acetone (1:1, by volume) using centrifugation to separate layers completely. In whole tissue the hexane (top) layer was taken as the base-neutral-extractable fraction and analyzed for parent organic compounds. In the case of cytosolic pools, the hexane layer was discarded, because parent organic compounds were always determined in total tissue. The remaining aqueous phase was heated to 90°C for 30 min to hydrolyze possible conjugates (Gingell and Wallcave, 1974; Gold et al., 1981; Brown et al., 1982a). The solution was allowed to cool, adjusted to pH 1 with 6 *N* HCl, extracted three times with 50 ml of methylene chloride, and centrifuged to separate layers, if necessary. The methylene chloride (bottom) layer was rotovaporized to dryness. Dried samples were methylated with a solution of 5 mg of 3-methyl-1-*p*-tolytriazene and 1 ml of diethyl ether by swirling in a round-bottomed flask; the samples were then dried again under a stream of nitrogen. The sample was redissolved in 100 ml of methanol and retained as the acid-extractable fraction for analysis of oxygenated metabolites (Brown et al., 1982a). Parent organic compounds and oxygenated metabolites were analyzed by

Table 6.1. Stations and Collection Methods for Organisms, Sediment, and Effluent Samples

Sample	Collection Date	Station	Method of Collection
Shrimp[a] *(Sicyonia ingentis)*	October 1982	PV 7–3	Otter trawl
Scorpion fish[a] *(Scorpaena guttata)*	October 1982 March 1983	PV 7–3 Catalina Island[b]	Otter trawl
Croakers *(Genyonemus lineatus)*	September 1981	PV 7–3 Dana Point[b]	Otter trawl
Dover sole[a,c] *(Microstomus pacificus)*	March 1982	PV 7–3	Otter trawl
Mussels *(Mytilus californianus)*	September 1981	Intertial zones adjacent to PV 7–3	By hand
Sediment[d]	October 1982	PV 7–3	Van Veen grab
Effluent composite[e]	November 1980 August 1981	Los Angeles County wastewater treatment plant	By hand

[a] Sample collected at a depth of 60 m.
[b] Relatively uncontaminated site.
[c] Dover sole with and without fin erosion.
[d] Top 2 cm of sediment collected.
[e] Results not reported in this chapter.

using gas chromatographs equipped with packed and capillary columns and both electron-capture and flame-ionization detectors (Brown et al., 1982a; Gossett et al., 1983). Metabolites have previously been confirmed through the use of a gas chromatograph–mass spectrometer equipped with a capillary column (Brown et al.,1982a). Analysis of parent organic compounds and metabolites in sediments was accomplished by the methods described by Gossett et al. (1984).

6.3. RESULTS

6.3.1. Sediments, Shrimp, and Fish

In sediments from station PV 7–3, the oxygenated metabolites DDA and DDOH accounted for 80% of the total measured forms of DDT, including the parent compound DDT, its primary derivatives DDE and DDD, and their oxygenated metabolites DDA and DDOH (Table 6.2). In shrimp muscle and scorpion fish liver from this station, the oxygenated metabolites of DDT accounted for 85 and 49%, respectively of all measured forms of DDT (Table 6.2). When livers of scorpion fish from near Catalina Island were analyzed, it was found that concentrations of DDT, DDE, DDD, and their oxygenated metabolites were lower, relative to Palos Verdes fish livers. However, oxygenated metabolites accounted for a higher proportion (77%) of the total of all measured forms of DDT (Table 6.2). In croaker livers from relatively uncontaminated Dana Point, oxygenated metabolites accounted for 96% of the total of all measured forms (Table 6.2).

6.3.2. Mussels

In mussel whole soft parts, the oxygenated metabolites of both DDT and PCB accounted for 90% or more of the total of all measured forms of these contaminants, whether taken from relatively uncontaminated Point Dume or from more contaminated Redondo Beach and Palos Verdes (Table 6.3). Both DDT parent compounds and metabolites were highest in mussels from Palos Verdes compared to those from Redondo Beach and Point Dume (Table 6.3). The PCB parent compounds were highest in mussels from Redondo Beach, although metabolites (PCBols) were highest in mussels from Palos Verdes.

6.3.3. Dover Sole

When the cytosolic distribution of DDA and DDOH were determined in livers of Dover sole with and without fin erosion from Palos Verdes, the following results were obtained: in Dover sole with fin erosion there was a 2.6-fold higher concentration of these metabolites in cytosol relative to controls without fin erosion; there was a 2-fold higher concen-

Table 6.2. Mean Concentrations (mg kg^{-1}, dry wt) of DDT, Its Derivatives, and Oxygenated Metabolites in Organisms and Sediment[a]

Station[b]	Sample	Number of Samples Analyzed	DDT and Its Derivatives (DDE, DDD)	Oxygenated Metabolites (DDA, DDOH)
Dana Point	Croaker liver	2	0.49 ± 0.22 (4%)	9.2 ± 0.8 (96%)
Catalina Island	Scorpion fish liver	5	1.73 ± 0.12 (23%)	4.7 ± 3.4 (77%)
Palos Verdes	Croaker liver	2	28.0 ± 10.9 (52%)	25.3 ± 0.1 (48%)
	Scorpion fish liver	5	24.9 ± 6.6 (51%)	24.3 ± 8.8 (49%)
	Shrimp muscle	5	0.87 ± 0.38 (15%)	4.83 ± 1.70 (85%)
	[illegible]	5	3.6 ± 2.7 (20%)	25.9 ± 14.3 (80%)

[illegible] parentheses is the percent of total dry wt.

[illegible] see [illegible]

[illegible]centrations (mg kg^{-1}, wet wt) of DDT and PCB Compounds and Oxygenated Metabolites in Whole Mussels[a]

	DDTs		PCBs	
Station[b]	Compounds DDT, DDE, DDD	Oxygenated Metabolites DDA, DDOH	Compounds Aroclors 1242, 1254	Oxygenated Metabolites PCBols[c]
Point Dume	0.015 (0.3%)	4.30 (99.7%)	0.005 (1.5%)	0.036 (98.5%)
Redondo Beach	0.035 (0.2%)	17.3 (99.8%)	0.040 (0.6%)	0.378 (90.4%)
Palos Verdes	0.080 (0.3%)	27.2 (99.7%)	0.018 (4.1%)	0.426 (95.9%)

[a] Soft parts only. Composite of 10 organisms.
[b] For station locations see Fig. 6.1.
[c] PCBols include compounds with chlorine atoms at 3,4, 5′ + 2′, 3,3, 4, 3, 4, and 5,5′.

tration of metabolites in the GSH pool relative to controls; there was a 25-fold higher concentration of metabolites in the MT pool relative to controls; and there was a 2.2-fold higher concentration of metabolites in the ENZ pool relative to controls (Fig. 6.2a).

In cytosolic pools of these same Dover sole, Zn was slightly higher in the ENZ pool of sole with fin erosion and more than double in the MT pool relative to controls; Cd and Cu were lower by one-half and one-third, respectively, in the ENZ pool of sole with fin erosion relative to controls; Cu and Cd were both one-third lower in the MT pool of sole with fin erosion relative to controls; and Cu was much higher in the GSH pool of sole with fin erosion (Fig. 6.2b).

6.3.4. Octanol-to-Water Partition Coefficients

The log octanol-to-water partition coefficients (log K_{ow}) (Gossett et al., 1983) of some contaminants in Los Angeles municipal wastewater effluent are given in Table 6.4; these range from 1.5 for dichloroethane to 6.0 for PCBs. Concentrations of contaminants in this effluent ranged from 0.90 μg liter^{-1} for total DDT (DDT, DDE, and DDD) to 210 μg liter^{-1} for toluene. In sediments collected from the immediate discharge zone of this outfall, contaminant concentrations ranged from <0.0005 mg kg^{-1} (wet wt) for vinylchloride and dichloroethane to 25.9 mg kg^{-1} (wet wt) for DDA. In

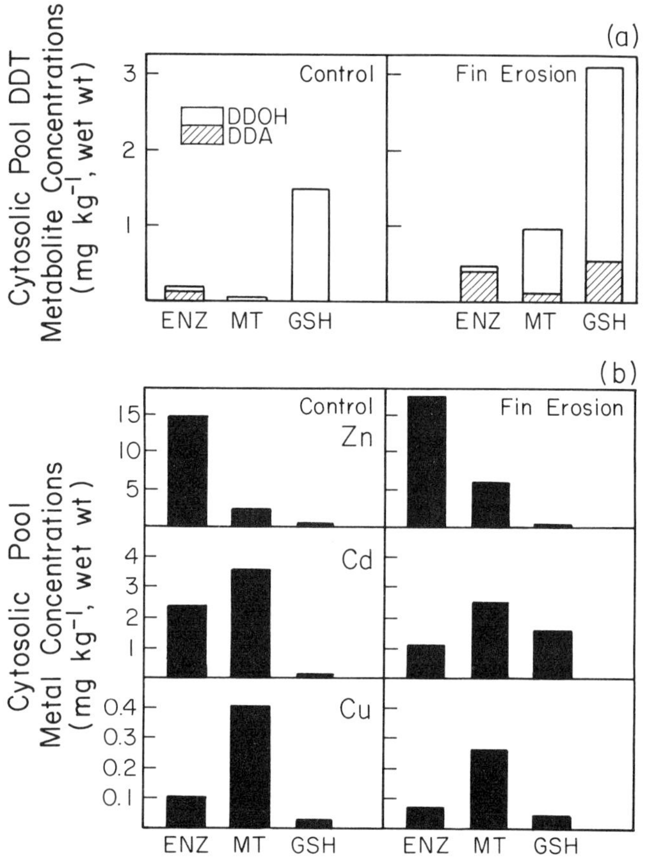

Figure 6.2. Concentrations of (a) DDT oxygenated metabolites (DDA and DDOH) and (b) Zn, Cd, and Cu in the enzyme (ENZ), metallothionein-containing (MT), and glutathione-containing (GSH) pools in liver cytosol of Palos Verdes Dover sole controls and Dover sole with fin erosion (composites of ten samples analysed).

livers of scorpion fish collected from this same area, contaminant concentrations ranged from <0.0003 mg kg^{-1} (wet wt) for vinylchloride and dichloroethane to 24.9 mg kg^{-1} (wet wt) for total DDT.

6.4. DISCUSSION AND CONCLUSIONS

Results from this study indicate that oxygenated metabolites may comprise the major form of contamination of the marine environment by some xenobiotic compounds. Previous investigators have reported on the presence of these metabolites from laboratory studies but not from samples collected from the marine environment; however, there is no reason to suppose that the results might differ. Laboratory studies using radiolabeled organic compounds have shown that they can be readily metabolized to oxygenated forms in sediments (Lee and Ryan, 1978, 1983; Lee et al., 1982), invertebrates (Lee, 1975; Metcalf et al., 1975; Sanborn and Malins, 1977), and fish (Metcalf et al., 1975; Roubal et al., 1977; Varanasi et al., 1979). In most of these studies, most of the recoverable dose was present as oxygenated metabolites several days after initial exposure to the parent organic compound.

In the present study, we found high concentrations of oxygenated DDT metabolites in fish, although previous investigators reported that fish cannot metabolize these compounds or DDE (Addison et al., 1977; Franklin et al., 1980). In addition, we noted the occurrence of high concentrations of oxygenated metabolites of DDT and PCBs in mussels, even though they are thought to metabolize xenobiotic compounds very slowly or not al all (Bayne et al., 1980). This apparent contradiction may be resolved by the fact that sediments can serve as significant reservoirs of metabolites. In sediments from Palos Verdes, the oxygenated metabolites of DDT represented 80% of the total DDT, its primary derivatives, DDE and DDD, and their oxygenated metabolites DDA and DDOH.

Microbial processes in sediments may serve as a partial source of the oxygenated metabolites found in samples from the marine environment. Lee and Ryan (1978) have shown that the half-life of DDE in marine sediments from near Savannah, Georgia, is approximately 3 y. Most of the so-called total DDT in the marine environment of southern California was discharged as DDE prior to 1972 (Schafer, 1982). Calculations based on Lee and Ryan's (1978) 3-y half-life value for DDE indicate that microorganisms in marine sediments had the potential to degrade at least 80% of the DDE by 1982.

In addition, oxygenated metabolites may have octanol-to-water partition coefficients that fall within the same range as those of organic compounds, which are readily bioaccumulated from sediments. Results in Table 6.4 and by Gossett et al. (1983) show that the accumulation of parent compounds in sediments and organisms is proportional to the compound's octanol-to-water partition coefficient rather than to the effluent concentration. This is because (1) compounds with higher partition coefficients are concentrated in particulate matter that serves as the base of food webs and (2) they are more likely to be retained in tissues. In general, compounds with octanol-to-water partition coefficients <3.5 are not accumulated in sediments or tissues. In sediments, DDA is readily detectable, and it would have an octanol-to-water partition coefficient in the stomach and upper portions of the duodenum that would allow bioaccumulation (Table 6.4). DDOH, which has a partition coefficient of 4.3, was not detected in sediments, but was present in scorpion fish livers. This may indicate production of this metabolite within the fish or absorption of an undetected source of DDOH.

In addition to microbial degradation of xenobiotic organic compounds in sediments, it has been demonstrated, at least in the case of DDT, that these compounds are readily taken up by phytoplankton (Cox, 1972) and metabolized to hydroxylated forms (Patil et al., 1972). Metabolism of DDT by phytoplankton resulted in the production of DDOH as a major hydroxylated product (Patil et al., 1972). This suggests that phytoplankton may also serve as a source of hydroxylated metabolites to the base of food webs and may account for their high accumulation in fish or filter-feeding mussels, which have little or no ability to metabolize some xenobiotic compounds.

The discovery of high concentrations of oxygenated metabolites in the marine environment raises questions regarding their potential to harm marine life or humans who consume seafood. In terms of acute toxicity we have been able to find only one published value, an LD_{50} value of 600 mg *p,p'*-DDA kg^{-1} (U.S. Department of Health and Human Services, 1980). This value is similar to that for the parent compound *p,p'*-DDE (880 mg *p,p'*-DDE kg^{-1}), which comprises 95% of the total DDT, DDE, and DDD in coastal southern California. Further studies of acute toxicity are needed; however, as suggested previously, acute effects of xenobiotic compounds are unlikely to occur in marine organisms at present exposure levels (Brown et al., 1982b).

Most previous investigators have suggested that parent compounds are, in fact, pretoxicants or precarcinogens that must be bioactivated by mixed-function oxygenase enzymes before they can cause toxic effects (Miller, 1970; Brodie et al., 1971; Bend et al., 1977). This is because hydrophobic parent xenobiotic compounds are partitioned into lipid pools (Allen et al., 1974, 1975) away from sensitive macromolecules, and are therefore unable to cause direct toxic effects. However, if these xenobiotic compounds are oxygenated by the mixed function oxygenases, highly reactive (bioactivated) epoxide intermediates are formed (Brodie et al., 1971; Jerina and Daley, 1974; Brown et al., 1982a). These epoxides can be rendered nontoxic by conjugation with glutathione or other similar conjugating agents such as glucuronic acid (Gillette et al., 1974). If the capacity of these conjugating agents is exceeded, however, the epoxides are then free to

Table 6.4. Partitioning of Parent Compounds and Oxygenated Metabolites in Wastewater, Sediment, and Scorpion Fish

Compound	Log K_{ow}[a]	Los Angeles County Municipal Wastewater Concentration (μg $liter^{-1}$)[b]	Palos Verdes Sediment Concentration (mg kg^{-1}, wet wt)[b]	Palos Verdes Scorpion Fish Livers Concentration (mg kg^{-1}, wet wt)[c]
Total PCB (1242, 1254)	6.0	0.99	0.93	2.2
Total DDT (DDT, DDE, DDD)	5.7	0.90	6.3	24.9
DDOH	4.3	NA[e]	<0.1	14.6
DDA	≤4.3[d]	NA[e]	25.9	9.7
Trichlorophenol	3.7	2.4	0.005	0.164
Naphthalene	3.5	15	0.005	<0.02
Toluene	2.7	210	<0.001	<0.001
Vinylchloride	1.5	6.2	<0.0005	<0.0003
Dichloroethane	1.5	44	<0.0005	<0.0003

[a] Log K_{ow} is the partition coefficient. (Hansch and Leo, 1979).
[b] Two samples analyzed.
[c] Five samples analyzed.
[d] DDA would have a log K_{ow} of 4.3 when fully protonated (below pH of 4) such as might be encountered in a fish stomach or upper portions of the duodenum, but could have log K_{ow} values as low as 1.6 when completely ionized (as could occur with pH significantly greater than 4) (A. Leo, personal communication).
[e] Not analyzed.

affect sites of toxic action, including DNA, enzymes, and metallothionein (Reid and Krishna, 1973; Brown et al., 1982a, 1982b).

In the present study, we have found oxygenated metabolites conjugated to substances in the GSH, MT, and ENZ pools even though the source of the metabolites may have been exogenous. This may indicate that oxygenated metabolites have properties that allow them to bind to substances even after deactivation (Brown et al.,1982a) from the highly reactive epoxide intermediate stage.

Results in the present study show an apparent association between the occurrence of oxygenated metabolites in the ENZ and MT pools, and pathological effects in fish. It was found that in livers of Dover sole without fin erosion almost all of the measured oxygenated metabolites occurred in the GSH pool, partitioned away from sensitive proteins. In livers of Dover sole with fin erosion, however, there was a large increase of oxygenated DDT metabolites in the GSH pool, which caused an apparent overloading of this pool, with spillover of metabolites into the ENZ and MT pools. In livers of croakers from Palos Verdes, the highest portions of oxygenated DDT metabolites spilled into the MT pool relative to the ENZ pool (Brown et al., 1982a). This may occur because of the high sulfhydryl content of MT (Brown et al., 1982a).

The present study further supports the proposal (Brown et al., 1982a) that reduced metal retention occurs with the binding of oxygenated metabolites of xenobiotic compounds to MT. In croakers with oxygenated DDT metabolites in the MT pool, there were lower levels of Cu and Cd in this pool than in controls. Young and Jan (1979) noted an inverse relationship between concentrations of xenobiotic compounds and metals in organisms from southern California coastal waters. Several studies have suggested that reduced metal levels are involved in the etiology of disease processes such as cancer (Fong et al., 1978; Brown et al., 1980). Further studies are needed to assess the specific effects of oxygenated metabolites in areas of fin erosion of Dover sole.

These preliminary results tend to support the proposal that oxygenated metabolites may affect sensitive cellular processes. Since oxygenated metabolites appear to be the major form of organic contamination in the marine environment, their effects should be further investigated.

ACKNOWLEDGMENTS

The authors are grateful to Valerie Raco for technical assistance with some portions of this study and to Scherrie Maidel and Peter Szalay for assistance in preparation of this manuscript. Research was funded by the Office of Marine Pollution Assessment, U.S. National Oceanic and Atmospheric Administration, grants NA80RAD00040 and NA82RAD 00003.

REFERENCES

Addison, R. F., M. E. Zinck, and D. E. Willis. 1977. Mixed function oxidase enzymes in trout (*Salvelinus frontinalis)* liver: absence of induction following feeding of *p,p'*-DDT or *p,p'*-DDE. *Comparative Biochemistry and Physiology,* **57**, 39–43.

Allen, J. R., L. A. Cartens, and D. A. Barsotti. 1974. Residual effects of short-term, low level exposure of nonhuman primates to polychlorinated biphenyls. *Toxicology and Applied Pharmacology,* **30**, 440–451.

Allen, J. R., L. A. Cartens, L. J. Abrahamson, and R. J. Marlar. 1975. Responses of rats and nonhuman primates to 2,5,2',5'-tetrachlorobiphenyl. *Environmental Research,* **9**, 265–273.

Bayne, B. L., D. A. Brown, F. Harrison, and P. D. Yevich. 1980. Mussel health. *In:* The International Mussel Watch. National Research Council, U.S. National Academy of Sciences, Washington, D. C., pp. 163–248.

Bend, J. R., M. O. James, and P. M. Dansette. 1977. *In vitro* metabolism of xenobiotics in some marine animals. *Annals of the New York Academy of Science,* **298**, 505–521.

Brodie, B. B., W. D. Reid, A. K. Cho, G. Sipes, G. Krishna, and J. R. Gillette. 1971. Possible mechanism of liver necrosis caused by aromatic organic compounds. *Proceedings of the National Academy of Sciences,* **68**, 160–164.

Brown, D. A., and T. R. Parsons. 1978. Relationship between cytoplasmic distribution of mercury and toxic effects to zooplankton and chum salmon (*Oncorhynchus keta)* exposed to mercury in a controlled ecosystem. *Journal of the Fisheries Research Board of Canada,* **35**, 880–884.

Brown, D. A., C.A. Bawden, K.W. Chatel, and T. R. Parsons. 1977. The wildlife community of Iona Island Jetty, Vancouver, B.C., and heavy-metal pollution effects. *Environmental Conservation,* **4**, 213–216.

Brown, D. A., K. W. Chatel, A. Y. Chan, and B. Knight. 1980. Cytosolic levels and distribution of cadmium, copper, and zinc in pretumorous livers from diethylnitrosamine exposed mice in non-cancerous kidneys from cancer patients. *Chemico–Biological Interactions,* **32**, 13–27.

Brown, D. A., R. W. Gossett, and K. D. Jenkins. 1982a. Contaminants in white croakers *Genyonemus lineatus* (Ayres, 1855) from the southern California Bight. II: Chlorinated hydrocarbon detoxification/toxification. *In:* Physiological Mechanisms of Marine Pollution Toxicity, W. B. Vernberg, A. Calabrese, F. P. Thurberg, and F. J. Vernberg (Eds.). Academic Press, New York, pp. 197–213.

Brown, D. A., K. D. Jenkins, E. M. Perkins, R. W. Gossett, and G. P. Hershelman. 1982b. Detoxification of metals and organic compounds in white croakers. *In:* Biennial Report, 1981–1982, Southern California Coastal Water Research Project, W. Bascom (Ed.). Southern California Water Research Project, Long Beach, Calif., pp. 157–172.

Cox, J. L. 1972. DDT residues in marine phytoplankton. *Residue Reviews,* **44**, 23–38.

Fong, L. Y., A. Sivak, and P. M. Newberne. 1978. Zinc deficiency and methylbenzyl-nitrosamine-induced esophageal cancer in rats. *Journal of the National Cancer Institute,* **61**, 145–150.

Franklin, R. B., C. R. Elcombe, M. J. Vodicnik, and J. J. Lech. 1980. Comparative aspects of the disposition and metabolism of xenobiotics in fish and mammals. *Federation Proceedings,* **39**, 3144–3149.

Gillette, J. R., J. R. Mitchell, and B. B. Brodie. 1974. Biochemical mechanisms of drug toxicity. *Annual Review of Pharmacology,* **14,** 271–288.

Gingell, R., and L. Wallcave. 1974. Species differences in the acute toxicity and tissue distribution of DDT in mice and hamsters. *Toxicology and Applied Pharmacology,* **28,** 285–394.

Gold, B., T. Leuschen, G. Brunk, and R. Gingell. 1981. Metabolism of a DDT metabolite via a chloroepoxide. *Chemico–Biological Interactions,* **35,** 159–178.

Gossett, R. W., D. A. Brown, and D. R. Young. 1983. Predicting the bioaccumulation of organic compounds in marine organisms using octanol/water partition coefficients. *Marine Pollution Bulletin,* **14,** 387–392.

Gossett, R. W., D. A. Brown, S. R. McHugh, and A. M. Westcott. 1984. Measuring the oxygenated metabolites of chlorinated hydrocarbons. *In:* Biennial Report 1983–1984, Southern California Coastal Water Research Project, W. Bascom (Ed.). Southern California Water Research Project, Long Beach, California, pp. 155–169.

Hansch, C., and A. Leo. 1979. Substituent Constants for Correlation Analysis in Chemistry and Biology. John Wiley & Sons, New York, 616 pp.

Hinga, K. R., M. E. Q. Pilson, R. F. Lee, J. W. Farrington, K. Tjessem, and A. C. Davis. 1980. Biogeochemistry of benzanthracene in an enclosed marine ecosystem. *Environmental Science and Technology,* **14,** 1136–1143.

Jenkins, K. D., and D. A. Brown. 1985. Determining the biological significance of contaminant bioaccumulation. *In:* Proceedings of the Workshop on Meaningful Measures of Marine Pollution Effects (April 1982), Pensacola, Florida, H. White (Ed.). University of Maryland Sea Grant Program, College Park, Maryland, in press.

Jenkins, K. D., D. A. Brown, G. P. Hershelman, and W. C. Meyer. 1982. Contaminants in white croakers *Genyonemus lineatus* (Ayres, 1855) from the southern California Bight. I: Trace metal detoxification/toxification. *In:* Physiological Mechanism of Marine Pollutant Toxicity, W. B. Vernberg, A. Calabrese, F. P. Thurberg, and F. J. Vernberg (Eds.). Academic Press, New York, pp. 177–196.

Jerina, D. M., and J. W. Daley. 1974. Arene oxides: A new aspect of drug metabolism. *Science,* **185,** 573–582.

Lee, R. F. 1975. Fate of petroleum hydrocarbons in marine zooplankton. *In:* Proceedings of Conference on Prevention and Control of Oil Pollution. American Petroleum Institute, Washington, D.C., pp. 549–553.

Lee, R. F., and C. Ryan. 1978. Microbial degradation of organochlorine compounds in estuarine waters and sediments. *In:* Proceedings of the Workshop: Microbial Degradation of Pollutants in Marine Environments (9–14 April 1978), Pensacola Beach, Florida, A. W. Bourquim and P. H. Pritchard (Eds.). Environmental Research Laboratory, U. S. Environmental Protection Agency, Gulf Breeze, Florida, pp. 443–450.

Lee, R. F., and C. Ryan. 1983. Microbial and photochemical degradation of polycyclic aromatic hydrocarbons in estuarine waters and sediments. *Canadian Journal of Fisheries and Aquatic Science,* **40**(Supplement 2), 86–94.

Lee, R. F., W. S. Gardner, J. W. Anderson, J. W. Blaylock, and J. Barwell-Clarke. 1978. Fate of polycyclic aromatic hydrocarbons in controlled ecosystem enclosures. *Environmental Science and Technology,* **12,** 832–838.

Lee, R. F., K. Hinga, and G. Almquist. 1982. Fate of radiolabelled polycyclic aromatic hydrocarbons and pentachlorophenol in enclosed marine ecosystems. *In:* Marine Mesocosms, G. D. Grice and M. R. Reeve (Eds.). Springer-Verlag, New York, pp. 123–135.

Malins, D. C. 1977. Metabolism of aromatic hydrocarbons in marine organisms. *Annals of the New York Academy of Science,* **298,** 482–496.

Metcalf, R. L., J. R. Sanborn, P.-Y. Lu, and D. Nye. 1975. Laboratory model ecosystem studies of the degradation and fate of radiolabelled tri-, tetra-, and pentachlorobiphenyl compared with DDE. *Archives of Environmental Contamination and Toxicology,* **3,** 151–165.

Miller, J. A. 1970. Carcinogensis by chemicals: an overview. *Cancer Research,* **30,** 559–576.

Oshida, P. S. 1982. Optimum temperature for tissue cytosol storage. *In:* Biennial Report 1981–1982, Southern California Coastal Research Project, W. Bascom (Ed.). Southern California Coastal Research Project, Long Beach, California, pp. 201–203.

Patil, K. C., F. Matsumura, and G. M. Boush. 1972. Metabolic transformation of DDT, Dieldrin, Aldrin and Endrin by marine microorganisms. *Environmental Science and Technology,* **6,** 629–632.

Peterson, R. E., and P. D. Guiney. 1979. Disposition of polychlorinated biphenyls in fish. *In:* Pesticide and Xenobiotic Metabolism in Aquatic Organisms, M. A. Q. Khan, J. J. Lech, and J. J. Menn (Eds.). American Chemical Society, Washington, D.C., pp. 21–36.

Reid, W. D., and G. Krishna. 1973. Centrolobular hepatic necrosis related to covalent binding of metabolites of halogenated aromatic hydrocarbons. *Experimental and Molecular Pathology,* **18,** 80–99.

Roubal, W. T., T. K. Collier, and D. C. Malins. 1977. Accumulation and metabolism of carbon-14 labeled benzene, naphthalene, and anthracene by young coho salmon (*Oncorhynchus kisutch*). *Archives of Environmental Contamination and Toxicology,* **5,** 513–529.

Sanborn, H. R., and D. C. Malins. 1977. Toxicity and metabolism of naphthalene: A study with marine larval invertebrates. *Proceedings of the Society for Experimental Biology and Medicine,* **154,** 151–155.

Schafer, H. A. 1982. Characteristics of municipal wastewaters 1980–1981. *In:* Biennial Report 1981–1982, Southern California Coastal Water Research Project, W. Bascom (Ed.). Southern California Coastal Water Research Project, Long Beach, California, pp. 201–203.

U.S. Department of Health and Human Services. 1980. Registry of Toxic Effects of Chemical Substances, Vol. 1. Public Health Service, Center for Disease Control, National Institute for Occupational Safety and Health, Cincinnati, Ohio 828 pp.

Varanasi, U., and D. C. Malins. 1977. Metabolism of petroleum hydrocarbons: accumulation and biotranformation in marine organisms. *In:* Effects of Petroleum on Arctic and Subarctic Marine Environments and Organisms, Vol. 11, D. C. Malins (Ed.). Academic Press, New York, pp. 175–200.

Varanasi, U., D. G. Gmur, and P. A. Treseler. 1979. Influence of time and mode of exposure on biotransformation of naphthalene by juvenile starry flounder (*Platichthys stellatus*) and rock sole (*Lepiodopsetta bilineata*). *Archives of Environmental Contamination and Toxicology,* **8,** 673–692.

Young, D. R., and T.-K. Jan. 1979. Trace metal contamination of the rock scallop *Hinnites giganteus,* near a large southern California municipal outfall. *Fisheries Bulletin,* **76,** 936–939.

Chapter 7

Persistent Pollutants and Nitrogen Fixation and Denitrification in Coastal Marine Sediments

Jennifer Slater and Douglas G. Capone

Marine Sciences Research Center
State University of New York
Stony Brook, New York

ABSTRACT

The effects of several synthetic organic compounds on nitrogen fixation and denitrification were examined in slurries of sediments from *Spartina alterniflora* communities. At 1000 $\mu g\ g^{-1}$, heptachlor, phenanthrene, pentachlorophenol (PCP), DDT, and endrin caused significant inhibition of ethylene production. Toxaphene stimulated ethylene production to more than twice the amount produced in the controls. The effects of Aroclor 1254, lindane, DDT, endrin, and PCP were different for long-term than for short-term experiments. Nitrous oxide production was inhibited by lindane, phenanthrene, toxaphene, naphthalene, and PCP, and stimulated slightly by Aroclor 1254.

7.1. INTRODUCTION

In many marine ecosystems the availability of nitrogen limits the rate of primary production, and sediments play an important role in regulating this productivity (Rowe et al., 1975; Nixon, 1981). Coastal marine sediments are often rich in organic material and have high microbial activity. They generally contain abundant nutrients, mostly formed by microbial degradation of organic material. Other microbial transformations of nitrogen occur in the sediments. In anaerobic areas, both denitrification (Sorensen, 1978) and nitrogen fixation (Capone, 1983) may be important.

Nitrogen fixation is the microbial reduction of N_2 to NH_4^+ and its subsequent conversion to organic form; this reduction is catalyzed by the enzyme nitrogenase. Estimates of nitrogen fixation in estuarine sediments range from 0.58 μmole N $m^{-2}\ h^{-1}$ in sediments of Tokyo Bay (Maruyama et al., 1974) to 4.7 μmole N $m^{-2}\ h^{-1}$ in Hawaii (Hanson and Gundersen, 1977). Higher values are associated with saltmarsh sediments. Whitney et al. (1975) reported values up to 20 μmole N $m^{-2}\ h^{-1}$ in the rhizosphere of tall *Spartina alterniflora* in Flax Pond saltmarsh, Long Island, New York. In saltmarsh sediment taken from the same salt marsh, we have obtained rates of nitrogenase activity between 0.18 and 0.81 nmole ethylene $g^{-1}\ h^{-1}$ (between 17 and 80 μmole N $m^{-2}\ h^{-1}$), depending on the time of year (unpublished data).

Denitrification is the dissimilatory reduction of nitrate to a gaseous end product, either N_2O or N_2. It is carried out by heterotrophic bacteria that use NO_3^- as a terminal electron acceptor in place of oxygen. The reduction occurs in several stages, each catalyzed by a different reductase. The reduction sequence is $NO_3^- \rightarrow NO_2^- \rightarrow [NO] \rightarrow N_2O \rightarrow N_2$ (Payne, 1981). Coastal sediments, with high organic input and low oxygen concentrations, are ideal sites for denitrification, and high rates have been reported. Sorensen (1978) measured rates of 560 μmole N m^{-2} h^{-1} in sediments of Limfjord, Denmark, and Nishio et al. (1982) found up to 790 μmole N m^{-2} h^{-1} in Tama estuary, Japan. Collins (1981) found denitrification rates of 2.6 nmoles N_2O g^{-1} h^{-1} (1200 μmole N m^{-2} h^{-1}) in saltmarsh sediments of Flax Pond, New York. Thus, coastal sediments may be sites of appreciable nitrogen loss; it has been suggested that saltmarshes could be used as a repository for nitrogenous wastes because of this dissimilatory capacity (Van Raalte et al., 1974).

Although nitrogen fixation may not be a major source of combined nitrogen in most sediment systems, it is very important because all other organisms need to assimilate nitrogen in a combined form and are therefore ultimately dependent on nitrogen-fixing bacteria (Burns and Hardy, 1975). Denitrification is certainly one of the major processes that decreases the supply of available nitrogen. The rates of nitrogen fixation and denitrification occurring naturally in the sediments will therefore affect the availability of combined nitrogen. This is especially important in estuarine environments, where conditions are ideal for denitrification and where processes occurring in the sediments may rapidly and directly affect the water column. Rowe et al. (1975) calculate that nearshore sediments could supply sufficient nitrogen to support all of the primary production of the New York Bight.

Marine sediments are also sites of accumulation of many of the anthropogenic pollutants entering the ocean. The increasing occurrence of chlorinated hydrocarbons (Pavlou and Hom, 1976) and polycyclic aromatic hydrocarbons (PAHs) (Hites et al., 1977) in marine sediments has been documented. O'Conner and Stanford (1979) listed pollutants of concern in the New York Bight; this list includes chlorinated pesticides, polychlorinated biphenyls (PCBs), polycyclic aromatic hydrocarbons (PAHs), and chlorophenols. The effects of these toxicants on microbial activity in marine sediments are not well known. Any disruption of the natural balance of nitrogen-transforming reactions may have an impact on productivity in the sediments and the water column.

In this chapter we will discuss the results of preliminary experiments designed to assess whether some pollutants affect microbial nitrogen fixation and denitrification.

7.2. EXPERIMENTAL

The organic compounds tested for their effects on denitrification potential and nitrogenase activity in saltmarsh sediments were heptachlor, Aroclor 1254 (Monsanto Co., St. Louis, Missouri), phenanthrene, naphthalene, pentachlorophenol (PCP), 1,1,1-trichloro-2,2-*bis*(chlorophenyl)ethane (DDT), endrin, chlordane, lindane, and toxaphene.

7.2.1. Evaluation of Methods

Nitrogenase catalyzes the reduction of acetylene to ethylene, which is easily measured by gas chromatography. This is the basis of the acetylene reduction assay for nitrogenase activity (Hardy et al., 1968), which has been extensively used in many environments (Taylor, 1983). The method is easy, sensitive, precise, cheap, and still regarded as one of the best methods available (Payne, 1984). If used as a measure of actual rates of nitrogen fixation, however, attention must be paid to the possible effects of acetylene on other microbial processes, the production of ethylene from other sources, and the stimulation of nitrogenase activity by acetylene leading to an overestimation of nitrogen fixation rates (Payne, 1984). Some of these possibilities can be ruled out by careful use of acetylene- and sediment-free controls, and by determination of short-term time courses.

Acetylene has also been shown to inhibit the last stage of denitrification, reduction of nitrous oxide to nitrogen, leading to accumulation of nitrous oxide, which is also measurable by gas chromatography. This method of estimating denitrification has been applied in pure cultures (Yoshinari and Knowles, 1976), in soil samples (Yoshinari et al., 1977), and in estuarine sediments (Sorensen, 1978).

Simultaneous estimates of nitrogen fixation and denitrification in a sediment system may be possible by addition of acetylene. In our experiments, for actual rate determinations both methods were calibrated with ^{15}N determinations. In these experiments, only a comparative measure of activity in control and amended incubations was needed.

In order to determine the effects of the pollutants, all other conditions varied as little as possible among the replicates. In this inhomogeneous system, preliminary experiments showed that to keep the coefficient of variation among triplicates below 10% it was necessary to use screened slurries of sediment. During these preliminary experiments, rates of denitrification were extremely variable, ranging from 0 to 1155 μmole N m^{-2} h^{-1}. In the experiments reported in this chapter, nitrate amendment was used to ensure reliable nitrous oxide production. Preliminary experiments on the effects of different amounts of added nitrate showed that 100 μM of nitrate in the slurry guaranteed nitrous oxide production but never reduced ethylene production by more than 19%.

7.2.2. Sediment Preparation

Sediment from Flax Pond, a relatively pristine saltmarsh on Long Island, New York, was obtained in core tubes 2.5 cm in diameter and brought immediately to the laboratory. The top 10 cm of the core was homogenized with deoxygenated filtered Flax Pond water (salinity 28‰) at a ratio of 2:1 (seawater to sediment, by vol), and screened through a 2-mm sieve. Because the aerobic layer of Flax Pond is a few millimeters, the slurry was continuously purged with oxygen-free nitrogen. Fifty-milliliter aliquots of the slurry were pipetted into 125-ml Erlenmeyer flasks. The flasks were immediately stoppered and purged for 60 s with oxygen-free nitrogen. Oxygen concentrations in the flasks were routinely measured by gas chromatography, and were <1 nmole ml^{-1} at the beginning of each experiment. All additions were made with a syringe through the stoppers. The flasks were shaken gently in a water bath in the dark at 25°C (± 1°), the mid-range of temperatures for Flax Pond in summer and fall. Experiments performed in winter used sediment maintained in a seawater table at 18°C to minimize thermal shock.

Nitrogenase activity was assessed by measuring ethylene production from acetylene (Capone and Taylor, 1980), and denitrification potential was assessed by measuring the accumulation of nitrous oxide in the presence of acetylene (Sorensen, 1978). Acetylene was added to the gas phase of the incubations to a final concentration of 20% by volume. Potassium nitrate was routinely added to the slurry to a concentration of 100 μM. When added, the organic compounds being tested were dissolved in acetone. Concentrations were calculated so that addition of 0.2 ml of the solution would give a concentration of 1000 μg of the organic compound per g of dry sediment. Two series of controls were run for each experiment, one set amended with an equal volume of deionized distilled water and one set amended with acetone. In each experiment, three replicates of each treatment were used and the experiments were run for at least 3 weeks. The volume of the gas phase in each flask was determined by adding an equal volume of water at the end of each experiment, and the sediment was dried at 105°C to a constant weight. Control experiments were run periodically with filtered seawater only, with autoclaved homogenates, and without acetylene.

7.2.3. Gas Chromatography

Ethylene and nitrous oxide concentrations were measured by gas chromatography on a Perkin-Elmer Sigma 2B gas chromatograph (Perkin-Elmer Corp., Norwalk, Connecticut). Samples were withdrawn from the gas phase periodically with a 100-μl syringe and injected onto a 3-m, ⅛-inch (outside diameter) column of 80–100 mesh Porapak-R® (Waters Associates, Milford, Massachusetts). Nitrous oxide was measured by thermal conductivity and ethylene by flame ionization, with the detectors connected in series. The carrier gas was helium at 30 ml min^{-1} and the column temperature was 60°C. Gas standards close to the concentrations in the incubation flasks were checked before and after each sample run. Concentrations of ethylene and nitrous oxide in the gas phase were corrected for solubility.

7.2.4. Data Presentation

The initial rate of gas production (nmole C_2H_4 or N_2O per g of dry sediment per h) in the controls was taken as the regression slope of the first linear portion of the time course of gas production. The initial production rate of each amended flask was then taken as the rate over the same period used in the control incubations. The rates for each flask were calculated separately and then the mean and standard error were calculated for each set of three replicates. Differences between the means of control and treated incubations were assessed for significance using a *t*-test. Similar calculations were done for total gas production (nmole gas per g of dry sediment) at various times in the course of the experiment. The production of ethylene or nitrous oxide in the amended incubations was then expressed as a ratio of the production in the control incubations.

7.3. CONTROL INCUBATIONS: RESULTS AND DISCUSSION

The experiments reported in this chapter were designed to test the hypothesis that addition of potential environmental contaminants to sediments could alter rates of nitrogen fixation and denitrification. The methods used were designed to give a comparative measure and not to assess *in situ* rates of nitrogenase activity or denitrification potential. It is noteworthy, however, that control rates measured for nitrogen fixation and denitrification in saltmarsh sediment throughout this investigation were within the ranges reported in other studies (Haines et al., 1977; Hattori, 1983).

Ethylene production in controls increased linearly until 8–23 h and then decreased (Fig. 7.1). In no case did ethylene production cease, however, even after 4-week incubations. The initial rate observed varied from 0.32 to 1.0 nmole ethylene per g of dry sediment per h. Production at 72 h was between 14 and 34 nmole of ethylene per g of dry sediment.

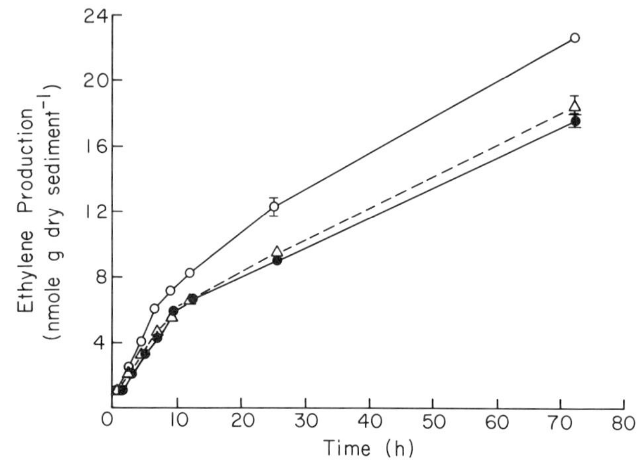

Figure 7.1. Ethylene production in sediment slurries: (○) control, (△) DDT added, and (●) endrin added (18 January 1983). Points are means of triplicate samples, with standard error shown.

A constant rate of production of nitrous oxide was seen for 5–12 h. Nitrous oxide concentration remained constant for a further 5–16 h and then decreased rather rapidly (Fig. 7.2). The initial rates observed were from 25 to 53 nmole of N_2O per g of dry sediment per h, and maximum production was from 212 to 296 nmole N_2O per g of dry sediment. The linearity of production of both ethylene and nitrous oxide from time zero suggests that the system was not greatly disrupted by the setup procedure.

No measurable nitrous oxide or ethylene production occurred in incubations with autoclaved homogenates or with seawater only. These control experiments ruled out the possibility of chemical gas production, or gas production by bacteria not associated with the sediment. Incubations without acetylene performed on many occasions also showed no production of ethylene or nitrous oxide. This suggests that the ethylene observed in the experimental incubations was produced from the acetylene.

Acetylene is known to affect other microbial processes (Payne, 1984). In our experiments, the concentration of acetylene was monitored throughout. In no case was there any large acetylene consumption. The possibility of ethylene consumption by methane oxidizers, demonstrated in lake sediments (Yeomans and Beauchamp, 1982) was not tested here.

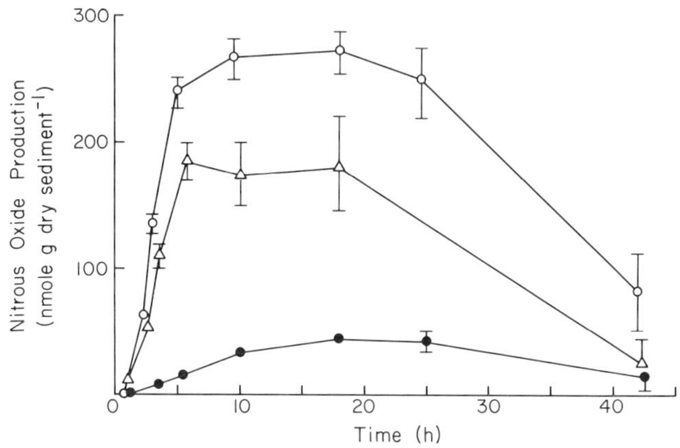

Figure 7.2. Nitrous oxide production in sediment slurries; (○) control, (△) toxaphene added, and (●) PCP added (12 January 1983). Points are means of triplicate samples, with standard error shown.

The disappearance of nitrous oxide after several days during acetylene blockage experiments was noted by Yeomans and Beauchamp (1978). In our experiments, the nitrous oxide disappeared rapidly, and the concentrations measured were regarded as a balance between its production by denitrifiers and its consumption, either by acetylene-resistant denitrifiers or by some other process. The mechanism of the nitrous oxide disappearance is under study now. It is not caused by nitrate depletion, because addition of more nitrate after the nitrous oxide concentration began to decrease had no effect.

In six experiments, no significant differences (at the 95% confidence level) could be detected in ethylene production between the acetone and water controls. In one experiment (30 November 1982), ethylene production in the acetone controls diverged from that in the water controls after 76 h and became twice as great in 2 weeks. After 76 h, results from this experiment were not used. Nitrous oxide production in acetone controls never differed from that in water controls at the 95% confidence level. Thus, acetone appears to have had no effect on acetylene reduction or nitrous oxide accumulation. These results contrast with those reported by Stratton et al. (1980). They found that acetone alone stimulated acetylene reduction in a cyanobacterium, and that the effects of acetone and permethrin together were not necessarily additive but could be either synergistic or antagonistic.

7.4. EFFECTS OF ORGANIC AMENDMENTS

7.4.1. Results

Tables 7.1 and 7.2 summarize the effects of organic compounds on ethylene and nitrous oxide production at 1000 μg per g of dry sediment. The mean and standard error for the initial rate of the acetone control incubations in each experiment are given. The ratio of the mean of the experimental incubations to the mean of the control incubations is listed for the initial rates of gas production, and for gas production (nmole per g of dry sediment) at various time points in the experiments. The significance of the difference between the mean control value and the mean treatment value in each experiment is shown. The controls containing acetone were used in these calculations.

Heptachlor, phenanthrene, PCP, DDT, and endrin caused significiant ($p<0.05$) inhibition of both initial rate and 72-h production of ethylene. Toxaphene, in contrast, stimulated nitrogenase activity. Chlordane caused significant inhibition of the intial rate of ethylene production, but the effect disappeared by 48 h. Aroclor 1254, naphthalene, and lindane had no significant effect at the 95% probability level on initial rate or on 72-h production of ethylene.

Table 7.1. Effects of Organic Contaminants on Ethylene Production

Compound	Ethylene Production: Initial Rate[a] (nmole g^{-1} h^{-1})	Experimental : Control: Initial Rate	Experimental : Control: 72-h	Experimental : Control: 3-week
Heptachlor	0.32 ± 0.019	0.79[b]	0.83[b]	0.81[b]
Aroclor 1254	0.32 ± 0.019	0.84[c]	0.92[c]	1.3[d]
Phenanthrene	0.41 ± 0.025	0.89[b]	0.90[b]	—[e]
	0.58 ± 0.049	0.86[d]	0.85[d]	0.86[d]
Naphthalene	0.41 ± 0.025	0.84[c]	0.81[c]	—
PCP	1.0 ± 0.16	0.93[b]	0.84[d]	0.62[b]
	0.58 ± 0.049	0.92[b]	0.87[b]	0.68[b]
Toxaphene	1.0 ± 0.16	1.2	2.5[d]	2.2[b]
	0.67 ± 0.061	1.1	2.3[b]	2.4[b]
DDT	0.88 ± 0.026	0.60[b]	0.85[b]	0.91[b]
Endrin	0.88 ± 0.026	0.73[d]	0.80[d]	0.82[d]
Chlordane	0.47 ± 0.011	0.85[b]	1.0[c]	0.97[c]
Lindane	0.47 ± 0.011	0.99[c]	1.1[c]	0.78[d]
	0.67 ± 0.061	1.0[c]	0.96[c]	0.70[d]

[a] Initial rate is given as the mean of three replicates ± standard error.
[b] Significant at $p < 0.05$.
[c] Not significant.
[d] Significant at $p < 0.01$.
[e] No data available.

Table 7.2. Effects of Organic Contaminants on Nitrous Oxide Production[a]

Compounds	Nitrous Oxide Production: Initial Rate[a] (nmole g^{-1} h^{-1})	Experimental : Control: Initial Rate	Experimental : Control: Maximum Production
Heptachlor	40 ± 1.1	1.0[b]	1.1[b]
Aroclor 1254	40 ± 1.1	0.98[b]	1.1[c]
Phenanthrene	37 ± 2.6	0.84[c]	0.87
	36 ± 3.1	0.83[c]	0.82[c]
Naphthalene	37 ± 2.6	0.59[d]	0.77[d]
PCP	53 ± 8.5	0.067[d]	0.18[c]
	36 ± 3.1	0.15[c]	0.20[c]
Toxaphene	53 ± 8.5	0.73[b]	0.69[b]
	41 ± 4.5	0.81[c]	0.74[c]
DDT	54 ± 2.7	1.0[b]	1.1[b]
Endrin	54 ± 2.7	1.0[b]	1.0[b]
Chlordane	25 ± 1.1	0.94[b]	0.99[b]
Lindane	25 ± 1.1	0.92[c]	0.90[c]
	41 ± 4.5	0.90	0.89[c]

[a] Initial rate is given as the mean of three replicates ± standard error.
[b] Not significant.
[c] Significant at $p < 0.05$.
[d] Significant at $p < 0.01$.

In these experiments, effects on acetylene reduction after longer incubation were sometimes quite different from effects at 72 h. For both Aroclor 1254 and lindane, the results at 3 weeks were markedly different from the results at 72 h. Aroclor 1254 stimulated ethylene production by 30% over controls at 3 weeks (compared with an insignificant 8% inhibition at 72 h). Lindane inhibited ethylene production at 3 weeks by 22% in one experiment and by 30% in another (compared with an insignificant 10% stimulation at 72 h in the first and 4% inhibition in the second experiment). The inhibition caused by PCP increased progressively over time; it was 16% at 72 h, 31% at 1 week, and 38% at 3 weeks in one experiment, and 13% at 72 h, 32% at 3 weeks, and 42% at 4 weeks in the other experiment. For DDT, however, there appeared to be some recovery after 3 weeks (9% inhibition at 3 weeks compared with 15% at 72 h). There was some indication that this might also be true for endrin. For the other compounds tested, 3- to 4-week and 72-h results were the same.

Because nitrous oxide disappeared rapidly in the denitrification experiments, initial rate and maximum production were the only parameters compared. Heptachlor and Aroclor 1254 had no effect on the initial rate of nitrous oxide production but slightly stimulated maximum production, although the stimulation by heptachlor was significant only at $p<0.1$.

Lindane slightly but significantly ($p<0.05$) depressed both initial rate and maximum production of nitrous oxide. Phenanthrene, toxaphene, and especially naphthalene and PCP inhibited nitrous oxide production, although the results for toxaphene were statistically significant at $p<0.05$ in only one of the two experiments.

7.4.2. Discussion

There are few reports on the effects of organic pesticides on nitrogen fixation in marine sediments. Knowles and Wishart (1977) found that oil or any petroleum fraction (except for 1,2,4-trimethylbenzene, which caused complete inhibition) had no effect on nitrogen fixation in Arctic sediment. Two studies showed decreases in both nitrogen fixation and denitrification in subarctic marine sediments exposed to oil (Haines et al., 1981; Griffiths et al., 1982). Some reports of effects of pesticides on nitrogen fixation in soils exist, and pesticide metabolism in marine sediments may be qualitatively similar to that in soils (Ware and Roan, 1970).

Many of the organic compounds tested in this work had a slight inhibitory effect on acetylene reduction. Similar results in soil were shown by Tu (1978) for eight carbamate and organophosphorus insecticides, including chlordane, which in our experiments was inhibitory only for the first 48 h. Tam and Trevors (1981a, 1981b) also reported inhibition of acetylene reduction in pure culture and in soil by PCP; this is similar to our results. In contrast, Nayak and Rao (1980, 1982) found that several pesticides stimulated nitrogen fixation in flooded soil; however, they did not test toxaphene, the only stimulatory compound in our study. Hexachlorocyclohexane (lindane) was stimulatory in three types of soil and inhibitory in two. In our experiments, lindane had no effect initially, but later it was inhibitory.

Grant and Payne (1982) found that, of the six pesticides they studied, only dalapon had a clearly inhibitory effect on denitrification activity in saltmarsh sediment. The compounds tested in our study were not the same, but we saw inhibition of nitrous oxide production by lindane, phenanthrene, toxaphene, naphthalene, and PCP.

Some of the compounds we tested (PCP, phenanthrene, naphthalene) were inhibitory to both nitrogenase activity and denitrification potential. However, there were several compounds that had opposite effects on the two processes studied: toxaphene stimulated nitrogenase activity and inhibited denitrification, and heptachlor and Aroclor 1254 inhibited nitrogenase and stimulated denitrification. A similar pattern was observed in the effect of some metals on these activities (Slater and Capone, 1984): nickel stimulated nitrogenase and inhibited denitrification potential, whereas Pb, Cu, Zn, and Cr inhibited nitrogenase and stimulated denitrification. Such compounds would affect the pool of combined nitrogen in the sediment twice over, affecting a source and a sink in opposite directions.

Long-term effects of oil on denitrification and nitrogen fixation were noted by Griffiths et al. (1982). The observed inhibition persisted, and in some cases increased in severity, for at least 1.5 y. There is also some indication in the results of our experiments that long-term effects of organic pollutants on acetylene reduction may be different from immediate responses. Thus, the inhibition caused by PCP progressively increased over a 4-week incubation. For DDT, on the other hand, some recovery from the initial inhibition was apparent by 3 weeks. For Aroclor 1254 and lindane, effects at 3 d were not significant, but by 2 weeks Aroclor 1254 caused significant stimulation and lindane caused significant inhibition of acetylene reduction. Once established, these effects appeared to continue for at least 4 weeks. These effects are worth further study in chronic-dosing microcosm experiments, where sediment, under more natural conditions, is exposed for longer time periods to low concentrations of a compound.

It is possible that the change in effects over time may be related to biodegradation of the compound studied. In general, anaerobic mineralization of synthetic compounds is negligible compared to aerobic mineralization (Kobayashi and Rittman, 1982). Although reductive dechlorination may occur for chlorinated pesticides, the products formed are often persistent and may themselves be toxic. Both DDT and endrin can undergo reductive dechlorination (Lal and Saxena, 1982). It may be that the products, probably 2,2-*bis*(chlorophenyl)-1,1-dichloroethane (DDD) and ketoendrin, are less inhibitory to nitrogenase than are DDT and endrin themselves; this would account for the gradual decrease in severity of inhibition that we observed. The time scale of conversion of DDT to DDD (15–30 d) observed by Castro and Yoshida (1971) in flooded soil is similar to the length of our experiments. For lindane, there is also evidence of anaerobic degradation, first to either β-pentachlorocyclohexane or γ-tetrachlorocyclohexane (γ-TCCH) (Lal and Saxena, 1982). *Clostridium* spp. in flooded soils have been shown to degrade γ-TCCH to CO_2 in days to weeks (Sethunathan et al., 1969). Because in our experiments lindane became more, rather than less, toxic it may be that its degradation to γ-TCCH occurred, but it did not completely mineralize. For PCP and PCBs, however, biodegradation in the absence of oxygen appears slight (Liu et al., 1981; DeLaune et al., 1983) and probably does not account for the apparent changes in activity over time.

The effects observed in our experiments with organic compounds were generally not as extreme as the effects seen in our previous experiments with metal (Slater and Capone, 1984), even at the high concentrations used. However, estuarine sediments, particularly near dredged-material and sewage dumpsites, contain appreciable concentrations of many organic compounds (McLeod et al., 1981). These experiments suggest that there might be some interference in the processes of nitrogen fixation and denitrification. Recovery to the control level was not observed within 4 weeks. In some cases (lindane and Aroclor 1254) the indications are that long-term exposure may have increasingly disruptive effects.

ACKNOWLEDGMENTS

We thank John Budin, Dwight Reese, and Ron Kiene for their assistance on various aspects of this study. Financial assistance was provided by the Jessie Smith Noyes Foundation, the Office of Marine Pollution Assessment, the U.S. National Oceanic and Atmospheric Administration (Grant NA-80-RAD-00057), the U.S. Environmental Protection Agency (Grant R-809475-01-0), the U.S. National Science Foundation (Grant OCE-82-00157), and the Hudson River Foundation (Grant 14-83B-12).

REFERENCES

Burns, R. C., and R. W. F. Hardy. 1975. Nitrogen Fixation in Bacteria and Higher Plants. Springer-Verlag, New York, 189 pp.

Capone, D. G. 1983. Benthic nitrogen fixation. *In:* Nitrogen in the Marine Environment, E. J. Carpenter and D. G. Capone (Eds.). Academic Press, New York, pp. 153–161.

Capone, D. G., and B. F. Taylor. 1980. Microbial nitrogen cycling in a seagrass community. *In:* Estuarine Perspectives, V. S. Kennedy (Ed.). Academic Press, New York, pp. 153–161.

Castro, T. F., and T. Yoshida. 1971. Degradation of organochlorine insecticides in flooded soils in the Philippines. *Journal of Agricultural and Food Chemistry,* **19,** 1168–1170.

Collins, S. 1981. Measurement of Denitrification in Undisturbed Salt Marsh Sediments. M.S. Thesis, State University of New York, Stony Brook, 27 pp.

DeLaune, R. D., R. P. Gambrell, and K. S. Reddy. 1983. Fate of pentachlorophenol in estuarine sediment. *Environmental Pollution,* **6**(B), 279–308.

Grant, M. A., and W. J. Payne. 1982. Effects of pesticides on denitrifying activity in salt marsh sediments. *Journal of Environmental Quality,* **11,** 369–372.

Griffiths, R. P., B. A. Caldwell, W. A. Broich, and R. Y. Morita. 1982. The long-term effects of crude oil on microbial processes in subarctic marine sediments. *Estuarine and Coastal Shelf Science,* **15,** 183–198.

Haines, E., A. Chalmers, R. Hanson, and B. Sherr. 1977. Nitrogen pools and fluxes in a Georgia salt marsh. *In:* Estuarine Processes, Vol. 2, M. Wiley (Ed.). Academic Press, New York, pp. 241–254.

Haines, J. R., R. M. Atlas, R. P. Griffiths, and R. Y. Morita. 1981. Denitrification and nitrogen fixation in Alaskan Continental Shelf sediments. *Applied and Environmental Microbiology,* **41,** 412–421.

Hanson, R. B., and K. Gundersen. 1977. Relationship between nitrogen fixation (acetylene reduction) and the C:N ratio in a polluted coral reef ecosystem, Kaneohe Bay, Hawaii. *Estuarine and Coastal Marine Science,* **5,** 437–444.

Hardy, R. W. F., R. D. Holsten, E. K. Jackson, and R. C. Burns. 1968. The acetylene–ethylene assay for N_2 fixation: laboratory and field evaluation. *Plant Physiology,* **43,** 1185–1207.

Hattori, A. 1983. Denitrification and dissimilatory nitrate reduction. *In:* Nitrogen in the Marine Environment, E. J. Carpenter and D. G. Capone (Eds.). Academic Press, New York, pp. 191–231.

Hites, R. A., R. E. Laflamme, and J. W. Farrington. 1977. Sedimentary polycyclic aromatic hydrocarbons: the historical record. *Science,* **198,** 829–831.

Knowles, R., and C. Wishart. 1977. Nitrogen fixation in arctic marine sediments: effect of oil and hydrocarbon fractions. *Environmental Pollution,* **13,** 133–149.

Kobayashi, H., and B. E. Rittman. 1982. Microbial removal of hazardous organic compounds. *Environmental Science and Technology,* **16,** 170A–183A.

Lal, R., and D. M. Saxena. 1982. Accumulation, metabolism, and effects of organochlorine insecticides on microorganisms. *Microbiological Reviews,* **46,** 95–127.

Liu, D., K. Thomson, and W. M. J. Strachan. 1981. Biodegradation of pentachlorophenol in a simulated aquatic environment. *Bulletin of Environmental Contamination and Toxicology,* **26,** 85–90.

Maruyama, Y., T. Suzuki, and K. Otobe. 1974. Nitrogen fixation in the marine environment: The effect of organic substrates on acetylene reduction. *In:* Effect of the Ocean Environment on Microbial Activities, R. R. Colwell and R. Y. Morita (Eds.). University Press, Baltimore, pp. 341–353.

McLeod, W. D., Jr., L. S. Ramos, A. J. Friedman, D. G. Burrows, P. G. Prohaska, D. L. Fisher, and D. W. Brown. 1981. Analysis of Residual Chlorinated Hydrocarbons, Aromatic Hydrocarbons and Related Compounds in Selected Sources, Sinks and Biota of the New York Bight. Technical Memorandum OMPA-6, National Oceanic and Atmospheric Administration, Boulder, Colorado, 128 pp.

Nayak, D. N., and V. R. Rao. 1980. Pesticides and heterotrophic nitrogen fixation in paddy soils. *Soil Biology and Biochemistry,* **12,** 1–4.

Nayak, D. N., and V. R. Rao. 1982. Pesticides and nitrogen fixation in a paddy soil. *Soil Biology and Biochemistry,* **14,** 207–210.

Nishio, T., I. Koike, and A. Hattori. 1982. Denitrification, nitrate reduction, and oxygen consumption in coastal and estuarine sediments. *Applied and Environmental Microbiology,* **43,** 648–653.

Nixon, S. W. 1981. Remineralization and nutrient cycling in coastal marine ecosystems. *In:* Estuaries and Nutrients, B. J. Neilson and L. E. Cronin (Eds.). Humana Press, Clifton, New Jersey, pp. 111–138.

O'Conner, J. S., and H. M. Stanford (Eds.). 1979. Chemical Pollutants of the New York Bight: Priorities for Research. National Oceanic and Atmospheric Administration, Boulder, Colorado, 217 pp.

Pavlou, S. P., and W. Hom. 1976. Interlaboratory calibration results from chlorinated hydrocarbon analysis in marine sediments. *Marine Chemistry,* **4,** 155–163.

Payne, W. J. 1981. Denitrification. Wiley-Interscience, New York, 214 pp.

Payne, W. J. 1984. Influence of acetylene on microbial and enzymatic assays. *Journal of Microbiological Methods*, **2,** 117–133.

Rowe, G. T., C. H. Clifford, and K. L. Smith, Jr. 1975. Benthic nutrient regeneration and its coupling to primary productivity in coastal waters. *Nature*, **255,** 215–217.

Sethunathan, N., E. M. Bautista, and T. Yoshida. 1969. Degradation of benzene hexachloride by soil bacterium. *Canadian Journal of Microbiology*, **15,** 1349–1354.

Slater, J., and D. G. Capone. 1984. Effects of metals on nitrogen fixation and denitrification in slurries of anoxic saltmarsh sediment. *Marine Ecology–Progress Series*, **18,** 89–95.

Sorensen, J. 1978. Denitrification rates in a marine sediment as measured by the acetylene inhibition technique. *Applied and Environmental Microbiology*, **36,** 139–143.

Stratton, G. W., R. E. Burrell, M. L. Kurp, and C. T. Corke. 1980. Interactions between the solvent acetone and the pyrethroid insecticide permethrin on activities of the blue-green alga *Anabaena*. *Bulletin of Environmental Contamination and Toxicology*, **24,** 562–569.

Tam, T.-Y., and J. T. Trevors. 1981a. Toxicity of pentachlorophenol to *Azotobacter vinelandii*. *Bulletin of Environmental Contamination and Toxicology*, **27,** 230–234.

Tam, T.-Y., and J. T. Trevors. 1981b. Effects of pentachlorophenol on asymbiotic nitrogen fixation in soil. *Water, Air, and Soil Pollution*, **16,** 409–414.

Taylor, B. F. 1983. Assays of microbial nitrogen transformations. *In:* Nitrogen in the Marine Environment, E. J. Carpenter and D. G. Capone (Eds.). Academic Press, New York, pp. 809–837.

Tu, C. M. 1978. Effect of pesticides on acetylene reduction and microorganisms in a sandy loam. *Soil Biology and Biochemistry*, **19,** 451–456.

Van Raalte, C. D., I. Valiela, E. J. Carpenter, and J. M. Teal. 1974. Inhibition of nitrogen fixation in salt marshes measured by acetylene reduction. *Estuarine and Coastal Marine Science*, **2,** 301–305.

Ware, G. W., and C. C. Roan. 1970. Interaction of pesticides with aquatic microorganisms and plankton. *Residue Review*, **33,** 15–45.

Whitney, D. E., G. M. Woodwell, and R. W. Howarth. 1975. Nitrogen fixation in Flax Pond: a Long Island salt marsh. *Limnology and Oceanography*, **20,** 640–643.

Yeomans, J. C., and E. G. Beauchamp. 1978. Limited inhibition of nitrous oxide reduction in soil in the presence of acetylene. *Soil Biology and Biochemistry*, **10,** 517–519.

Yeomans, J. C., and E. G. Beauchamp. 1982. Acetylene as a possible substrate in the denitrification process. *Canadian Journal of Soil Science*, **62,** 139–144.

Yoshinari, T., and R. Knowles. 1976. Acetylene inhibition of nitrous oxide reduction by denitrifying bacteria. *Biochemical and Biophysical Research Communications*, **69,** 705–710.

Yoshinari, T., R. Hynes, and R. Knowles. 1977. Acetylene inhibition of nitrous oxide reduction and measurement of denitrification and nitrogen fixation in soil. *Soil Biology and Biochemistry*, **9,** 177–183.

Chapter 8

Biodegradation Rates of Xenobiotic Compounds in Plankton Communities

Jan Kuiper and Arnbjörn O. Hanstveit

Laboratory for Applied Marine Research
Division of Technology for Society TNO
Den Helder, The Netherlands

ABSTRACT

There is little experimental basis for the extrapolation of results of laboratory biodegradation tests to more natural conditions. For this reason several experiments were performed during 1977–1981 with North Sea plankton communities. The plankton were enclosed in plastic bags with selected organic compounds. Biodegradation tests in the laboratory were performed simultaneously with model plankton ecosystem (MOP) experiments using water from the enclosed plankton communities. The compounds tested included phenol, 4-chlorophenol (4CP), 2,4-dichlorophenol (DCP), 3,4-dichloroaniline (DCA), 4-nitrophenol (4NP), 4,6-dinitroorthocresol (DNOC), tetrapropylenebenzenesulphonate (TPBS) and two commercial PCB mixtures (Aroclors 1221, 1254). The polychlorinated biphenyls (PCBs), DNOC and DCA, were not degraded in either test system. Phenol and chlorinated phenols were degraded in both systems, but differences were found in the degradation rates. The degradation rate of phenol and 4CP in the MOPs was at times much lower than that in the laboratory die-away test with the same seawater. This difference between the two test methods was even larger for 4NP and TPBS. These compounds were easily degraded in the laboratory tests but appeared to be resistant to biodegradation in the MOPs. Factors at the ecosystem level can limit degradation rates under natural conditions. These factors probably include (1) competition for inorganic nutrients or other organic substrates between bacteria capable of degrading a specific compound and other bacteria, (2) competition between bacteria and phytoplankton for inorganic nutrients, and (3) predation on bacteria by microzooplankton. Standard laboratory die-away tests only show whether a compound can be degraded by microorganisms. The results of these tests cannot be used for extrapolation to degradation rates actually existing in the sea. Use of natural water for the die-away tests is a better approach. Model ecosystems such as those employed in the present study can be used to bridge the gap between the laboratory and the field when information on the behavior of chemicals under more natural conditions is necessary.

8.1. INTRODUCTION

Information on the biodegradability of specific chemicals or chemical wastes is required in the framework of many international conventions on dumping of chemicals and wastes in the sea. There are also regulations for the introduction of new chemicals [Anonymous, 1979; Organisation for Economic Co-operation and Development (OECD), 1981; Oslo

Commission, 1984]. This information is generally derived from laboratory biodegradation tests (OECD, 1981; Oslo Commission, 1984). Although there is little experimental basis for the extrapolation of these test results to more natural conditions, it is generally considered that a compound classified as easily degradable by the OECD (1981) will be repidly decomposed by microorganisms in the natural environment.

The experiments described in this chapter, carried out from 1977 to 1981, compare the biodegradation of chemicals in a natural environment to relevant laboratory tests. The natural environment experiments were performed in model plankton ecosystems (MOPs) consisting of natural North Sea plankton communities enclosed by large plastic bags, in which the fate and effects of selected organic compounds were studied. The merits of this method as a model for marine pelagic ecosystems have been discussed extensively by Grice and Reeve (1982) and by Kuiper (1982a, 1982b).

Simultaneous biodegradation tests were performed in the laboratory using water from the MOPs. The model compounds tested differed widely in their degradability and included the following chemicals: phenol, 4-chlorophenol (4CP), 2,4-dichlorophenol (DCP), 3,4-dichloroaniline (DCA), 4-nitrophenol (4NP), 4,6-dinitroorthocresol (DNOC), tetrapropylenebenzenesulphonate (TPBS), and two commercial PCB mixtures [Aroclors 1221 and 1254 (Monsanto, St. Louis, Missouri)]. Some of the results with phenol, 4CP, DCP, and DCA have been published by De Kock and Kuiper (1981), Kuiper and Hanstveit (1982a, 1982b) and Kuiper (1982a). In this chapter the discussion will focus on the comparison between laboratory and MOP experiments.

8.2. EXPERIMENTAL

8.2.1. Model Plankton Ecosystem Experiments

In each experiment, 8 plastic bags (3 m long, 75 cm in diameter) were filled with 1.5 m^3 of seawater collected a few miles offshore of Den Helder, The Netherlands. The bags were anchored near a raft in the harbor of Den Helder where they were exposed to a natural light and temperature regime. The development of the enclosed community (bacteria, phytoplankton, and zooplankton) and a set of physicochemical parameters (pH, salinity, and light intensity, and concentrations of dissolved oxygen, phosphate, silicate, nitrate, and ammonia) were monitored for 4–8 weeks (Kuiper, 1981, 1982a).

The test compounds were generally added as single doses at relatively low concentrations at the start of the experiments. In some experiments, additional doses were given in order to study the possible adaptation of the community. Table 8.1 shows the compounds tested and their concentrations. The fate of the added contaminants in the enclosures was determined by measuring the concentrations of the compound in the water and in the sediment at the bottom of the bags. The disappearance of the parent compound was used as a measure of degradation. Possible absorption and diffusion of the test compounds through the plastic foil of the bags was studied in the laboratory (Kuiper and Hanstveit, 1982a, 1982b).

8.2.2. Laboratory Experiments

A die-away test (Hanstveit and De Kreuk, 1980; Kuiper and Hanstveit, 1982a), was used to study biodegradation in the laboratory. Seawater was collected from the MOPs and transported to the laboratory the same day. One-hundred and fifty-milliliter samples of this seawater were transferred to 300-ml conical flasks of either glass or polycarbonate. The test compound was then added to each set of flasks at concentrations similar to those in the MOPs (Table 8.1). In many experiments, part of the added test compound was radioactively labeled (Table 8.1). All experiments included sterilized controls.

The flasks were incubated in the dark in a Psycotherm shaking apparatus (New Brunswick Scientific Co. Inc., New Brunswick, New Jersey) at the same temperature as that in the bags. In order to follow degradation, concentrations of the parent compound were measured in flasks in which radioactive tracers were not added.

In the experiments with radioactively labeled compounds, loss of $^{14}CO_2$ and possible assimilation of ^{14}C in bacteria were used as indicators of biodegradation. Experiments lasted up to 57 d. Laboratory tests using a number of compounds were initiated during the MOP experiments in order to study possible adaptation of the bacterial community to the test compound. In some die-away tests, the inorganic nutrients $KHPO_4$ (3 or 10μM), NH_4NO_3 (180μM), and KNO_3 (10μM) were added to study the influence of nutrients on biodegradation.

8.2.3. Analytical Methods

The analytical methods used to measure the different parameters are described by Kuiper (1981, 1982a) and by Kuiper and Hanstveit (1982a, 1982b). The phenols DCA and DNOC were analyzed by high performance liquid chromatography (HPLC) (Kuiper and Hanstveit, 1982a, 1982b), and TPBS was analyzed by the methylene blue method (OECD, 1971). The PCBs were measured using gas chromatography after perchloration to decachlorobiphenyl (Kuiper, 1982c).

To measure the $^{14}CO_2$ production in the laboratory die-away tests with labeled compounds, 5-ml samples were acidified with 0.1-ml 2 N H_2SO_4 and flushed with air to remove the $^{14}CO_2$. One ml of the acidified sample was added to 15 ml of Lumagel® scintillation liquid (Lumac Systems AG, Basel, Switzerland) and counted in a Tricarb Model 2650 scintillation counter (Packard Instrument Company, Inc., Warrenville, Illinois).

The assimiliation of labeled compounds by bacteria was measured by filtration of 4 ml of the acidified sample through 0.2-μm pore diameter membrane filters, which were then washed with 5 ml of filtered seawater and counted as described above. Adsorption of labeled compounds to the filter was measured on separate filters and the results were used to correct the bacterial assimilation figures.

8.3. RESULTS

No biodegradation of the PCBs, DNOC, or DCA could be detected in the laboratory die-away tests or the MOP tests used in this study. However, in the MOPs concentrations of PCBs and DCA decreased due to adsorption and subsequent settling of suspended particles. Phenol, 4CP, and DCP were degraded in both test systems, but differences were sometimes observed in the degradation rates and patterns.

Figure 8.1 shows the percentage of degradation of phenol in two MOP experiments and one laboratory test. The phenol was degraded rapidly by bacteria in the laboratory test. A higher degradation rate was observed in one MOP experiment; in the other MOP experiment, carried out simultaneously with the laboratory test, phenol was degraded more slowly and, instead of showing an exponential decrease, phenol concentrations decreased linearly. This indicates that a factor other than phenol was limiting bacterial growth rates.

Figure 8.2 shows the degradation of 4CP in the May 1978 experiments. Here, direct comparison is more difficult because $^{14}CO_2$ (the end product of degradation) production was used as the indicator of biodegradation in the laboratory test, whereas the disappearance of the parent compound was used in the MOP experiment (the start of the degradation route was measured). The length of the lag phases, however, can be compared directly. After the initial addition of 1 mg 4CP

Table 8.1. Compounds Tested in Laboratory Die-Away Tests and Plankton Ecosystems

Test Compound	Initial Concentration (mg liter^{-1})	Date		Laboratory Die-Away Tests[a]			
				Duration (d)	Nutrients[b] Added	No Nutrients Added	Radioactively Labeled Compound Added
Phenol	9.0	August	1977	8			
Phenol	3.8	September	1978	12			x
4-Chlorophenol	0.1–1.0	May	1978	42	x(2)		x
4-Chlorophenol	0.3–1.0	August	1978	28	x(2)		x
4-Chlorophenol	0.3–1.0	August	1980	28	x(2)	x	x
2,4-Dichlorophenol	0.1–1.0	May	1978	42	x(2)		x
2,4-Dichlorophenol	0.3–1.0	August	1978	28	x(2)		x
2,4-Dichlorophenol	0.3–1.0	August	1980	28	x(2)	x	x
4-Nitrophenol	0.1–1.0	August	1979	49	x(1)	x	
4-Nitrophenol	0.1–1.0	April	1980	46		x	
4-Nitrophenol	0.3–1.0	August	1980	37	x(2)	x	x
4,6-Dinitroorthocresol	0.1–1.0	August	1979	57	x(1)	x	
4,6-Dinitroorthocresol	0.1–1.0	April	1980	38		x	
Tetrapropylenebenzenesulphonate	0.1–1.0	April	1981	38	x(2)		
3,4-Dichloroaniline	0.000–0.025	August	1977	35			
3,4-Dichloroaniline	0.1–1.0	May	1978	42	x(1)	x	x
Aroclor 1221	0.001–0.010	April	1981	38	x(2)		
Aroclor 1254	0.0005–0.005	April	1981	38	x(2)		

[a] The symbol "x" indicates that the test was performed simultaneously in the field with the Model Plankton Ecosystems.

[b] For nutrients added, (1) indicates a nutrient mixture of 15 μM KNO_3, 3μM K_2HPO_4; and (2) indicates a nutrient mixture of 180μM NH_4NO_3, 10μM K_2HPO_4.

liter^{-1} to the MOP, a lag phase of 2 weeks occurred, after which 4CP disappeared rapidly. A similar phenomenon occurred in the laboratory tests (Fig. 8.2a), although lag phases in the die-away tests started on day zero, they were much longer and showed a large variation (Fig. 8.2b). After the first dose of 0.1 mg 4CP liter^{-1} (not shown in Fig. 8.2), lag phases in the laboratory also varied but were usually shorter (16 d). After the second addition of 4CP to the MOPs and the laboratory tests, degradation started immediately, indicating that an adapted bacterial community was present. In the MOP experiments, 4CP concentations decreased linearly, whereas in the laboratory tests with nutrients added the results showed an exponential decrease. As in the case of phenol (Fig. 8.1), the resulting degradation rate of 4CP in the MOP was lower than that in the laboratory test with the same water.

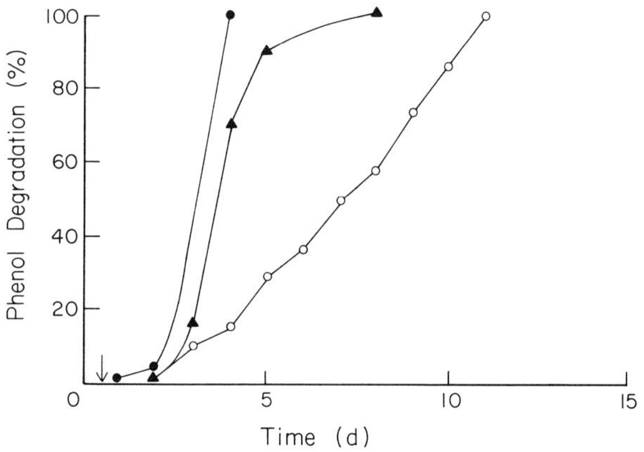

Figure 8.1. Degradation of phenol in two MOP experiments and one laboratory experiment using water from the MOPs: (●) MOP experiment, August 1977; (▲) laboratory experiment, September 1978; and (○) MOP experiment, September 1978. The arrow indicates the addition of phenol to the experiments.

Figure 8.3 shows the results of similar tests carried out in August 1980. Comparable results were obtained in the laboratory (Fig. 8.3a) and the MOP test (Fig. 8.3b). Although initially 4CP degraded somewhat slower in the laboratory test than in the MOP experiment, repeated additions produced very similar results. Repeated additions again resulted in shorter lag phases and more rapid degradation, indicating an adapted bacterial community. This adaptation was also shown by the number of colony forming units (CFUs) grown on medium containing 4CP (Fig. 8.4). The differences found in complete degradation between the laboratory and MOP tests (Figs. 8.2 and 8.3) were partly due to the analytical method; complete degradation was not reached in the laboratory tests because of the formation of stable intermediates.

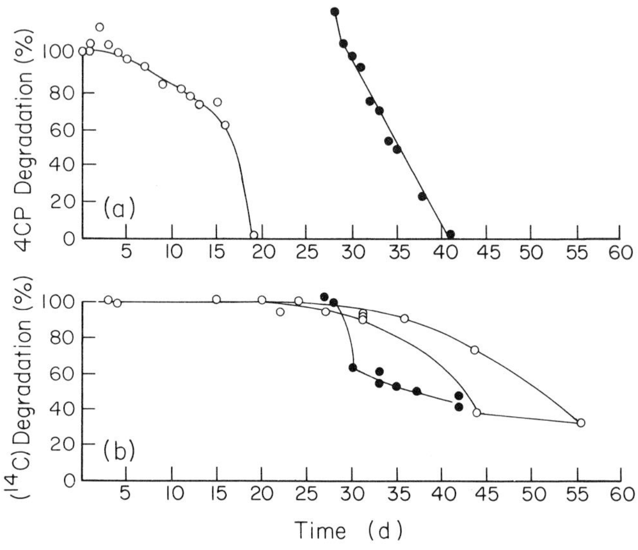

Figure 8.2. Degradation of 4CP in (a) MOPs and (b) laboratory experiments (May 1978): (○) first dose of 1 mg liter^{-1} 4CP and (●) second dose of 1 mg liter^{-1} 4CP. Laboratory tests were performed in duplicate.

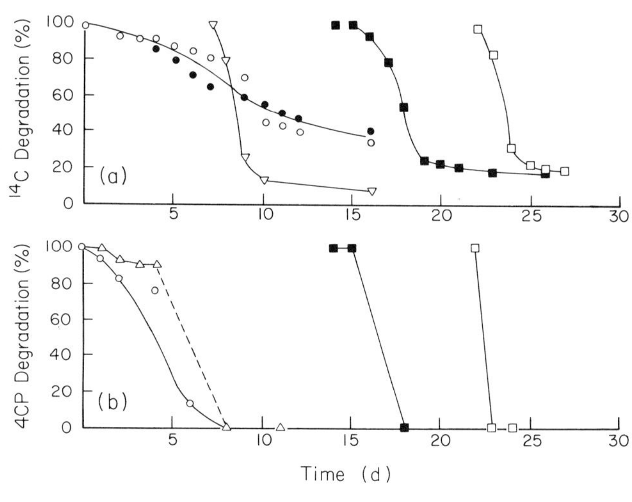

Figure 8.3. The degradation of 4CP in (a) laboratory and (b) MOP experiments (August 1980): (△) first dose of 0.3 mg liter^{-1} 4CP, (○) first dose of 1 mg liter^{-1} 4CP, (●) first dose of 1 mg liter^{-1} 4CP, (▽) second dose of 1 mg liter^{-1} 4CP (laboratory test), (■) third dose 1 mg liter^{-1} 4CP, and (□) fourth dose of 1 mg liter^{-1} 4CP.

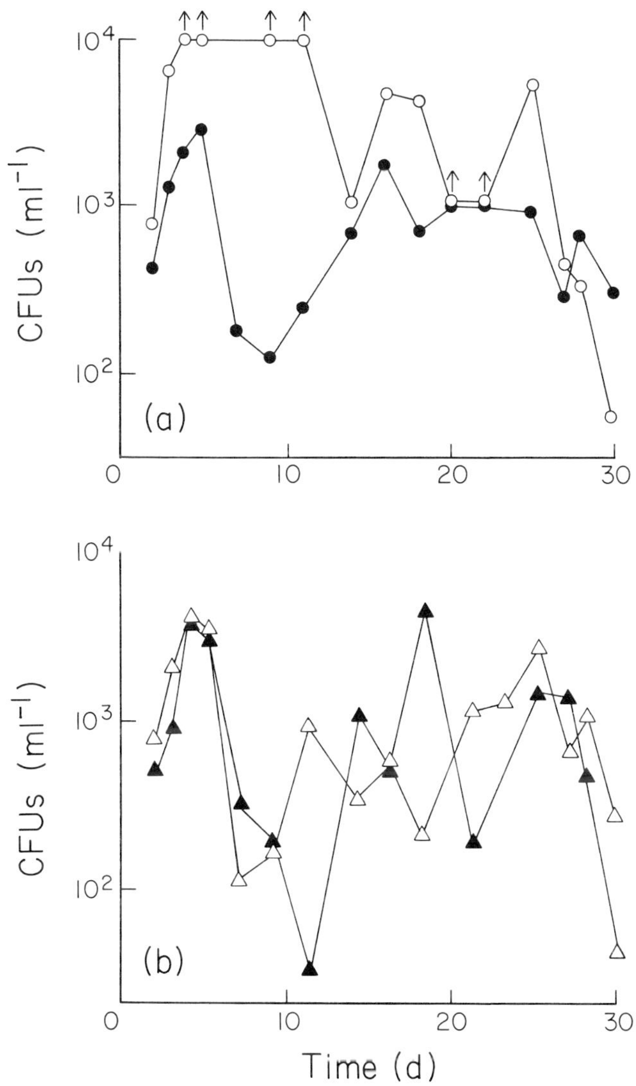

Figure 8.4. Number of colony forming units (CFUs) in water (from different MOPs) incubated on medium containing 4CP (August 1980): (a) 4CP added to water at concentrations of (○) 0.3 mg liter^{-1} and (●) 1.0 mg liter^{-1}, and (b) no 4CP added to water; (△, ▲) represents duplicate control runs.

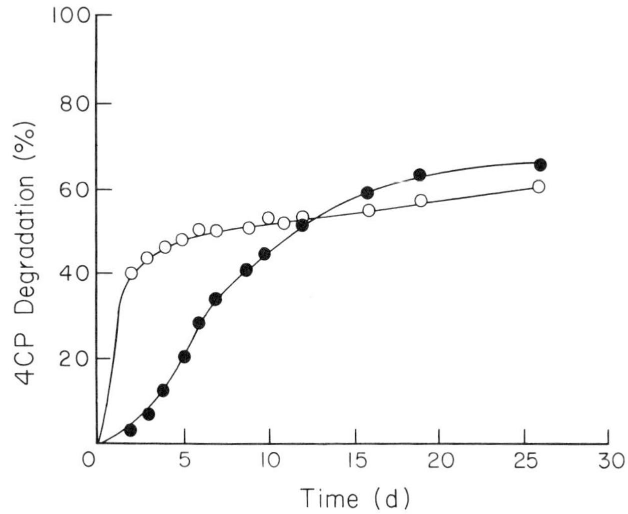

Figure 8.5. The degradation of 4CP in laboratory tests with (○) nutrients added and (●) no nutrients added.

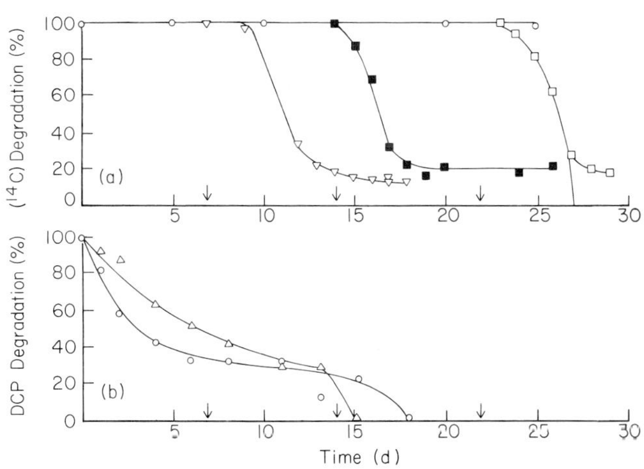

Figure 8.6. The degradation of DCP in (a) laboratory and (b) MOP experiments (August 1980): (△) first test, 0.3 mg liter^{-1} DCP; (○) first test, 1 mg liter^{-1} DCP; (▽) second test, 1 mg liter^{-1} DCP; (■) third test, 1 mg liter^{-1} DCP; and (□) fourth test, 1 mg liter^{-1} DCP. In the laboratory experiments, the arrows indicate the commencement of the second, third, and fourth tests, respectively, and in the MOP experiments they indicate days on which water was collected for laboratory experiments.

Several tests were carried out with and without addition of nitrate and phosphate in order to test the influence of nutrients on the degradation of 4CP. In all cases, addition of nutrients stimulated the biodegradation of 4CP; an example of a strong stimulation is given in Fig. 8.5. Figure 8.6 shows the results of an experiment with DCP. The variation in the results from laboratory tests (Fig. 8.6a), initiated at the moment of the first addition of DCP to the MOPs, was very large in all experiments. In the first laboratory tests, no degradation was found after 28 d. The results of laboratory tests using water removed from the MOPs (Fig. 8.6b) 7, 14, and 22 d later showed much better reproducibility and faster degradation. In the MOP, DCP degraded in 16–20 d, which was faster than in the laboratory tests started the same day but slower than in laboratory tests started later. Experiments

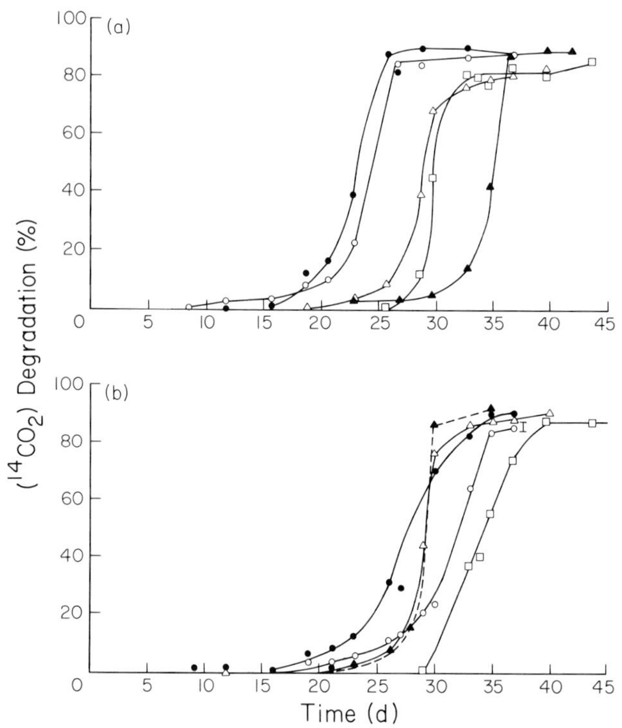

Figure 8.7. The degradation of 4NP in August 1980 tests with water from MOPs after (a) an initial dose, 0.3 mg liter^{-1} 4NP; and (b) an addition of 1.0 mg liter^{-1} 4NP. Residence time in MOPs: (○, ●) 0 d, (△) 7 d, (▲) 14 d, and (□) 21 d.

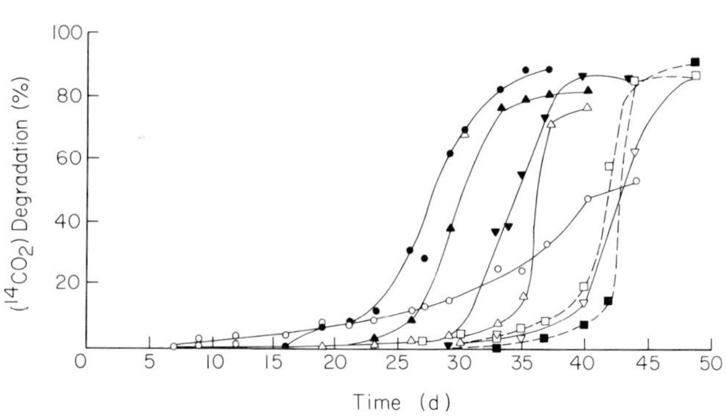

Figure 8.8. The degradation of 4NP in laboratory experiments with water from MOPs (August 1980). Open symbols represent experiments with added nutrients, solid symbols represent experiments with no added nutrients. Residence time in MOPs: (○, ●) 0 d, (△, ▲) 7 d, (□, ■) 14 d, and (▼, ▽) 21 d.

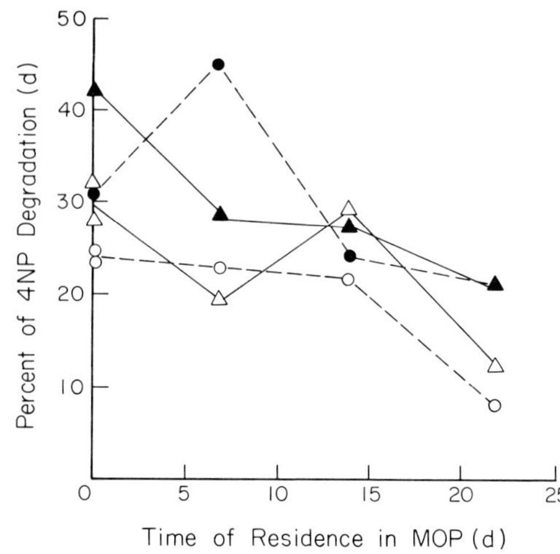

Figure 8.9. Time necessary to reach 50% degradation of 4NP in laboratory experiments, as a function of the residence time of the water in the MOPs, at initial doses of: (●) 0.3 mg liter^{-1} 4NP, with nutrients; (○) 0.3 mg liter^{-1} 4NP, with nutrients; without nutrients; (▲) 1.0 mg liter^{-1} 4NP, with nutrients; and (△) 1.0 mg liter^{-1} 4NP, without nutrients.

with added inorganic nutrients were also carried out with DCP. These tests showed the same results as those without addition of nutrients.

Figure 8.7 shows the percentage of degradation of 4NP in die-away tests carried out in 1980. No degradation of 4NP was observed in the simultaneous MOP experiment. In the laboratory, 4NP was degraded easily and degradation rates in water removed from the MOP increased as residence times increased, indicating adaptation of bacteria to the presence of 4NP in the MOPs. Experiments with added nutrients were also carried out with 4NP. Some results of these tests are shown in Fig. 8.8. Addition of large amounts of nutrients (Table 8.1) generally resulted in lower degradation rates. Results from tests with 4NP in the 1981 experiments are shown in Fig. 8.9.

During the MOP experiments with 4NP in 1981, water samples were plated out on media containing 4NP that selected for 4NP-degrading bacteria; the numbers of CFUs did not increase markedly in the MOPs during the experiment. Growth rates on plates containing 4NP were very low compared with rates measured on similar 4CP containing plates (Fig. 8.4); by week 3, 90% of the total CFUs were present on the media containing 4CP. On the media containing 4NP, however, only 30–40% of the total CFUs were present after 3 weeks.

Similar results were obtained with 4NP in two earlier experiments. In the first experiment (August 1979), 4NP disappeared from the water in the laboratory test after 9 d. When relatively small amounts of nutrients were added (15 μM KNO_3 and 3μM K_2HPO_4), 4NP disappeared after 7 d. In the second experiment (April 1980), 4NP could not be traced 14 d after the start of laboratory tests. In the two MOP experiments, lasting 49 and 46 d, respectively, degradation of 4NP was not observed. The 4NP was used as a model compound in other degradation studies in our laboratory. Using enriched media (OECD, 1981), 4NP disappeared rapidly in these experiments (complete degradation in 18–22 d, initial concentrations of 2–10 mg 4NP liter^{-1}).

Comparable results were obtained from tests with TPBS. Figure 8.10 shows the degradation of TPBS in the MOPs and the die-away tests. No significant degradation was found in the MOPs with 0.1 mg liter^{-1} TPBS; at a concentration of 1 mg liter^{-1}, 15–20% had degraded after 38 d. In the laboratory tests with the MOP water, 60–75% degradation was found after 39 d, whereas TPBS was degraded even faster (65–70% degradation within 28 d) in the OECD (1981) routine test.

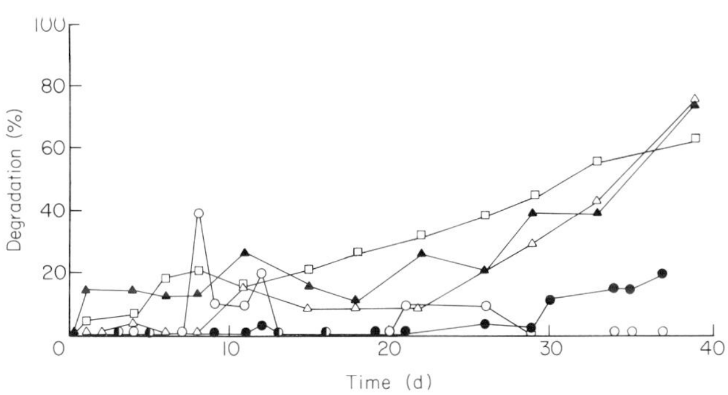

Figure 8.10. The degradation of TPBS in MOPs and laboratory experiments (April 1981): (○) MOP with 0.1 mg liter^{-1} TPBS, (●) MOP with 1 mg liter^{-1} TPBS, (□) laboratory experiment with 1 mg liter^{-1} TPBS, and (△, ▲) laboratory experiments with 2 mg liter^{-1} TPBS.

8.4. DISCUSSION

Laboratory die-away tests and experiments with natural marine plankton communities produced similar results with respect to biodegradation of PCBs, DNOC, and DCA. The relative persistence of these chemicals in the present study is in agreement with results reported by Hutzinger et al. (1974), Still and Herrett (1976), West and Hatcher (1980), Liu (1981), and Tabak et al. (1981). In most experiments, both test systems showed that phenol and chlorinated phenols were degraded at comparable rates (Kuiper and Hanstveit, 1982b), and 4CP disappeared faster than DCP. This is in agreement with laboratory results obtained by DiGeronimo et al. (1979), Lee and Ryan (1979), and Baker and Mayfield (1980).

The results of replicate die-away tests with 4CP and DCP initiated at the start of an MOP experiment sometimes showed large variation, particularly during the lag phase. Such large variations are often found when relatively persistent chemicals are tested with small inocula of bacteria. However, the bacterial community became adapted to the presence of the chemicals. Adaptation was shown by increasing degradation rates after repeated addition of 4CP to the MOPs and by repeated laboratory tests with 4CP, DCP, and 4NP, using water previously exposed to the same compound. This adaptation was also shown by the increased number of CFUs when water samples with 4CP resistant bacteria were plated on medium containing 4CP. Similar adaptation has also been shown for other compounds and other media (DiGeronimo et al., 1978; Sayler et al., 1978; Vela and Ralston, 1978; Baker and Mayfield, 1980; Fournier, 1980).

The degradation rates of phenol and 4CP in the MOP were occasionally much lower than those found in laboratory die-away tests with the same water. In these cases decreasing concentrations followed a linear pattern, and not the expected rapid exponential decrease found in simultaneous laboratory experiments with the same water. These linear decreases in concentration indicate that a factor other than 4CP or phenol limited the growth of the bacteria degrading these compounds. The very low phosphate (<0.1 μM) or nitrate (<0.1 μM) concentrations during these experiments probably limited the growth of the bacteria. Addition of these nutrients in the laboratory tests during the third experiment with 4CP resulted in increased degradation rates; this indicates that these nutrients limited the degradation rates in the MOPs.

Limitation of biodegradation rates by low concentrations of inorganic nutrients has been reported by Atlas and Bartha (1972), Olivieri et al. (1978), and Lee and Ryan (1979). In natural oligotrophic waters, bacteria capable of degrading a compound compete with other bacteria and phytoplankton for these nutrients. Goldman et al. (1979) showed that in oligotrophic oceanic waters phytoplankton can grow at maximum rates, indicating that phytoplankton have the ability to compete with bacteria at very low nutrient concentrations. Davis et al. (1977) found that the biodegradation rates of various oils depended mainly on the nutrient status of the seawater sample used in the experiment. Degradation rates of 4CP were higher in winter than in summer (when nutrient concentrations were low) in water samples from a station in the North Sea, 30 km off the Dutch coast (De Kreuk and Hanstveit, 1981).

The fact that the degradation rate of a compound can be limited by low concentrations of mineral nutrients such as phosphate and nitrate is important for extrapolation of results from laboratory biodegradation tests to the field. Laboratory tests are often performed in the dark, so that competition for nutrients between bacteria and algae does not occur. Moreover, nutrients are often added to laboratory cultures in unrealistically large amounts (OECD, 1981).

The results of the laboratory experiments with 4CP indicated that stable intermediates were formed; this was also found by De Kreuk and Hanstveit (1981). This was probably also the case in the MOPs, because the water was tinged yellow after repeated additions of 4CP. Using chromatography on samples from laboratory experiments, Kuiper and Hanstveit (1982b) reported that this yellow color was probably caused by the presence of 2-hydroxy-5-chloromuconic-acid-semi-aldehyde. This compound is a final metabolite formed from 4CP by co-metabolism and meta-cleavage of the ring structure. This formation of stable metabolites also explains the seemingly different results between the laboratory and MOP tests with 4CP, when using complete degradation as a criterion (Figs. 8.2 and 8.3).

The differences between the two test methods were larger in the cases of 4NP and TPBS. These compounds were degraded in the laboratory tests but appeared to be resistant to biodegradation in the MOP test. In the case of 4NP, this marked difference was found in three separate MOP experiments conducted during different seasons and lasting between 5 and 8 weeks. Surprisingly, increased biodegradation rates were found in laboratory tests using water from the MOPs that had been exposed to 4NP for 1–3 weeks. This indicates adaptation of the bacteria in the MOP to 4NP. Nevertheless, no degradation of 4NP was observed in the MOPs, and relative numbers of 4NP resistant bacteria, counted on medium containing 4NP, did not increase during the MOP experiment. The slower degradation of 4NP after addition of high concentrations of nutrients in laboratory experiments was unexpected, because nutrient additions stimulated 4CP degradation and did not influence DCP degradation. The compound 4NP is often used in biodegradation tests (Raymont and Alexander, 1971; Munnecke and Hsieh, 1974, 1976; Sudhakar-Barik et al., 1976, 1978). Complete degradation is generally found after 5–30 d. The lower degradation rates of 4NP and TPBS under near-natural conditions may be explained by different mechanisms. First, rapid adaptation of bacterial flora to 4NP may be inhibited by high concentrations of easily degradable natural organic substrate, favoring the development of other bacteria. Chou and Bohonos (1979) found that the presence of readily assimilated substrate inhibited the formation and maintenance of a microbial flora which would decompose the more difficult compounds. An indication that such diauxic effects are important in the degradation of 4NP may be found in biodegradation studies with seawater in chemostats, which show that 4NP is relatively stable and that degradation occurs only when the concentrations of other organic carbon sources are very low (Van den Berg, 1982). A second factor that might cause the difference between laboratory and MOP experiments is the competition from other bacteria and phytoplankton for nutrients. Shortage of ammonia, nitrate, or phosphate probably did not occur in the present study, but another nutrient not measured in the experiments may have been limiting. A third possibility, also related to the dynamics of the plankton community, is that there may be predation by protozoa or other microzooplankton on all bacteria in the MOP, including those capable of degrading 4NP. Bacteria that grow slowly have a disadvantage over those that grow rapidly. Results of the CFU counts on the media containing 4NP, and the long residence times in the chemostat experiments mentioned above, indicate that these 4NP-degrading bacteria grow slowly.

Standard laboratory biodegradation tests are generally carried out under conditions which exclude competitive interrelations. Predators are excluded or die, phytoplankton growth and production of easily assimilated organic substrate is inhibited by performing tests in the dark, large inocula that disrupt the original bacterial community are introduced, and inorganic nutrients are added, thus excluding competition for these nutrients. These competitive interrelations should not be excluded, because overestimated degradation rates may result.

8.5. CONCLUSIONS

Results of laboratory die-away tests using natural seawater and experiments with natural marine plankton communities are generally in agreement, although large differences are sometimes found for some compounds, for which laboratory tests to a large extent overestimate the degradation rate under more natural conditions. We therefore conclude that laboratory die-away tests can only show the potential of a compound for biodegradation. The results of these tests cannot be extrapolated to degradation rates actually occurring in the sea. The results with 4NP in particular show that a compound that would be classified as easily degradable in a standardized laboratory test may persist under more natural conditions. Model ecosystems such as those employed in the present study can be used to bridge the gap between the laboratory and field if information on the behavior of chemicals under more natural conditions is necessary.

ACKNOWLEDGMENTS

We acknowledge the financial support of the Commission of the European Communities [Contract Nos. ENV–461–N and 277–77–ENV(N)], the Umwelt Bundes Amt, Berlin (Contract No. 25-97 406/28), and the Ministry of Transport and Public Works, The Hague. We would also like to thank many colleagues from our laboratories in Delft and Den Helder, who carried out the various experiments, and P. B. Davis, who corrected the English.

REFERENCES

Anonymous. 1979. 6th Amendment EEC. *Official Journal of the European Communities*, **L259**, 10–28.

Atlas, R. M., and R. Bartha. 1972. Degradation and mineralization of petroleum in sea water: limitation by nitrogen and phosphorus. *Biotechnology and Bioengineering*, **14**, 309–318.

Baker, M. D., and C. I. Mayfield. 1980. Microbial and non-biological decomposition of chlorphenols and phenol in soil. *Water, Air and Soil Pollution*, **13**, 411–424.

Chou, T., and N. Bohonos. 1979. Diauxic and cometabolic phenomena in biodegradation evaluations. *In:* Proceedings of a Workshop on Microbial Degradation of Pollutants in Marine Environments, A. W. Bourquin and P. H. Pritchard (Eds.). U.S. Environmental Protection Agency, Gulf Breeze, Florida, pp. 76–88.

Davis, S. J., C. F. Gibbs, and K. B. Pugh. 1977. Quantitative studies on marine biodegradation of oil. III. Comparison of different crude oil residues and effects of sea water source. *Environmental Pollution*, **13**, 203–215.

De Kock, W. C., and J. Kuiper. 1981. Possibilities for marine pollution research at the ecosystem level. *Chemosphere*, **10**, 575–603.

De Kreuk, J. F., and A. O. Hanstveit. 1981. Determination of the biodegradability of the organic fraction of chemical wastes. *Chemosphere*, **10**, 561–573.

DiGeronimo, M. J., M. Nikaido, and M. Alexander. 1978. Most-probable-number technique for the enumeration of aromatic degraders in natural environments. *Microbial Ecology*, **4**, 263–266.

DiGeronimo, M. J., R. S. Boethling, and M. Alexander. 1979. Effect of chemical structure and concentration on microbial degradation in model ecosystems. *In:* Proceedings of a Workshop on Microbial Degradation of Pollutants in Marine Environments, A. W. Bourquin and P. H. Pritchard (Eds.). U. S. Environmental Protection Agency, Gulf Breeze, Florida, pp. 154–166.

Fournier, J. C. 1980. Enumeration of the soil microorganisms able to degrade 2,4-D by metabolism or co-metabolism. *Chemosphere*, **9**, 169–171.

Goldman, J. C., J. J. McCarthy, and D. G. Peavey. 1979. Growth rate influence on the chemical composition of phytoplankton in oceanic waters. *Nature*, **279**, 210–215.

Grice, G. D., and M. R. Reeve (Eds.). 1982. Marine Mesocosms. Springer-Verlag, New York, 430 pp.

Hanstveit, A. O., and J. F. De Kreuk. 1980. Determination of degradability. *In:* Degradability, Ecotoxicity and Bioaccumulation, H. J. Hueckled (Ed.). Dutch Government Publishing Office, The Hague, pp. 111–148.

Hutzinger, D., S. Safe, and V. Zitko. 1974. The Chemistry of PCBs. CRC Press, Cleveland, Ohio, 674 pp.

Kuiper, J. 1981. Fate and effects of cadmium in marine plankton communities in experimental enclosures. *Marine Ecological Progress Series*, **6**, 161–174.

Kuiper, J. 1982a. The Use of Enclosed Plankton Communities in Aquatic Ecotoxicology. Thesis, Agricultural University, Wageningen, The Netherlands, 256 pp.

Kuiper, J. 1982b. Ecotoxicological experiments with marine plankton communities in plastic bags. *In:* Marine Mesocosms, G. D. Grice and M. R. Reeve (Eds.). Springer-Verlag, New York, pp. 181–193.

Kuiper, J. 1982c. The Use of Model Ecosystems for the Validation of Screening Tests for Biodegradation and Acute Toxicity. Report No. CL 82/01, Division of Technology for Society TNO Den Helder, The Netherlands, 134 pp.

Kuiper, J., and A. O. Hanstveit. 1982a. Fate and effects of 3,4-dichloroaniline (DCA) in marine plankton communities in experimental enclosures. *Ecotoxicological Environmental Safety*, **8**, 34–54.

Kuiper, J., and A. O. Hanstveit. 1982b. Fate and effects of 4-chlorophenol (4CP) and 2,4-dichlorophenol (DCP) in marine plankton communities in experimental enclosures. *Ecotoxicological Environmental Safety*, **8**, 15–33.

Lee, R. F., and C. Ryan. 1979. Microbial degradation of organochlorine compounds in estuarine waters and sediments. *In:* Proceedings of a Workshop on Microbial Degradation of Pollutants in Marine Environments, A. W. Bourquin and P. H. Pritchard (Eds.). U. S. Environmental Protection Agency, Gulf Breeze, Florida, pp. 443–449.

Liu, D. 1981. Biodegradation of Aroclor 1221 type PCBs in sewage wastewater. *Bulletin of Environmental Contamination and Toxicology*, **27**, 695–703.

Munnecke, D. M., and D. P. H. Hsieh. 1974. Microbial decontamination of parathion and p-nitrophenol in aquous media. *Applied Microbiology*, **28**, 212–217.

Munnecke, D. M., and D. P. H. Hsieh. 1976. Pathways of microbiological metabolism of parathion. *Applied Environmental Microbiology*, **31**, 63–69.

Organisation for Economic Co-Operation and Development (OECD). 1971. Pollution by Detergents. Determinations of the Biodegradability of Anionic Synthetic Surface Active Agents/Pollution par les Detergents. Determination de la Biogradabilite des Agents de Surface Synthetiques Anioniques. Organisation for Economic Co-Operation and Development, Paris, 62 pp.

OECD. 1981. OECD Guidelines for Testing of Chemicals. Organisation for Economic Co-Operation and Development, Paris, 332 pp + appendices.

Olivieri, R., A. Robertiello, and L. Degan. 1978. Enhancement of microbial degradation of oil pollutants using lipophilic fertilizers. *Marine Pollution Bulletin*, **9**, 217–220.

Oslo Commission. 1984. General Guidelines for the Implementation of Test Procedures Under the Prior Consultative Procedure. Seventh Annual Report on the Activities of the Oslo Commission, Annex 3.Chameleon Press, London, 51 pp.

Raymont, D. G. M., and M. Alexander. 1971. Microbial metabolism and cometabolism of nitrophenols. *Pesticide Biochemical Physiology*, **1**, 123–130.

Sayler, G. S., R. Thomas, and R. R. Colwell. 1978. Polychlorinated biphenyl (PCB) degrading bacteria and PCB in estuarine and marine environments. *Estuarine and Coastal Marine Science*, **6**, 553–567.

Still, G. G., and R. A. Herrett. 1976. Methylcarbamates, carbanilates and acylanilides. *In:* Herbicides Chemistry, Degradation and Mode of Action, Vol. 2, P. C. Kearney and D. D. Kaufmann (Eds.). Marcel Dekker, New York, pp. 610–664.

Sudhakar-Barik, R., R. Sidderamppa, and N. Sethunathan. 1976. Metabolism of nitrophenols by bacteria isolated from flooded soil. *Antonie van Leeuwenhoek*, **42**, 461–470.

Sudhakar-Barik, R., R. Sidderamppa, P. H. Wahid, and N. Sethunathan. 1978. Conversion of p-nitrophenol to 4-nitro catechol by a *Pseudomonas* sp. *Antonie van Leeuwenhoek*, **44**, 171–176.

Tabak, H. H., S. A. Quave, C. I. Mashni, and E. P. Barth. 1981. Biodegradability studies with organic priority pollutants compounds. *Journal Water Pollution Control Federation*, **53**, 1503–1508.

Van den Berg, R. 1982. Biodegradability in the marine environment. The chemostat as a model system for the aquatic environment. *In:* Proceedings of the 5th International Congress of Pesticide Chemistry (IUPAC), J. Miyamoto (Ed.). Pergamon Press, Oxford, pp. V-10.

Vela, G. R., and J. R. Ralston. 1978. The effect of temperature of phenol degradation in wastewater. *Canadian Journal of Microbiology*, **24**, 1366–1370.

West, R. H., and P. G. Hatcher. 1980. Polychlorinated biphenyls in sewage sludge and sediments of the New York Bight. *Marine Pollution Bulletin*, **11**, 126–129.

Chapter 9

Lysosomal and Microsomal Responses to Oil-Derived Hydrocarbons in *Littorina littorea*

Michael N. Moore, Richard K. Pipe, Suzanne V. Farrar, Sarah Thomson, and Peter Donkin

Natural Environment Research Council
Institute for Marine Environmental Research
Plymouth, England

ABSTRACT

A cytochemical investigation of the hepatopancreatic digestive cells of the marine gastropod *Littorina littorea* from the vicinity of the Sullom Voe Oil Terminal in Shetland (Scotland) revealed reductions in latency of lysosomal β-glucuronidase, increases in total β-glucuronidase activity, and enlargement of lysosomes in comparison with *Littorina* from a nearby clean reference site. Snails from a contaminated reference site and many of the other test sites showed similar lysosomal responses. Cytochemically determined activity of blood cell NADPH-neotetrazolium reductase, a xenobiotic-inducible component of microsomal detoxification systems, was significantly elevated at four sites, including the contaminated reference site. Similar responses had been measured previously at several of these sites. Animals chosen from one of the test sites showed fourfold higher levels of phenanthrenes and tenfold higher levels of naphthalenes than did those from the clean reference site. Experimental treatment of *L. littorina* with 400 μg liter^{-1} of phenanthrene for 5 d induced lysosomal and microsomal responses similar to those described for some of the field animals. These responses were all shown to be reversible following 8 d in clean seawater. These findings are consistent with the hypothesis that the effects observed in some of the field samples are induced by polycyclic aromatic hydrocarbons (PAHs). The lysosomal responses are typical of enhanced autophagy, also inducible by starvation, and it is proposed that an autophagic process is involved in cell injury induced by PAHs. These results demonstrate the sensitivity of a cytochemical approach to the detection of cellular reactions to injury by environmental agents; this applies particularly to disturbance of lysosomal function.

9.1. INTRODUCTION

The use of cytochemistry or cellular biochemistry in the investigation of cellular function can increase the sensitivity of measurement to many thousands of times greater than that of conventional biochemistry (Chayen, 1978). In general, cytochemical measurements of components of single identifiable cell types have provided high sensitivity, often with low variability. This enables one to overcome the problem of having to deal with biochemical homogenates consisting of subcellular components derived from the many cell types comprising a particular tissue or organ (Chayen, 1978; Moore, 1980).

Cytochemistry has proven to be extremely sensitive in the determination of adaptive responses to environmental factors and cell injury in some marine organisms (Moore, 1980). The basis for the development of indexes of stress for use in environmental monitoring is to enable the detection of deleterious effects at an early and, hopefully, reversible stage. Although cytochemical tests can often provide the extreme sensitivity required, these measurements should be linked to the health of the organisms (Bayne et al., 1979).

Lysosomal cytochemistry has proven to be extremely useful both in mammals (Bitensky and Chayen, 1979) and in marine organisms (Moore, 1980), because lysosomal disturbances associated with cell injury are frequently related to deleterious consequences for the individual organism. Previous cytochemical investigation of lysosomal and microsomal responses in the snail *Littorina littorea* from the vicinity of Sullom Voe Oil Terminal in the Shetland Islands (Scotland) showed evidence of lysosomal disturbances in the digestive cells and induction of a component of the microsomal detoxification system in hemocytes of snails from four sites, including a site contaminated with hydrocarbons (Moore et al., 1982). These findings were interpreted as a probable response to the injurious presence of polycyclic aromatic hydrocarbons (PAHs), based on previous experimental data (Moore et al., 1978; Moore, 1979, 1980). Polycyclic aromatic hydrocarbons are also known to be concentrated in the lysosomal-vacuolar system (Allison, 1969).

The objectives of the present investigation were: (1) to determine if the effects described previously (Moore et al., 1982) had persisted for the intervening period of 10 months, (2) to determine if comparable effects could be induced by the experimental exposure of snails to phenanthrene, (3) to determine if such effects are reversible, (4) to elucidate the lysosomal mechanism involved in the cell injury, and (5) to test the hypothesis that the lysosomal responses observed in snails from some Shetland sites and those treated with phenanthrene were analogous to autophagy induced by a non-xenobiotic means such as nutritional deprivation.

9.2. EXPERIMENTAL METHODS

9.2.1. Chemicals

All enzyme substrates, as well as fast violet *B*, NADPH-neotetrazolium chloride, POLYPEP P5115, and polyvinyl alcohol (Type II, molecular weight ~12,000) were obtained from the Sigma Chemical Co., Ltd. (Poole, Dorset, United Kingdom). Phenanthrene (≥98% pure) was obtained from the Aldrich Chemical Co., Ltd. (Gillingham, Dorset, United Kingdom).

9.2.2. Sample Collection

Littorina littorea, with a shell length of 20–25 mm, were collected at low tide from the nine sites (Fig. 9.1) in mid-July 1982. All littorinids sampled appeared to be actively migrating and feeding at the time of collection. There is no evidence of a correlation between phasic activity in the digestive gland and the tidal cycle (Merdsoy and Farley, 1973). Ronas Voe animals represented the clean site, whereas Mavis Grind animals represented a contaminated reference site (Moore et al., 1982). The latter were exposed to oily runoff from an asphalt production plant, and were collected ~15 m north of the runoff point of entry.

9.2.3. Phenanthrene Exposure

The *Littorina littorea* were acclimated in the laboratory for 3 weeks at 15°C and fed an unlimited supply of the alga *Ulva lactuca*. For experimental purposes, 20 snails were placed in 7-liter, seasoned (for 1 month, with seawater changed daily) polypropylene tanks containing filtered (0.7-μm) aerated seawater (15°C, 34‰ salinity) from the Eddystone area, near Plymouth, England. This seawater was changed at 24-h intervals throughout the experiment, and a supply of *U. lactuca* was available at all times. Phenanthrene dissolved in analar acetone (10 mg ml^{-1}) was added to the test tank daily to give a final concentration of 400 μg $liter^{-1}$. Acetone alone was added to the control tank daily. Snails were dosed with phenanthrene and acetone for 5 d, then two samples, each consisting of five animals, were taken from the control and the experimental conditions. The remaining snails were put in tanks with clean seawater for an additional 8 d; five snails were then taken from each condition.

9.2.4. Starvation

Snails were acclimated as described above. After acclimation control organisms were fed an unlimited supply of *Ulva lactuca*, while experimental organisms were not fed. Snails were sampled after 4 and 11 d, with each sample consisting of five snails.

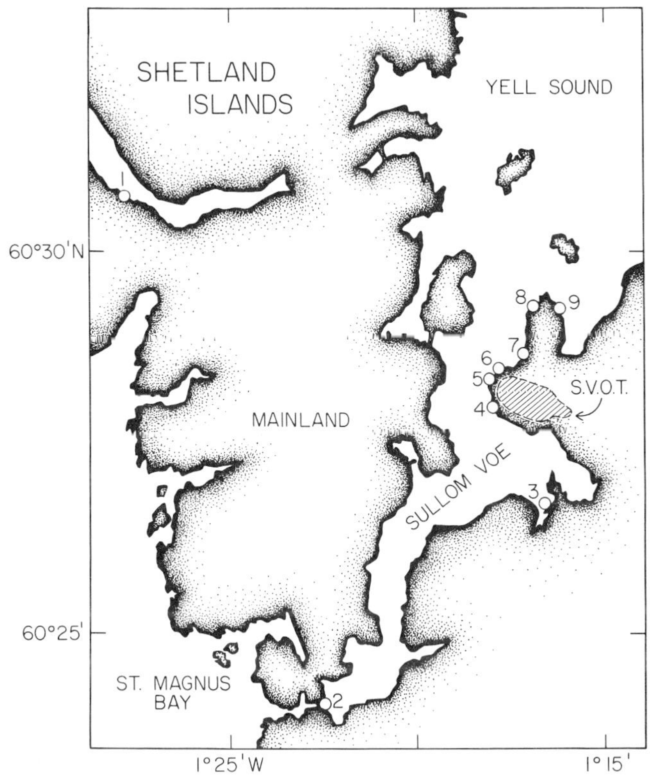

Figure 9.1. Location of *Littorina littorea* populations sampled: (1) Ronas Voe, clean sample site; (2) Mavis Grind, contaminated sample site; (3) Outer Houb of Scatsta; (4) Tanker Jetty 4; (5) Kames Tidal Pool; (6) The Kames; (7) Brei Wick; (8) Skaw Taing; and (9) Swarta Taing. The diagonal lines show the Sullom Voe Oil Terminal (S.V.O.T.).

9.2.5. Tissue Preparation for Enzyme Cytochemistry

For cytochemical examination, small pieces (5-mm^3) of freshly excised visceral mass (one portion from each of the five animals per sample) were placed on an aluminum cryostat chuck. The chuck was then placed for 1 min in a small bath of hexane (boiling range 67–70°C), which had been precooled to -70°C using a surrounding bath of liquid nitrogen (Moore, 1976). Serial sections (10-μm) were cut in a cryostat with motorized cutting (10 cutting strokes per minute); the cabinet temperature was kept below -25°C and the knife was cooled with crushed, solid CO_2. The sections were transferred to slides kept at 20°C, which effectively flash-dried them (Bitensky et al., 1973), and were stored in the cryostat for up to 4 h before use. All samples were coded prior to freezing and decoded only after all measurements were made.

9.2.6. Determination of Latency of Lysosomal Hydrolases

Latent lysosomal activity of the lysosomal enzyme β-glucuronidase (GLUC) was tested in tissue sections by use of naphthol AS-BI β-glucuronide and postcoupling with the diazonium salt, fast violet *B*, as described by Moore (1976). Latent activity of acid phosphatase (ACP) was assessed using naphthol AS-BI phosphate (8 mg:50 ml^{-1}) as substrate dissolved in 0.5 ml of dimethylformamide. Incubation with the substrate proceeded for 30 min at pH 5.5 and 37°C, and postcoupling was with fast violet *B* (Moore et al., 1982).

A simplified pretreatment sequence at pH 4.5 and 37°C was used in order to determine labilization period (pretreatment time required to activate latent enzyme; total activity = free activity + latent activity) (Moore, 1976, 1980). This sequence was as follows: 2, 5, 10, 15, 20, and 25 min; 2 min was used as the minimal pretreatment time because sections that have not undergone pretreatment may show stronger reactivity than do short-term pretreated sections (Moore, 1976). This enzymatic reaction product is believed to be due to a free enzyme resulting from damage to some of the larger lysosomes during the process of sectioning. The free enzyme can be lost by diffusion from the section when no polypeptide stabilizer is present, as is the case in the pretreatment medium (Moore, 1976). Due to this complication, a zero preincubation is usually omitted and the 2-min pretreatment is taken as representing the free lysosomal activity.

Labilization periods were determined by visual assessment of maximal enzymatic reaction-product intensity for each of the lysosomal hydrolases by examining the pretreatment sequences of sections from each animal (Moore, 1976, 1980). Data were statistically tested against the clean reference site (Ronas Voe) or experimental control animals by using the Mann-Whitney *U*-test (Siegal, 1956). When we applied this test to our data, based on the interval series of 2–25 min, there was sometimes a problem with ties; that is, many of the labilization periods for one sample (five animals) were the same, and sometimes there was overlap between the data sets being compared. When there were ties and overlap, a special table of tie vectors was used, which permitted the determination of the level of significance for critical values of the *U*-test.

9.2.7. Determination of β-Glucuronidase Activity

Total activity of lysosomal β-glucuronidase was determined in sections that showed maximal (free latent activity) enzymatic reaction product as described in the preceding section. Measurements were made using a Vickers M85 scanning microdensitometer (Vickers, Ltd., York, United Kingdom) at a wavelength of 560 nm with a measuring spot 0.5 μm in diameter and a mask area of 132.8 μm^2 (size A3) as described by Moore et al. (1978). Microdensitometric measurements (10 per section of individual digestive glands) of lysosomal enzyme activities (as relative absorption) were made in the mid-cytoplasmic regions of digestive cells. Statistical analysis of the data was made using the Mann-Whitney *U*-test.

9.2.8. Determination of NADPH-Neotetrazolium Reductase Activity

NADPH-neotetrazolium reductase (NADPH-NTR) was analyzed in 10-μm cryostat sections (in duplicate) as described by Moore (1979). Microdensitometric measurements, as relative absorption, were made using a Vickers M85 scanning microdensitometer at a wavelength of 585 nm with a measuring spot 0.5 μm in diameter and a mask area of 12.6 μm^2 (size A1) as described by Moore (1979). Statistical analysis of the data was made by using the Mann-Whitney *U*-test.

9.2.9. Hydrocarbon Analysis

Fifteen animals were collected from Ronas Voe, the clean reference site, and the site at Tanker Jetty 4 (Fig. 9.1). Each sample set was divided into two groups: one with seven snails and one with eight snails. The total wet tissue from the animals in each group was homogenized and extracted into hexane by steam distillation according to methods described by Widdows et al. (1982). The distillates were analyzed by high performance liquid chromatography (HPLC) on a 250-mm × 5-mm column of Partisil-10 PAC (polaramino-cyano) (Whatman Chemical Separations, Inc., Maidstone, United Kingdom) eluted with hexane. This column, and related amino phases, provides a separation of aromatic hydrocarbons based on the number of aromatic rings; alkyl substituents have little influence on compound elution time (Wise et al., 1977; Suatoni, 1979).

The PAHs were detected by means of a Perkin-Elmer LC 75 ultraviolet spectrophotometric detector (Perkin-Elmer Corp., Norwalk, Connecticut) set at 225 nm to selectively measure alkyl substituted naphthalenes and at 254 nm to selectively measure phenanthrene and its alkyl homologues. In order to quantify the two major PAH groups measured, 2,3-dimethylnaphthalene and 1-methylphenanthrene were chosen as representative standards. However, because each chromatographic peak derived from petroleum residues is composed of many components of differing molar extinction coefficients at the selected wavelengths, quantification is only approximate and likely to underestimate PAH content.

9.3. RESULTS

9.3.1. Shetland Samples

The digestive cells of snails from Mavis Grind, Tanker Jetty 4, The Kames Tidal Pool, and Skaw Taing showed frequent evidence of greatly enlarged lysosomes (Fig. 9.2a, b). Values for the latency of lysosomal β-glucuronidase (as labilization periods) in the digestive cells of *Littorina littorea* are summarized in Table 9.1. There were statistically significant reductions in latency of lysosomal β-glucuronidase at all sites when compared with the clean reference sample for Ronas Voe. The contaminated sample from Mavis Grind showed a value comparable to many of the other test samples, with the exception of the Outer Houb of Scatsta, The Kames, and Skaw Taing, which were all significantly higher ($p \leq 0.05$, *U*-test) than the Mavis Grind value. The mean latency value of 18 min for the Outer Houb of Scatsta sample is a considerable increase over the value of 5 min recorded in September 1981 ($p < 0.01$, *U*-test) (Moore et al., 1982).

Total activities of lysosomal β-glucuronidase in the digestive cells are summarized in Table 9.1. Except for the Outer Houb of Scatsta and Swarta Taing sites, these activities show significant elevation when compared to the clean reference sample from Ronas Voe. The contaminated reference samples from Mavis Grind showed significant elevation of β-glucuronidase activity ($p \leq 0.01$, *U*-test) compared to the clean reference sample from Ronas Voe.

Microdensitometric determinations of relative hemocytic NADPH-neotetrazolium reductase activity are given in Table 9.1. There was evidence of significant elevation in samples from four sites, including the contaminated reference sample from Mavis Grind, Tanker Jetty 4, The Kames Tidal Pool, and Brei Wick, compared to the sample from Ronas Voe.

Table 9.1. Latencies of Lysosomal β-Glucuronidase and Relative Activities of β-Glucuronidase and NADPH-Neotetrazolium Reductase in *Littorina* from Nine Sample Stations

Sample Site[a]	Mean Labilization Period of β-Glucuronidase (min)[b]	Mean Relative Activites (%) of Ronas Voe Values (± standard error)	
		β-Glucuronidase	NADPH-Neotetrazolium Reductase
Ronas Voe	24 (20, 25)	100.0 ± 2.0	100.0 ± 3.6
Mavis Grind	6 (5, 10)[c]	153.2 ± 12.6[c]	147.7 ± 8.4[c]
Outer Houb of Scatsta	18 (15, 20)[d]	88.4 ± 1.8[c]	105.9 ± 3.2
Tanker Jetty 4	9 (5, 10)[c]	149.1 ± 8.0[c]	130.7 ± 4.8[c]
The Kames Tidal Pool	5.4 (2, 10)[c]	167.4 ± 8.3[c]	117.5 ± 4.7[d]
The Kames	14 (10, 15)[c]	121.8 ± 7.6[c]	106.5 ± 3.1
Brei Wick	9 (5, 10)[c]	121.4 ± 3.7[c]	131.4 ± 6.8[c]
Skaw Taing	13 (10, 15)[c]	144.3 ± 9.5[c]	—[e]
Swarta Taing	9 (5, 10)[c]	107.0 ± 4.3	105.9 ± 4.9

[a] For sample site locations see Fig. 9.1.
[b] Range in parentheses.
[c] Mann-Whitney *U*-test, $p \leq 0.01$.
[d] Mann-Whitney *U*-test, $p \leq 0.05$.
[e] Dash denotes no data available.

Determinations of concentrations of PAHs in the tissues of snails from Ronas Voe and Tanker Jetty 4 are shown in Table 9.2. The results indicate that there was a fourfold increase in phenanthrenes and a tenfold increase in naphthalenes in the sample from Tanker Jetty 4.

It should also be noted that each snail contained food in its gut, indicating that it was actively feeding.

9.3.2. Phenanthrene Exposure

Exposure of snails to 400 μg liter^{-1} of phenanthrene for 5 d induced significant decreases in the latency of lysosomal β-glucuronidase and acid phosphatase and induced increases in total β-glucuronidase and NADPH-neotetrazolium reductase (Table 9.3). Lysosomes in the digestive cells of phenanthrene-treated snails were greatly enlarged (Fig. 9.2c).

The enlarged lysosomes of phenanthrene-treated snails returned to normal conditions after placement in clean seawater for 8 d (Table 9.3). These results indicate that the lysosomal and microsomal responses described above are reversible under these experimental conditions.

9.3.3. Starvation

The digestive cells of snails from sample sites other than Ronas Voe and Outer Houb of Scatsta, snails from both of which were in the healthy range for lysosomal latency measurement, often contained enlarged lysosomes. This condition was also observed in the snails treated with phenanthrene

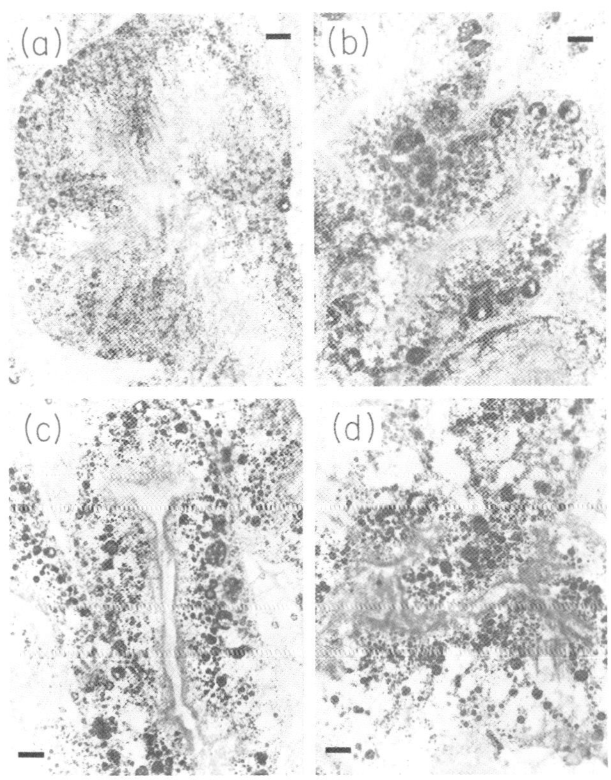

Figure 9.2. Cryostat sections through digestive tubules of *L. littorea* showing lysosome reaction to β-glucuronidase: (a) Ronas Voe, normal appearance of lysosomes; (b) The Kames Tidal Pool, enlarged lysosomes; (c) phenanthrene treated for 5 d, enlarged lysosomes; and (d) starved for 4 d, enlarged lysosomes. Scale bar ≈20 μm.

for 5 d and is characteristic of autophagy in molluscs (Bayne et al., 1978; Lowe et al., 1981; Moore and Clarke, 1982). *Littorina littorea* that underwent starvation showed significantly decreased latency of lysosomal β-glucuronidase (Table 9.4) as well as consistent evidence of lysosomal enlargement (Fig. 9.2d). This depression of latency was on the same order for both periods of starvation, indicating that a steady state had been attained, presumably of controlled autophagy. Total-activity measurements of β-glucuronidase showed that there was a signficant elevation of this enzyme after 4 d, but it returned to a much lower level after 11 d. This level was still significantly higher than was the control level (Table 9.4).

9.4. DISCUSSION AND CONCLUSIONS

The results show that there were significant reductions in latency of lysosomal β-glucuronidase in the digestive cells of *Littorina littorea* from some of the Shetland test sites, including the contaminated reference, in comparison with the clean reference site. These findings are in accord with our previous report (Moore et al., 1982), although there is clearly a recovery in the lysosomal hydrolase latency of snails from Outer Houb of Scatsta since the previous sampling in September 1981. The sample from this latter site also showed no evidence of lysosomal enlargement, elevation of β-glucuronidase or NADPH-neotetrazolium reductase activities, which is a further indication of recovery. Elevated activities of NADPH-neotetrazolium reductase in snails from the contaminated reference sample (Mavis Grind), Tanker Jetty 4, The Kames Tidal Pool, and Brei Wick are indicative of induction of this component of the microsomal detoxification system in the hemocytes. This phenomenon was recorded previously from Mavis Grind and Tanker Jetty 4 (Moore et al., 1982). Evidence of elevated levels of PAHs in tissues of snails from the Tanker Jetty 4 site is consistent with the hypothesis that reduced lysosomal latency in combination with induction of NADPH-neotetrazolium reductase is indicative of a response to these compounds.

Table 9.2. Concentrations ($\mu g\ g^{-1}$, wet wt) of Phenanthrene and Naphthalene in *Littorina* Whole-Body Tissues[a]

	Concentrations	
Sample Site[b]	**Phenanthrene**	**Naphthalene**
Ronas Voe	0.11	0.12
	0.11	0.10
Tanker Jetty 4	0.37	1.22
	0.41	1.03

[a] Duplicate samples of each treatment were analyzed.

[b] For sample site locations see Fig. 9.1.

Samples from the remaining three sites did not show any elevation of NADPH-neotetrazolium reductase, although their cellular condition was poor (as determined by measurement of lysosomal latency). Therefore, we cannot infer that these samples were affected by PAHs, although the snails were obviously stressed. The possiblity that these effects could be induced by natural factors is unlikely, since physical conditions at all sites were similar and food was available to all snails.

Experimental exposure to phenanthrene-induced lysosomal and microsomal effects within 5 d were comparable to the effects observed in the field. The return of these responses to the normal condition after a period of clean seawater indicates that the effects are reversible and that snails can recover, as is apparently the case with snails from the Houb of Scatsta (Moore et al., 1982).

Experimental starvation of snails induced lysosomal responses that were initially comparable to those observed in some of the Shetland samples and in the phenanthrene-treated snails. The snails, however, appeared to have established an altered steady-state condition after 11 d of starvation with a significant decrease in the initial elevation of lysosomal β-glucuronidase. This is indicative of a controlled autophagic response to nutritional stress: the lysosomal catabolism of cytoplasmic components may be functioning as an adaptive physiological maintenance mechanism (Ericsson, 1969). Increased specific activities of lysosomal enzymes are also known to accompany some autophagic processes induced by starvation in mammalian cells (Desai, 1971).

Augmented autophagy is frequently observed in mammalian cells subjected to sublethal injury (Ericsson, 1969). This response is characterized by enlargement of lysosomes (autolysosomes) and reduced hydrolase latency. We consistently found that an autophagic response was induced by experimental exposure to phenanthrene and by environmental factors in some samples from the Shetland sites. Three of the Shetland samples (The Kames, Skaw Taing, and Swarta Taing) showed possible evidence of autophagy, as indicated by reduced hydrolase latency; however, it may not have been due to oil-derived PAHs since there was no evidence of induction of NADPH-neotetrazolium reductase.

These findings show that there is a requirement for further, more detailed investigation of autophagic processes and PAH-induced cell injury in order to determine whether autophagy is induced directly or indirectly by these compounds. Direct induction could involve damage to the existing secondary lysosomes; indirect induction could involve either autoly-

Table 9.3. Latencies of Lysosomal β-Glucuronidase and Acid Phosphatase and Relative Activities of β-Glucuronidase and NAPDH-Neotetrazolium Reductase in *Littorina* Exposed to Phenanthrene (400 μg liter $^{-1}$)

	Mean Labilization Period (min)[a]		Relative Activites (%) of Control Values (± standard error)	
Experimental Treatment	**β-Glucuronidase**	**Acid Phosphatase**	**β-Glucuronidase**	**NADPH-Neotetrazolium Reductase**
Control (5 d)	22 (20, 25)[b]	22 (20, 25)	100.0 ± 0.5	100.0 ± 3.7
	22 (20, 25)[b]	22 (20, 25)	97.8 ± 0. 5	97.3 ± 5.5
Phenanthrene (5 d)	3 (2, 5)[b,c]	5.8 (2, 10)[c]	114.7 ± 2.2[c]	111.5 ± 2.2[d]
	7.4 (2, 10)[b,c]	6.8 (2, 10)[c]	113.6 ± 2.7[d]	113.4 ± 1.9[d]
Control (13 d)	22 (20, 25)	21 (20, 25)	100.0 ± 1.7	100.0 ± 4.7
Phenanthrene (5 d) and clean seawater (8 d)	20 (15, 25)	21 (20, 25)	97.1 ± 1.7	96.1 ± 6.0

[a] Range in parentheses.
[b] Data from duplicate samples of a particular treatment.
[c] Mann-Whitney *U*-test compared with control data, $p \leq 0.01$.
[d] Mann-Whitney *U*-test compared with control data, $p \leq 0.05$.

sosome formation as a mechanism for the isolation and removal of focal regions of hydrocarbon-damaged cytoplasm or impairment of heterophagic intracellular lysosomal digestion of food, leading to nutrient deprivation. The mechanism may be one or a combination of these processes. If a qualitative difference between autophagy induced by starvation and autophagy induced by PAHs could be discerned, the mechanisms could be better understood.

The findings described in this chapter provide further evidence for the use of lysosomal hydrolase latency and NADPH-neotetrazolium reductase as cytochemical indicators of cellular condition in marine organisms used for environmental monitoring purposes. Their use, coupled with experimental approaches, has enabled us to propose a probable autophagic lysosomal reaction to cell injury induced by PAHs. This interpretation does not exclude the possible interference of other environmental factors but is compatible with earlier findings for *Mytilus edulis* (Lowe et al., 1981; Moore and Clarke, 1982; Viarengo and Moore, 1982).

The use of this type of methodology demonstrates the sensitivity of the techniques. These cytochemical procedures are extremely rapid, easy to apply, and require only very small tissue samples from individual organisms.

ACKNOWLEDGMENTS

We are grateful to British Petroleum, Ltd., for allowing us to use their laboratory facilities at the Sullom Voe Oil Terminal. This work is part of the Experimental Ecology Programme of the Institute for Marine Environmental Research, a component of the Natural Environment Research Council. It was funded in part by the Shetland Oil Terminal Environmental Advisory Group. The authors thank Sheila Evans for her help in the analysis of tissues for polycyclic aromatic hydrocarbons.

Table 9.4. Latencies of Lysosomal β-Glucuronidase and Relative Activites of β-Glucuronidase in Starved *Littorina*

Experimental Treatment	Mean Labilization Period of β-Glucuronidase (min)[a]	Mean Relative Activities of β-Glucuronidase (%) of Control Values (± standard error)
Fed, control (4 d)	24 (20, 25)	100.0 ± 9.8
Starved (4 d)	15 (15, 15)[b]	165.6 ± 9.6[b]
Fed control (11 d)	25 (25, 25)	100.0 ± 2.0
Starved (11 d)	14 (10, 15)[b]	112.6 ± 2.3[c]

[a] Range in parentheses.
[b] Mann-Whitney *U*-test compared with control data, $p \leq 0.01$.
[c] Mann-Whitney *U*-test compared with control data, $p \leq 0.05$.

REFERENCES

Allison, A. C. 1969. Lysosomes and cancer. *In:* Lysosomes in Biology and Pathology, Vol. 2, J. T. Dingle and H. B. Fell (Eds.). North Holland–American Elsevier, Amsterdam, pp. 178–204.

Bayne, B. L., D. L. Holland, M. N. Moore, D. M. Lowe, and J. Widdows. 1978. Further studies on the effects of stress in the adult on eggs of *Mytilus edulis. Journal of the Marine Biological Association of the United Kingdom,* **58,** 825–841.

Bayne, B. L., M. N. Moore, J. Widdows, D. R. Livingstone, and P. Salkeld. 1979. Measurement of the responses of individuals to environmental stress and pollution: studies with bivalve molluscs. *Philosophical Transactions of the Royal Society of London: B., Biological Sciences,* **286,** 563–581.

Bitensky, L., and J. Chayen. 1979. Cytochemical methods for studying lysosomes. *In:* Quantitative Cytochemistry and Its Applications, J. R. Pattison, L. Bitensky, and J. Chayen (Eds.). Academic Press, London, pp. 49–61.

Bitensky, L., R. S. Butcher, and J. Chayen. 1973. Quantitative chytochemistry in the study of lysosomal function. *In:* Lysosomes in Biology and Pathology, Vol. 3, J. T. Dingle (Ed.). North Holland–American Elsevier, Amsterdam, pp. 465–510.

Chayen, J. 1978. Microdensitometry. *In:* Biochemical Mechanisms of Liver Injury, T. F. Slater (Ed.). Academic Press, London, pp. 259–292.

Desai, I. D. 1971. Regulation of lysosomal enzymes. II: Reversible adaptation of intestinal acid hydrolases during starvation and realimentation. *Canadian Journal of Biochemistry,* **49,** 170–176.

Ericsson, J. L. E. 1969. Mechanisms of cellular autophagy. *In:* Lysosomes in Biology and Pathology, Vol. 2, J. T. Dingle and H. B. Fell (Eds.). North Holland–American Elsevier, Amsterdam, pp. 345–394.

Lowe, D. M., M. N. Moore, and K. R. Clarke. 1981. Effects of oil on digestive cells in mussels: Quantitative alterations in cellular and lysosomal structure. *Aquatic Toxicology,* **1,** 213–226.

Merdsoy, B., and J. Farley. 1973. Phasic activity in the digestive gland cells of the marine prosobranch gastropod, *Littorina littorea* (L.). *Proceedings of the Malacological Society (London),* **40,** 473–482.

Moore, M. N. 1976. Cytochemical demonstration of latency of lysosomal hydrolases in digestive cells of the common mussel, *Mytilus edulis,* and changes induced by thermal stress. *Cell and Tissue Research,* **175,** 279–287.

Moore, M. N. 1979. Cellular responses to polycyclic aromatic hydrocarbons and phenobarbital in *Mytilus edulis. Marine Environmental Research,* **2,** 255–263.

Moore, M. N. 1980. Cytochemical determination of cellular responses to environmental stressors in marine organisms. *In:* Biological Effects of Pollution and Problems of Monitoring, A. D. McIntyre and J. B. Pearce (Eds.). Rapport et Procès Verbaux des Réunions, Vol. 179. Conseil International pour l'Exploration de la Mer, Copenhagen, pp. 7–15.

Moore, M. N., and K. R. Clarke. 1982. Use of microstereology and quantitative cytochemistry to determine the effects of crude oil-derived aromatic hydrocarbons on lysosomal structure and function in a marine bivalve mollusc, *Mytilus edulis. Histochemical Journal,* **14,** 713–718.

Moore, M. N., D. M. Lowe, and P. E. M. Fieth. 1978. Lysosomal responses to experimentally injected anthracene in the digestive cells of *Mytilus edulis. Marine Biology,* **48,** 297–302.

Moore, M. N., K. R. Pipe, and S. V. Farrar. 1982. Lysosomal and microsomal responses to environmental factors in *Littorina littorea* from Sullom Voe. *Marine Pollution Bulletin,* **13,** 340–345.

Siegel, S. 1956. Nonparametric Statistics for the Behavioral Sciences. McGraw-Hill Kogakusha, Tokyo, 312 pp.

Suatoni, J. C. 1979. Hydrocarbon group-type analysis by high performance liquid chromatography. *In:* Chromatography in Petroleum Analysis, K. H. Altgelt and T. H. Gouw (Eds.). Marcel Dekker, New York, pp. 121–136.

Viarengo, A., and M. N. Moore. 1982. Effects of aromatic hydrocarbons on the metabolism of the digestive gland of the mussel, *Mytilus edulis* L. *Comparative Biochemistry and Physiology,* **71**(C), 21–25.

Widdows, J., T. Bakke, B. L. Bayne, P. Donkin, D. R. Livingstone, D. M. Lowe, M. N. Moore, S. V. Evans, and S. L. Moore. 1982. Responses of *Mytilus edulis* on exposure to the water-accommodated fraction of North Sea oil. *Marine Biology,* **67,** 15–31.

Wise, S. A., S. N. Chesler, H. S. Hertz, L. R. Hilpert, and W. E. May. 1977. Chemically-bonded aminosilane stationary phase for the high-performance liquid chromatographic separation of polynuclear aromatic compounds. *Analytical Chemistry,* **49,** 2306–2310.

Chapter 10

Availability of PCBs and PAHs to *Mytilus edulis* from Artificially Resuspended Sediments

Richard J. Pruell and James G. Quinn

Graduate School of Oceanography
University of Rhode Island
Narragansett, Rhode Island

James L. Lake and Wayne R. Davis

Environmental Research Laboratory
U.S. Environmental Protection Agency
Narragansett, Rhode Island

ABSTRACT

Estuarine sediment containing high concentrations of polychlorinated biphenyls (PCBs) and polycyclic aromatic hydrocarbons (PAHs) was resuspended into a flowing seawater system. Analysis of the dissolved and particulate phases in the exposure tank indicated a release of some low-molecular-weight PCB and PAH compounds from the sediment into the dissolved phase. Blue mussels *(Mytilus edulis)* exposed to the resuspended sediment rapidly accumulated PCBs, reaching a steady-state concentration after 20 d of exposure. Exposed mussels took up PCB congeners with 3–7 chlorine atoms per molecule. The dissolved phase contained compounds with 3–7 chlorine atoms per molecule; however, the relative distributions of chromatogram peaks from the mussels and the dissolved phase differed. The exposure sediment contained compounds with 3–10 chlorine atoms. Exposed mussels also rapidly accumulated PAHs. The total PAH concentration reached a maximum concentration at 20 d, then decreased slightly. By the end of a 40-d exposure, the PAH distribution in the exposed mussels was similar to that found on the particles. Two and three-ring PAH compounds were more prevalent, however, in the particulate phase than in the mussels.

10.1. INTRODUCTION

Many laboratory studies (Neff et al., 1976; Langston, 1978) and field investigations (Young et al., 1976; Kveseth et al., 1982) have shown that filter-feeding bivalves can accumulate numerous organic contaminants. Also, several programs have successfully used both indigenous (Cowan, 1981; Knutzen and Sortland, 1982) and transplanted mussels (Phelps and Galloway, 1980; Farrington et al., 1982) to identify pollutant trends in coastal areas.

The organic compounds accumulated by aquatic organisms usually have low water solubilities and high partition coefficients (Chiou et al., 1977; Veith et al., 1980; Geyer et al., 1982). These properties also result in the association of a large percentage of these compounds with particles in aquatic environments (Steen et al., 1978; Karickhoff et al., 1979; Means et al., 1980). Therefore, sediments in areas receiving large inputs of organic pollutants often contain high concentrations of a variety of organic contaminants (Windsor and Hites, 1979; Lopez-Avila and Hites, 1980; Bopp et al., 1982).

Oviatt and Nixon (1975) reported that the material collected by sediment traps placed in Narragansett Bay was dominated by resuspended sediment particles, indicating that sediment resuspension may be a major contributor to particulate material in the water column. Therefore, bivalves in polluted estuaries or near oceanic disposal sites may be chronically exposed to resuspended sediment containing high concentrations of organic contaminants. Mussels efficiently filter particles from seawater (Bayne et al., 1976); however, little is known about the accumulation of dissolved and particle-bound organic contaminants by these organisms.

Upper Narragansett Bay receives large quantities of organic pollutants from a variety of sources, including urban runoff (Hoffman et al., 1983), sewage outfalls (Van Vleet and Quinn, 1977), and industrial discharges (Jungclaus et al., 1978). The distribution of several of these contaminants has been well documented in the sediments of this estuary (Paulson and Brown, 1978; Lake et al., 1979; Lopez-Alvila and Hites, 1980). In the present study, surface sediment from upper Narragansett Bay was collected and resuspended into a flowing seawater system containing blue mussels *(Mytilus edulis)*. Dissolved and particulate phases of the exposure-system seawater were analyzed for polychlorinated biphenyls (PCBs) and polycyclic aromatic hydrocarbons (PAHs). The concentrations and distributions of compounds in these phases were then compared with those detected in mussels maintained in the dosing system for up to 40 d.

10.2. METHODS

10.2.1. Sediment Collection

Sediment for the exposure system was collected from the Providence River (41°46′40″N, 71°23′5″W) at the head of the Narragansett Bay Estuary in Rhode Island. The water depth at the collection site was 5 m, and the salinity was 29‰. A Smith-McIntyre grab sampler was used to collect a surface sediment sample. The sediment was passed through a 2-mm-mesh screen to remove macrofauna and shell debris and then mixed well before freezing at −20°C. Particle-size analysis with a Coulter Counter® (Coulter Electronics, Hialeah, Florida) showed that the sediment particle-size distribution ranged from 4–64 μm in diameter with a maximum at ~15–20 μm; therefore, virtually all of the material could be classified as silt. The total organic carbon content of the sediment, determined in triplicate by using a Carlo Erba Model 1106 CHN Analyzer (Erba Instruments, Inc., Peabody, Massachusetts), was 43.1 ± 0.6 mg g^{-1}.

10.2.2. Dosing System

The sediment-dosing system, developed by Sinnett and Davis (1983), is outlined in Fig. 10.1. A sediment slurry was produced by mixing 40 g (wet weight) of sediment per liter of seawater in a separatory funnel. Air was bubbled into the bottom of the funnel to prevent settling; this quickly oxidized the originally anoxic sediment, changing its color from black to brown. Therefore, all of the slurry dosed into the 80-liter aquaria was in the oxidized state. The sediment slurry was replaced weekly.

A transmissometer was used to monitor particulate levels in the aquaria. An eight-bit microprocessor received measurements from the transmissometer every 2 min and controlled an air pump used to deliver sediment slurry to the aquaria and maintain constant particle levels in the system. The system was calibrated by filtering seawater samples through a pre-weighed 0.45-μm Millipore® filter (Millipore Corp., Bedford, Massachusetts), then drying and reweighing the filter. Particle concentrations measured in the exposure and control aquaria during the experiment were 14.6 ± 6.8 (n = 25) and 1.3 ± 0.6 (n = 9) mg $liter^{-1}$, respectively. Therefore, about 13 mg of exposure sediment was resuspended into each liter of seawater in the exposure system. The control system received seawater and no additional sediment.

Mussels were collected from a relatively uncontaminated intertidal area in Matunuck, Rhode Island, and ranged in length from 3.0–4.6 cm. They were acclimated to the exposure system for 3 d before the experiment was initiated. The sediment exposure lasted 40 d, from 5 April–15 May 1982. During this time, incoming seawater, which ranged in tem-

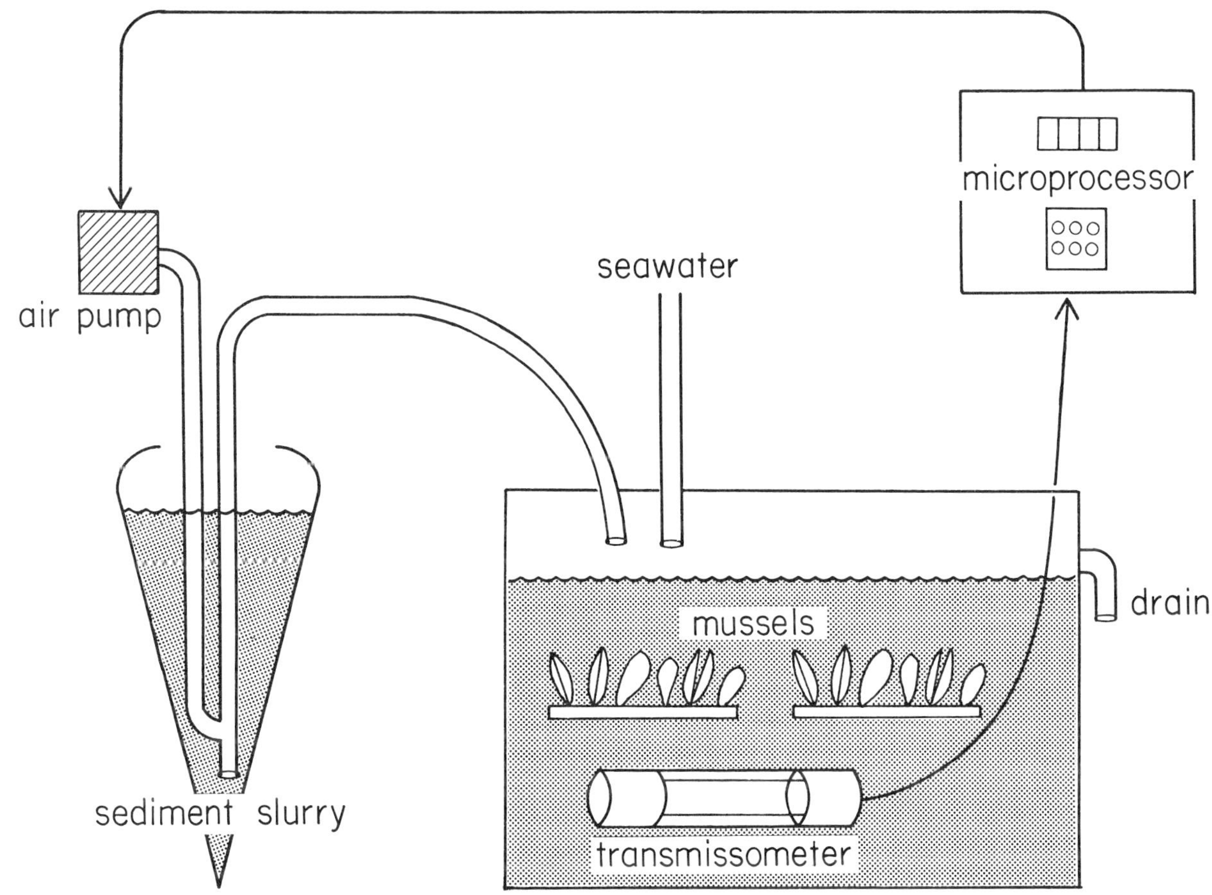

Figure 10.1. The sediment dosing system.

perature from 4–16°C, was heated to 15°C before entering the dosing and control systems. The salinity ranged between 27 and 31‰, and the pH varied between 8.0 and 8.3. Oxygen concentrations in the incoming seawater ranged from 8.0–10.6 mg liter^{-1}. Oxygen was not measured in the systems but was assumed to be at or above saturation, due to the use of aeration to prevent particle settling in the aquaria. The flow rate in both systems was maintained at 200 ml min^{-1}. Mortality of the mussels in both the control and exposure tanks was minimal throughout the experiment.

10.2.3. Sampling

Samples of the exposure sediment as well as fresh and 1-week-old slurried samples of the sediment were collected for analysis. Particulate and dissolved samples were taken from the exposure and control systems by using a method similar to that described by de Lappe et al. (1980). For this collection, the drain of each aquarium was connected to a filter holder containing a pre-ignited 27.3-cm (diameter) Gelman AE glass fiber filter (Gelman Sciences, Inc., Ann Arbor, Michigan). A metal cartridge containing two 4.5-cm (diameter) by 4.7-cm (long) polyurethane foam plugs, which were pre-cleaned by extraction with hexane and acetone, was located downstream from the filter. Particle analysis, using a Coulter Counter, indicated that the filter removed >95% of the particles >2 μm in diameter. The polyurethane foam plugs act as solid absorbants, retaining dissolved PCBs and PAHs (Gesser et al., 1971, Basu and Saxena, 1978). Spiking experiments conducted at a flow rate of ~200 ml min^{-1} showed that 80–90% of the dissolved PCBs and PAHs retained were in the first foam plug, indicating a very high efficiency for the two-plug cartridge. A 120-liter sample was obtained by allowing the outflow of the aquaria to flow through the system for 10 h at a rate of 200 ml min^{-1}. During the experiment, the exposure system was sampled four times and the control system was sampled three times.

The foam plugs and filters were used to provide information on dissolved and particle-bound PCBs and PAHs in the dosing system. However, many unknowns are involved with the separation of dissolved and particulate fractions and therefore the designations in this study should be considered as operationally defined parameters.

Mussels were collected from the system for analysis at the start of the exposure period and after 3, 10, 20, and 40 d. During several of the samplings, 50% of the mussels were frozen directly; the remainder were placed in flowing seawater for 6 h to allow depuration of their gut contents and then frozen. Subsequent chemical analyses showed no difference between these groups, indicating that the material in the gut had little influence on body-burden PCB and PAH measurements. A similar conclusion can be drawn from comparisons of the distributions of compounds detected in the particles and in the mussels. The results presented in this chapter are for mussels that were depurated for 6 h.

10.2.4. Extraction

Foam plugs were extracted three times with acetone, then three times with hexane. These extracts were combined in a separatory funnel along with distilled water and the sample was partitioned into the hexane. This partitioning step was repeated two times, and the extracts were combined.

After addition of methanol and an internal standard (*m*-terphenyl), sediment and filter samples were refluxed for 2 h. The samples were cooled, then filtered through a preignited Whatman GF/C glass fiber filter (Whatman, Inc., Clifton, New Jersey). The filtrate was added to an equal volume of water in a separatory funnel, and the sample was then partitioned against petroleum ether three times.

Mussels were shucked, frozen, and the contents of six organisms were combined. After thawing, the sample was homogenized in a stainless-steel blender. One to three grams of homogenate were removed and weighed, then dried at 100°C and reweighed to determine a wet weight:dry weight ratio. Methanol and *m*-terphenyl were added to a weighed portion of the remaining homogenate and the sample was refluxed for 2 h. After cooling, the sample was filtered through glass wool, and the filtrate was added to a separatory funnel. An equal volume of water was added and the sample was partitioned three times against petroleum ether. Ten percent of the combined extract was removed, dried, and weighed for the extractable lipid calculation.

The extracts from all samples were reduced in volume by rotary evaporation to ~5 ml, then charged to a 0.8-cm × 46-cm column of Bio-Sil® A silicic acid (Bio-Rad Laboratories, Richmond, Virginia). Nitrogen pressure was used to increase the flow rate through the column. The PCB fractions were eluted with 50 ml of petroleum ether. The PAH fractions were then eluted with 35 ml of an 80:20 petroleum ether–methylene chloride mixture.

Mussel PAH fractions contained large concentrations of polyolefin compounds, which interfered with PAH quantification; therefore, these samples were further purified by adsorption chromatography. The PAH fraction was evaporated to near dryness, and 0.5 ml of methanol was added. This solution was charged to a 1.0-cm × 50-cm column of Sephadex® LH-20 (Pharmacia, Inc., Piscataway, New Jersey) in methanol. Nitrogen pressure was used to produce a solvent flow rate of ~1–2 ml min^{-1}. The first 35 ml of methanol from the column contained the polyolefinic material and were discarded. The next 90 ml were collected as the PAH fraction. Ninety milliters of distilled water were added to the methanol in a separatory funnel, and the sample was extracted three times with petroleum ether.

10.2.5. Analysis

The PCB fractions from the sediment and filter samples were passed through activated copper powder to remove sulfur. All PCB fractions were reduced to 1 ml, then 1 μl was injected on a Hewlett Packard 5840 gas chromatograph (Hewlett Packard, Avondale, Pennsylvania) equipped with a 30-m DB-5 fused silica column (J & W Scientific, Inc., Rancho Cordo, California) and an electron capture detector. After injection, the column temperature was held at 80°C for 4 min and then heated to 290°C at a rate of 10°C min^{-1}, after which the temperature was held at 290°C for 20 min. The PCBs were quantified by comparison of peak heights using the most appropriate standard or mixture of standards, foam plugs and mussel samples were quantified using Aroclor 1254 (Monsanto Co., St. Louis, Missouri), sediment and filter samples were quantified against a mixture of Aroclors 1254 and 1260.

The PAH fractions from all samples were concentrated to ~100 μl, then 1–2 μl were injected into a Hewlett Packard 5710A gas chromatograph equipped with a 15-m DB-5 fused silica column and a flame ionization detector. The column was temperature-programmed from 40 to 290°C at a rate of 4°C min^{-1}. Quantification of the PAHs was against the *m*-terphenyl internal standard initially added to each sample, except for foam plug samples, which were quantified using *m*-terphenyl as an external standard. Peak areas were measured with a Hewlett Packard 3388A integrator.

The PCBs and PAHs were tentatively identified by comparing retention times with those of authentic standards. Identifications were checked in representative samples by gas chromatography–mass spectrometry (GC-MS). These samples were analyzed on a Finnigan 4531 GC-MS (Finnigan Corp., Sunnyvale, California) equipped with a 30-m DB-5 fused silica column and utilizing 70-eV electron impact ionization and an INCOS data system.

Blank analyses were conducted for each sample type. In all cases, results for the blanks showed only trace concentrations of PCBs; the specific PAH compounds reported here were not detected in the blanks. The recovery of standards averaged 86% for PCBs and from 52–95% for specific PAH compounds. The lower recoveries were for two-ring compounds; three- to six-ring compounds showed recoveries >90%. Reported concentrations were not corrected for procedural losses.

The precision of our analysis scheme for specific PAHs is indicated by the standard deviations for the triplicate analysis of the exposure sediment (Fig. 10.2). Relative standard deviations for these analyses ranged from 4% for benz(a)anthracene to 38% for acenaphthene. The standard deviations reported in Fig. 10.2 for the analysis of filter and foam plug samples include variability associated with the operation of the dosing system, the sampling methodology, and the analysis scheme. A mussel homogenate that had been used in an intercalibration for the Mussel Watch Program (Galloway et al., 1983) was analyzed to provide information on the precision of the mussel-analysis procedure. Triplicate analyses of this sample indicated relative standard deviations of 11, 12, and 8% for PCB wet, dry, and lipid weight calculations, respectively. The respective variations for total identified PAH analysis were 6, 5, and 13%.

10.3. RESULTS

Table 10.1 presents the results of PCB and PAH analyses of the exposure sediment and filter and foam plug samples.

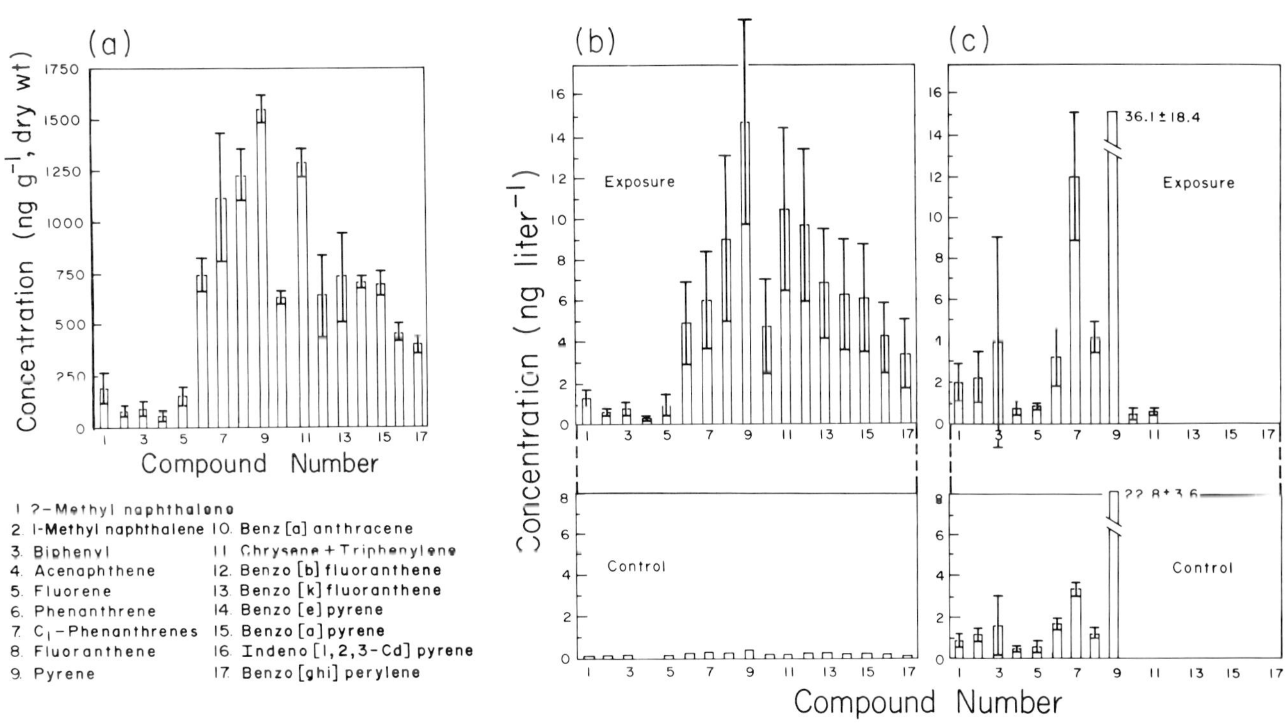

Figure 10.2. Concentrations of PAHs in (a) exposure sediment, (b) exposure and control tank filters, and (c) exposure and control tank foam plug samples. Error bars represent one standard deviation from the mean; $n = 3$ for PAHs in exposure sediment, control filter, and control foam plug samples, and $n = 4$ for exposure filter and exposure foam plug samples.

The PCB and PAH concentrations were also measured in freshly prepared and 1-week-old sediment slurry samples. The results of several analyses indicated, however, that no detectable changes occurred in either the concentrations or distributions of the PCB or PAH compounds in the slurry after 1 week in the separatory funnel.

The results show high concentations of both PCBs and PAHs in the exposure sediment (Table 10.1). Calculations based on the concentration of particulate material resuspended into the exposure tank (~13 mg liter^{-1}) and the exposure sediment PCB concentration (1070 ng g^{-1}) should produce a PCB concentration of ~13.9 ng liter^{-1} in the exposure tank. The measured increase (sum of the difference between exposure and control filter and foam plug samples) in the exposure tank was ~13.3 ng liter^{-1}. A similar analysis would predict an addition of 144.3 ng liter^{-1} for total identified PAHs. The measurements indicate a concentration of 119.6 ng liter^{-1}.

Table 10.1. Concentrations of PCBs and Total Identified PAHs in Exposure Sediment, Exposure and Control Filter, Samples and Foam Plug Samples[a]

Sample	PCBs	PAHs[b]
Exposure sediment (ng g^{-1},dry wt)	1070[c]	11,100 ± 330
Filter samples (ng liter^{-1})		
Exposure	12.0 ± 4.06	89.7 ± 34.8
Control	0.24 ± 0.02	1.72 ± 0.24
Foam plug samples (ng liter^{-1})		
Exposure	1.81 ± 0.19	64.9 ± 24.0
Control	0.32 ± 0.06	3.33 ± 4.4

[a] Mean ± standard deviation; n = 3 for PAHs in exposure sediment and control filter and foam plug samples, and n = 4 for exposure filter and foam plug samples.
[b] Total for the PAHs listed in Fig. 10.2.
[c] Only one analysis was done.

Figure 10.3 shows representative chromatograms of the PCB fractions of exposure sediment, filter, and foam plug samples. Exposure sediment contained a wide distribution of chlorobiphenyls containing from 3 to 10 chlorine atoms. A similar distribution was found in the exposure-tank filter samples. The foam plug samples only contained compounds with up to seven chlorine atoms. The highly chlorinated congeners (with 8, 9, and 10 chlorine atoms) were absent from these samples. Similar distributions of compounds were also observed in the control tank samples; however, the concentrations were lower (Table 10.1).

Figure 10.2 presents the concentrations of individual PAHs measured in the exposure sediment, filters, foam plugs, and controls. The sediment contained high levels of PAHs in a distribution dominated by parent PAH compounds. The exposure-tank filter samples show a similar PAH distribu-

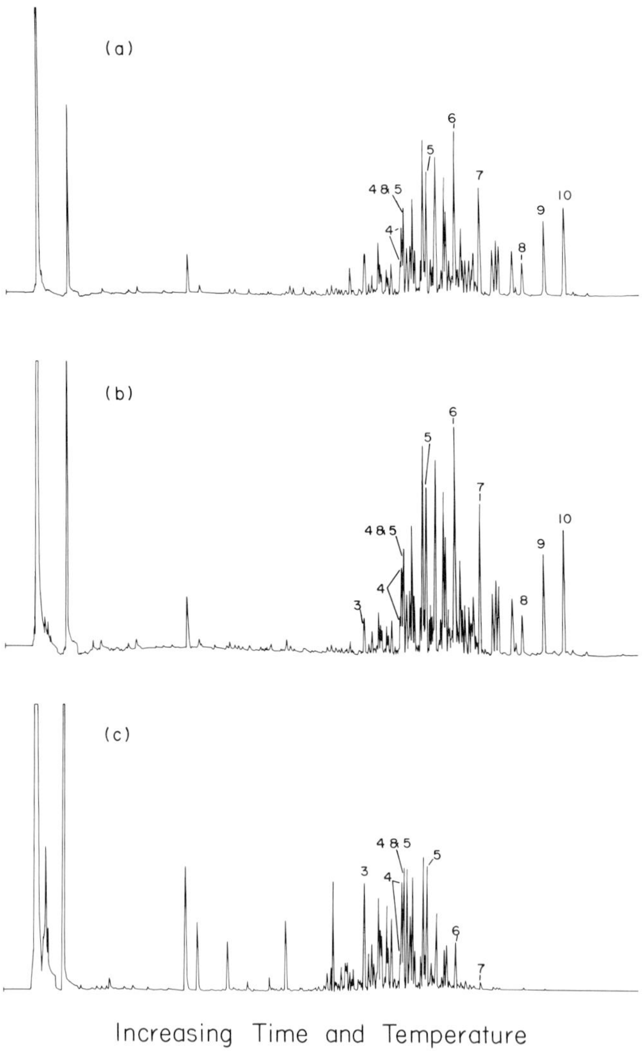

Figure 10.3. Polychlorinated biphenyl chromatograms of (a) exposure sediment, (b) exposure filter, and (c) exposure foam plug samples. The numbers refer to the number of chlorine atoms on the PCBs.

tion; nevertheless, there appears to be a depletion of some of the two-, three-, and four-ring PAH compounds in the filters as compared to the PAH distribution in the sediment. This is most pronounced for phenanthrene, C_1-phenanthrenes, and fluoranthene. These same compounds were found in high concentrations in the exposure foam plugs. This indicates a removal of these compounds from the particles to the dissolved phase.

The PAH distributions in both exposure tank and control tank foam plugs were very different from the sediment or filter samples (Fig. 10.2). The foam plugs contained a complex mixture of two-, three-, and four-ring compounds (Fig. 10.4). Also, very large concentrations of pyrene were detected in both exposure and control tank foam plugs. The reason for the anomalously high pyrene concentrations is unclear. This observation was, however, confirmed by GC-MS analysis.

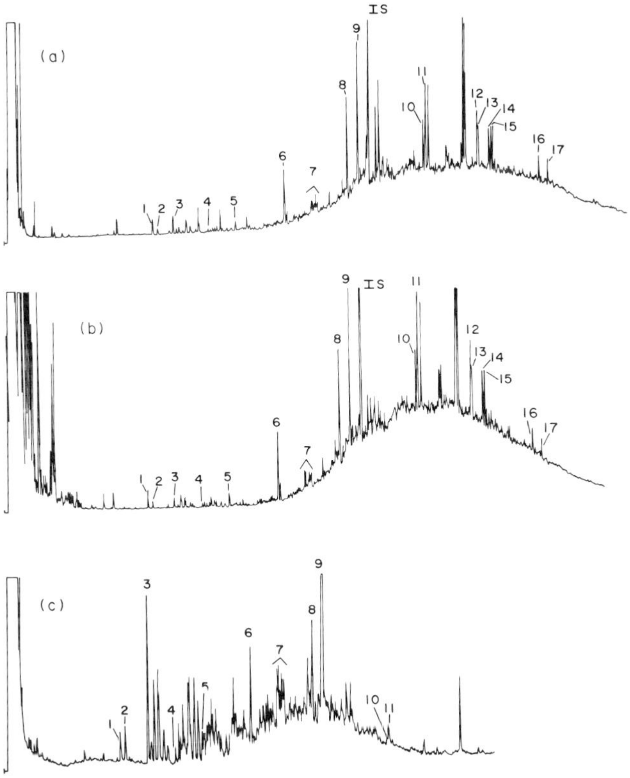

Figure 10.4. Polycyclic aromatic hydrocarbon chromatograms of (a) exposure sediment, (b) exposure filter, and (c) exposure foam plug samples. See Fig. 10.2 for identification of the peaks; IS stands for internal standard (m-terphenyl).

Mussel PCB concentrations are plotted in Fig. 10.5. These data are presented on a wet, dry, and extractable lipid weight basis to facilitate comparison with other studies. The wet weight:dry weight ratio was 13.7 ± 2.3, and the lipid content was 4.9 ± 1.2% (dry weight) for mussels analyzed in this study. There were no significant differences between exposed and control groups for these parameters.

Control mussels showed little change in PCB concentration throughout the experiment. PCB concentrations increased rapidly in the mussels dosed with the resuspended sediment, reaching a steady-state value after about 20 d, and the trend was similar for the three quantification methods. The PCB distribution detected in the mussels (Fig. 10.6) was depleted in the 8, 9, and 10 chlorobiphenyls compared to the exposure sediment. This was consistent for all mussel samples.

Total identified PAH concentrations remained constant or decreased slightly in control mussels during the 40-d experiment (Fig. 10.7). The exposed mussels rapidly accumulated PAHs, however, reaching maximum PAH levels at day 20, then decreasing slightly. The three quantification methods showed similar trends, as was the case for PCBs. Chromatograms of the PAH fraction of day-40 exposure and control mussels are shown in Fig. 10.8. The exposed mussels also contained an unresolved complex mixture (UCM) of hydrocarbons. The UCM had a narrower molecular-weight distribution (Fig. 10.8) than did that of the exposure sediment (Fig. 10.4). In addition, at day 40 the PAH distribution in the mussels showed low concentrations in two- and three-ring compounds relative to the distribution in the exposure sediment. Changes occurred in the distributions of specific PAHs in the mussels during the exposure period; these changes will not be discussed in this chapter.

10.4. DISCUSSION

10.4.1. Exposure Sediment

The PCB concentration measured in the exposure sediment (1070 ng g^{-1}) was similar to the value reported by Paulson and Brown (1978) for Providence River surface sediment (1110 ng g^{-1}). This sediment contained C1-3 to C1-10 PCBs in a distribution that indicated contributions of Aroclors 1254 and 1260.

Polycyclic aromatic hydrocarbon concentrations measured in the exposure sediment agreed with values reported by Lake et al. (1979) for surface sediments from upper Narragansett Bay. The PAH distribution was similar to that reported

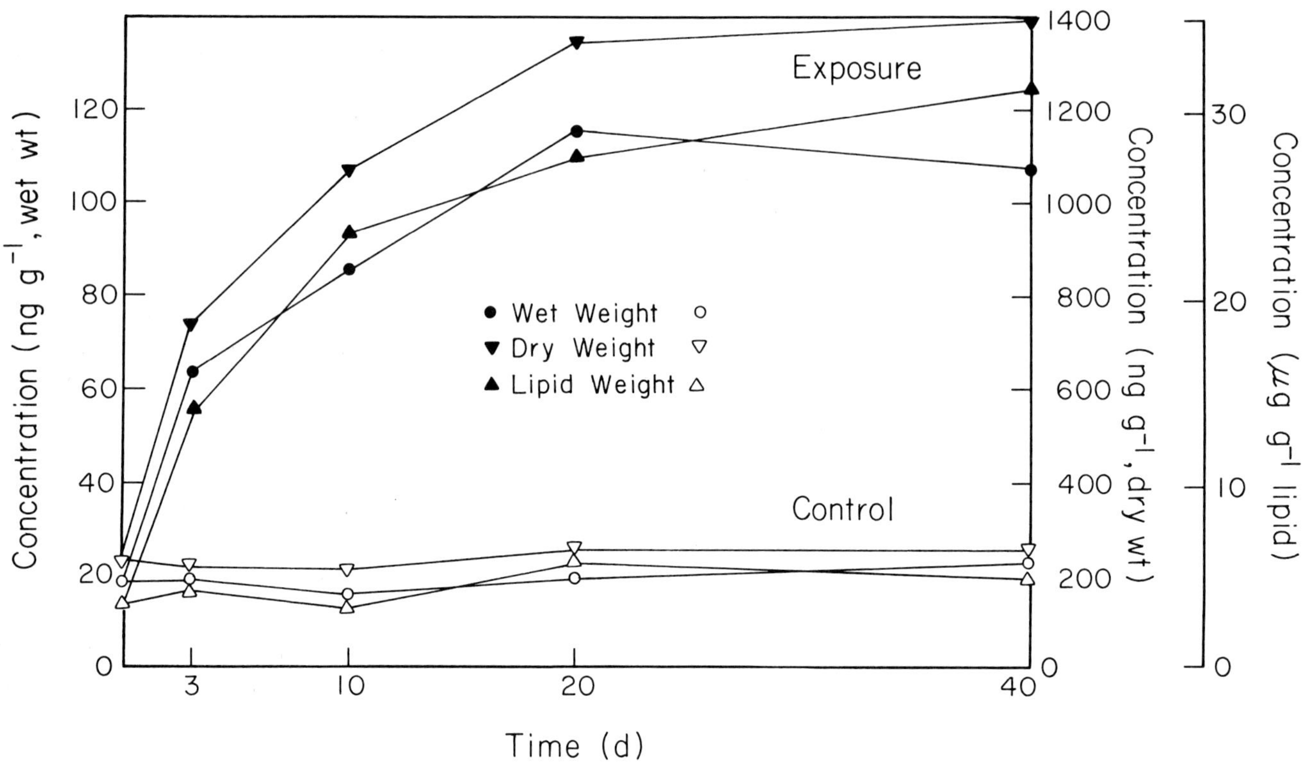

Figure 10.5. Polychlorinated biphenyls in mussels.

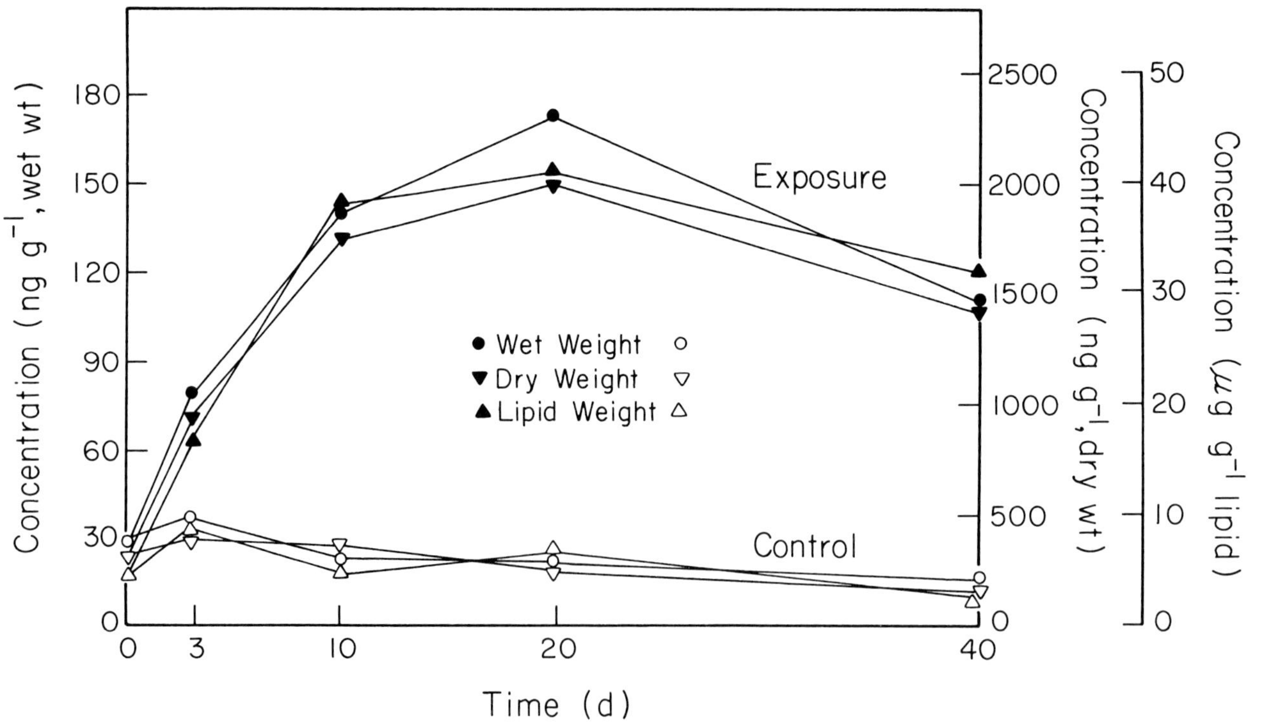

Figure 10.7. Total PCB concentrations in mussels.

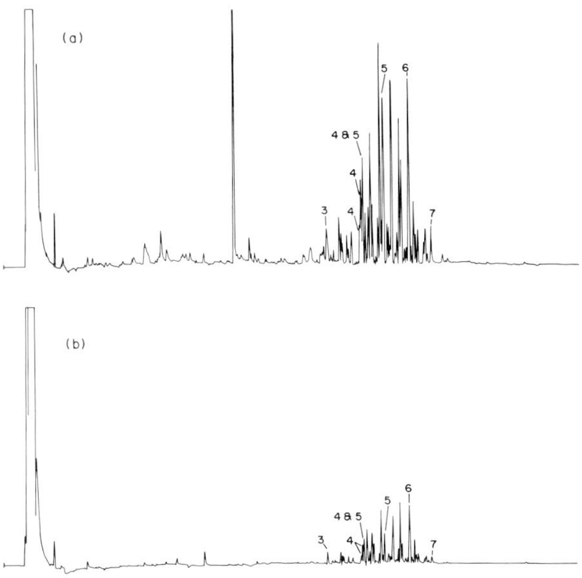

Figure 10.6. Chromatograms of the PCB fraction at day 40 for (a) exposure and (b) control mussel samples. The numbers refer to the number of chlorine atoms on the PCBs.

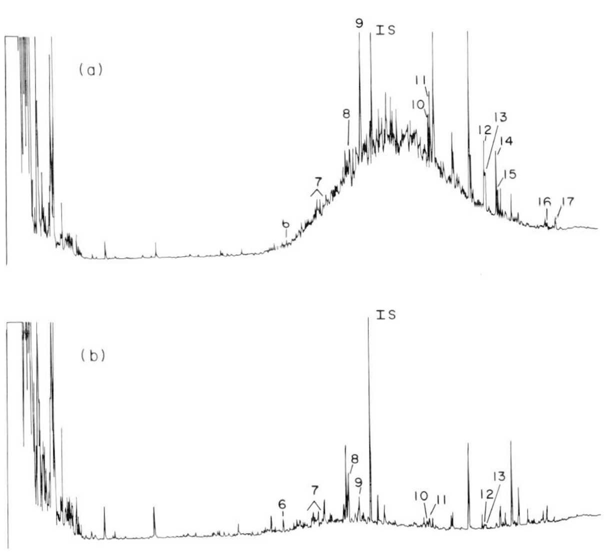

Figure 10.8. Chromatograms of the PAH fraction at day 40 for (a) exposure and (b) control mussel samples. See Fig. 10.2 for identification of the peaks; IS stands for internal standard (m-terphenyl).

for surface sediments collected worldwide (Laflamme and Hites, 1978). This distribution is believed to result from combustion processes (Youngblood and Blumer, 1975).

10.4.2. Exposure System

The concentrations of PCBs detected in the exposure tank (total of filter and foam plug samples) were consistent with the measured suspended-particle levels and exposure-sediment PCB concentrations. Also, the PCB peak pattern in the suspended particles of the exposure tank was almost identical to that of the exposure sediment (Fig. 10.3). The dissolved phase, however, contained a lower molecular-weight distribution of PCB compounds. Chlorobiphenyls with 8, 9, and 10 chlorine atoms, which were present in the exposure sediment, were not detected in the foam plug samples. Higher PCB levels were observed in exposed versus control foam plug samples. This indicates that some of the less-chlorinated PCBs may have desorbed from the sediment into the dissolved phase.

Low concentrations of numerous two-, three-, and four-ring PAH compounds were detected in the dissolved phase from the control tank (Fig. 10.2). We believe that this represents the background concentration of PAHs in the seawater entering the laboratory, which is located in the relatively uncontaminated lower portion of Narragansett Bay. The high concentrations of pyrene in both exposure and control tank foam plugs seem anomalous; this may be due to contamination, although contamination by a single PAH compound is unlikely.

Analysis of the exposure-tank filter samples for PAHs (Fig. 10.2) indicated a depletion of several compounds (phenanthrene, C_1-phenanthrenes, and fluoranthene) compared to the exposure sediment. Apparently, this depletion in the filter samples represents a partitioning of these compounds in the dissolved phase, as these compounds showed increased concentrations in the exposure-tank foam plugs. The variability observed for several of the other PAHs was large, making similar comparisons difficult for these compounds. Yet, the results indicated that as with the PCBs, some lower-molecular-weight compounds were desorbed from the exposure sediment into the dissolved phase.

The distribution of compounds observed in the particulate and dissolved phases of exposure tank seawater was consistent with partitioning models (Chiou et al., 1977; Mackay et al., 1980). These models predict that the lower-molecular-weight compounds, which have higher water solubilities, would be more readily desorbed from the sediment into the dissolved phase.

10.4.3. Mussels

Mussels exposed to contaminated sediment particles rapidly accumulated PCBs, reaching an apparent steady-state concentration after ~20 d of exposure (Fig. 10.5). The wet weight:dry weight ratio and lipid content of the organisms remained relatively constant during the experiment. Therefore, PCB accumulation trends were similar for the three quantification methods used.

In similar studies, Langston (1978) reported that cockles *(Cerastoderma edule)* exposed to Aroclor 1260 adsorbed onto aluminum oxide particles reached maximum PCB concentrations after a 20-d exposure. Vreeland (1974) showed that oysters exposed to six PCB compounds reached a steady-state concentration in ~1 month. However, Peddicord (1980) found that mussels dosed with very high concentrations (2–20 g liter^{-1}) of PCB-contaminated sediment for 25 d did not accumulate these compounds. The maximum PCB concentration observed in mussels from the present study was ~1400 ng g^{-1} (dry weight). This value is similar to levels reported for mussels *(Mytilus edulis)* collected from contaminated areas (Cowan, 1981) or caged near industrial centers (Farrington et al., 1982). It is about two times as high as concentrations reported for *Mytilus edulis* caged 1 m above the sediment in the Providence River (unpublished manuscript).

The distribution of PCBs accumulated by mussels in this study (Fig. 10.6) was depleted in higher-molecular-weight compounds relative to the exposure sediment (Fig. 10.3). This finding agrees with work done by Langston (1978), who reported that the accumulation of seven and eight chlorobiphenyls was reduced compared to other congeners of an Aroclor 1260 mixture in two bivalve species (*Cerastoderma edule* and *Macoma balthica).* Also, Courtney and Denton (1976) found that clams accumulated larger concentrations of the less-chlorinated compounds of Aroclor 1254 dosed on aluminum oxide particles. Similarly, it has been found that mussels caged in Narragansett Bay accumulated PCBs with 3–7 chlorine atoms per molecule, although surface sediment collected from the same location contained chlorobiphenyls with 3–10 chlorine atoms per molecule (unpublished manuscript).

In the present study, the PCB distribution found in exposed mussels (Fig. 10.6) was similar to that detected in the dissolved phase of the exposure tank (Fig. 10.3). The mussels did not contain several of the more highly chlorinated compounds that were present on the particles. This may indicate that the compounds accumulated by the mussels were desorbed from the particles into the dissolved phase before incorporation. Alternatively, these compounds may be more easily removed from particles by the mussels. These results are similar to those reported by Halter and Johnson (1977) from studies on the uptake of PCBs by fish. Other reasons such as molecular size or steric properties affecting membrane passage as described by Tulp and Hutzinger (1978) may also explain these observations.

Control mussels showed a slight decrease in total identified PAH levels during the experiment. However, PAHs were rapidly incorporated by exposed mussels, reaching a maximum concentration at day 20 (Fig. 10.7), then decreasing slightly. The PAH fraction of exposed mussels contained a large UCM and high concentrations of several PAH compounds. Unlike the PCBs, PAH distributions in the mussels were more like the distributions found in the particles than those detected in the dissolved phase. Compared to the particles, however, the UCMs showed a narrower distribution (Figs. 10.4 and 10.8), indicating a depletion of higher-molecular-weight unresolved components. This trend was also noted by Teal and Farrington (1977) for mussels and sediment collected from a salt marsh.

In the present study, exposed mussels did not bioaccumulate two- and three-ring compounds, but did accumulate high concentrations of several four- and five-ring compounds. These results agree with a study by Bjorseth et al. (1979), which compared PAHs in mussels and sediment collected from a contaminated Norwegian fjord. The reason for the low concentrations of the two- and three-ring compounds is not known, but it may be the result of their smaller partition coefficients or of metabolic processes in the mussels.

Concentration factors were calculated for PCBs and PAHs in the exposure mussels. This analysis was done for mussels sampled at day 20. At day 20, the PAH concentrations were highest, and between days 20 and 40 some of these compounds decreased in concentration. The reason for this decrease is not known, but it may have been due either to the induction of enzymes that metabolize these compounds or to toxic effects on mussels from the contaminated sediment. Therefore, it is thought that day-20 mussels would best represent the distribution of compounds taken up by the mussels.

The concentration factors for total PCBs in day-20 exposure mussels were 63,800, 9630, and 8370 (on a wet-weight basis) when calculated against the dissolved, particulate, and total concentrations, respectively. Benz(a)anthracene showed the largest concentration factors for the PAHs that could be measured both in the dissolved and in the particulate phases of the dosing system. The concentration factors for this PAH compound in mussels were 35,700, 2650, and 2470 for the dissolved, particulate, and total concentrations, respectively.

10.5. CONCLUSIONS

1. Analysis of filter and foam plug samples indicated that low-molecular-weight PCBs and PAHs were present in the dissolved phase of the exposure tank. The particulate phase contained PCBs and PAHs in distributions similar to those of the sediment.

2. Both PCBs and PAHs were accumulated by mussels exposed to contaminated resuspended sediment.

3. The PCBs reached a steady-state concentration in the mussels after 20 d of exposure. The PAHs reached a maximum concentration at day 20, then decreased slightly.

4. The PCB distributions of the exposed mussels were similar to those found in the dissolved phase.

5. After 40 d, PAH distributions in the mussels were generally similar to those detected in the particles, although when compared to the particle PAH distributions, mussels were depleted in low-molecular-weight compounds.

ACKNOWLEDGMENTS

We thank Michael Balboni, Robert Bowen, Curt Norwood, and Sharon Pavignano from the Narragansett Laboratory of the U.S. Environmental Protection Agency (EPA) and Gregg Douglas from the University of Rhode Island for their help. Although the information described in this article has been funded wholly or in part by the EPA under assistance agreement to J. G. Quinn, it has not been subjected to the agency's required peer and administrative review; therefore, does not necessarily reflect the views of the agency, and no official endorsement should be inferred. Use of product names does not constitute endorsement of that product by the EPA.

REFERENCES

Basu, D. K., and J. Saxena. 1978. Monitoring of polynuclear aromatic hydrocarbons in water. II: Extraction and recovery of six representative compounds with polyurethane foam plugs. *Environmental Science and Technology,* **12,** 791–795.

Bayne, B. L., R. J. Thompson, and J. Widdows. 1976. Physiology I. *In:* Marine Mussels: Their Ecology and Physiology, B. L. Bayne (Ed.). Cambridge University Press, New York, pp. 121–206.

Bjorseth, A., J. Knutzen, and J. Skei. 1979. Determination of polycyclic aromatic hydrocarbons in sediments and mussels from Saudafjord, W. Norway, by glass capillary gas chromatography. *The Science of the Total Environment,* **13,** 71–86.

Bopp, R. F., H. J. Simpson, C. R. Olsen, R. M. Trier, and N. Kostyk. 1982. Chlorinated hydrocarbons and radionuclide chronologies in sediments of the Hudson River and Estuary, New York. *Environmental Science and Technology,* **16,** 666–676.

Chiou, C. T., V. H. Freed, D. W. Schmedding, and R. L. Kohnert. 1977. Partition coefficient and bioaccumulation of selected organic chemicals. *Environmental Science and Technology,* **11,** 475–478.

Courtney, W. A. M., and G. R. Denton. 1976. Persistence of polychlorinated biphenyls in the hard-clam *(Mercenaria mercenaria)* and the effect upon the distribution of these pollutants in the estuarine environment. *Environmental Pollution,* **10,** 55–64.

Cowan, A. A. 1981. Organochlorine compounds in mussels from Scottish coastal waters. *Environmental Pollution (Series B),* **2,** 129–143.

de Lappe, B. W., R. W. Risebrough, A. M. Springer, T. T. Schmidt, J. C. Shropshire, E. F. Letterman, and J. R. Payne. 1980. The sampling and measurement of hydrocarbons in natural waters. *In:* Hydrocarbons and Halogenated Hydrocarbons in the Aquatic Environment, B. K. Afgan and D. Mackay (Eds.). Plenum Press, New York, pp. 29–68.

Farrington, J. W., R. W. Risebrough, P. L. Parker, A. C. Davis, B. de Lappe, J. K. Winters, D. Boatwright, and N. M. Frew. 1982. Hydrocarbons, Polychlorinated Biphenyls, and DDE in Mussels and Oysters from the U.S. Coast, 1976–1978. Technical Report WHOI-82-42, The Mussel Watch, Woods Hole Oceanographic Institution, Woods Hole, Massachusetts, 106 pp.

Galloway, W. B., J. L. Lake, D. K. Phelps, P. F. Rogerson, V. T. Bowen, J. W. Farrington, E. D. Goldberg, J. L. Laseter, G. C. Lawler, J. H. Martin, and R. W. Risebrough. 1983. The Mussel Watch: intercomparison of trace level constituent determinations. *Environmental Toxicology and Chemistry,* **2,** 395–410.

Gesser, H. D., A. Chow, F. C. Davis, J. F. Uthe, and J. Reinke. 1971. The extraction and recovery of polychlorinated biphenyls (PCB) using porous polyurethane foam. *Analytical Letters,* **4,** 883–886.

Geyer, H., P. Sheehan, D. Kotzias, D. Freitag, and F. Korte. 1982. Prediction of ecotoxicological behavior of chemicals: Relationship between physico-chemical properties and bioaccumulation of organic chemicals in the mussel *Mytilus edulis. Chemosphere,* **11,** 1121–1134.

Halter, M. T., and H. E. Johnson. 1977. A model system to study the desorption and biological availability of PCB in hydrosoils. *In:* Aquatic Toxicology and Hazard Evaluation, F. L. Mayer and J. L. Hamelink (Eds.). Special Technical Publication 634, American Society for Testing and Materials, Philadelphia, pp. 179–195.

Hoffman, E. J., G. L. Mills, J. S. Latimer, and J. G. Quinn. 1983. Annual input of petroleum hydrocarbons to the coastal environment via urban runoff. *Canadian Journal of Fisheries and Aquatic Science,* **40** (Supplement 2), 41–53.

Jungclaus, G. A., V. Lopez-Avila, and R. A. Hites. 1978. Organic compounds in an industrial wastewater: A case study of their environmental impact. *Environmental Science and Technology,* **12,** 88–96.

Karickhoff, S. W., D. S. Brown, and T. A. Scott. 1979. Sorption of hydrophobic pollutants on natural sediments. *Water Research,* **13,** 241–248.

Knutzen, J., and B. Sortland. 1982. Polycyclic aromatic hydrocarbons (PAH) in some algae and invertebrates from moderately polluted parts of the coast of Norway. *Water Research,* **16,** 421–428.

Kveseth, K., B. Sortland, and T. Bokn. 1982. Polycyclic aromatic hydrocarbons in sewage, mussels and tap water. *Chemosphere,* **11,** 623–639.

Laflamme, R. E., and R. A. Hites. 1978. The global distribution of polycyclic aromatic hydrocarbons in recent sediments. *Geochimica et Cosmochimica Acta,* **42,** 289–303.

Lake, J. L., C. Norwood, C. Dimock, and R. Bowen. 1979. Origins of polycyclic aromatic hydrocarbons in estuarine sediments. *Geochimica et Cosmochimica Acta,* **43,** 1847–1854.

Langston, W. J. 1978. Accumulation of polychlorinated biphenyls in the cockle *Cerastoderma edule* and the tellin *Macoma balthica. Marine Biology,* **45,** 265–272.

Lopez-Avila, V., and R. A. Hites. 1980. Organic compounds in an industrial wastewater. Their transport into sediments. *Environmental Science and Technology,* **14,** 1382–1390.

Mackay, D., A. Bobra, and W. Y. Shiu. 1980. Relationships between aqueous solubility and octanol–water partition coefficients. *Chemosphere,* **9,** 701–711.

Means, J. C., S. G. Wood, J. J. Hassett, and W. L. Banwart. 1980. Sorption of polynuclear aromatic hydrocarbons by sediments and soils. *Environmental Science and Technology,* **14,** 1524–1528.

Neff, J. M., B. A. Cox, D. Dixit, and J. W. Anderson. 1976. Accumulation and release of petroleum-derived aromatic hydrocarbons by four species of marine animals. *Marine Biology,* **38,** 279–289.

Oviatt, C. A., and S. W. Nixon. 1975. Sediment resuspension and deposition in Narragansett Bay. *Estuarine and Coastal Marine Science,* **3,** 201–217.

Paulson, A. J., and D. T. Brown. 1978. PCBs: Their Environmental Significance and Distribution in Rhode Island. Technical Report 68, Coastal Resources Center, University of Rhode Island, Kingston, 25 pp.

Peddicord, R. K. 1980. Direct effect of suspended sediments on aquatic organisms. *In:* Contaminants and Sediments, Vol. 1, R. A. Baker (Ed.). Ann Arbor Science Publishers, Ann Arbor, Michigan, pp. 501–536.

Phelps, D. K., and W. B. Galloway. 1980. A report on the Coastal Environmental Assessment Station (CEAS) Program. *In:* Biological Effects of Marine Pollution and the Problems of Monitoring, A. D. McIntyre and J. B. Pearce (Eds.). Rapports et Procès-Verbaux des Réunions, Vol. 179. Conseil International Pour l'Exploration de la Mer, Copenhagen, pp. 76–81.

Sinnet, J. C., and W. R. Davis. 1983. A programmable turbidistat for suspended particles in laboratory aquaria. *Journal of Experimental Marine Biology and Ecology,* **73,** 167–174.

Steen, W. C., D. F. Paris, and G. L. Baughman. 1978. Partitioning of selected polychlorinated biphenyls to natural sediments. *Water Research,* **12,** 655–657.

Teal, J. M., and J. W. Farrington. 1977. A comparison of hydrocarbons in animals and their benthic habitats. *In:* Petroleum Hydrocarbons in the Marine Environment, A. D. McIntyre and K. J. White (Eds.). Rapports et Procès-Verbaux des Réunions, Vol. 171. Conseil International Pour l'Exploration de la Mer, Copenhagen, pp. 79–83.

Tulp, M. T. M., and D. Hutzinger. 1978. Some thoughts on aqueous solubilities and partition coefficient of PCB, and the mathematical correlation between bioaccumulation and physico-chemical properties. *Chemosphere,* **7,** 849–860.

Van Vleet, E. S., and J. G. Quinn. 1977. Input and fate of petroleum hydrocarbons entering the Providence River and upper Narragansett Bay from wastewater effluents. *Environmental Science and Technology,* **11,** 1086–1092.

Veith, G. D., K. J. Macek, S. R. Petrocelli, and J. Carrol. 1980. An evaluation of using partition coefficients and water solubility to estimate bioconcentration factors for organic chemicals in fish. *In:* Aquatic Toxicology, J. G. Eaton, P. R. Parrish, and A. C. Hendricks (Eds.). Special Technical Publication 707, American Society for Testing and Materials, Philadelphia, pp. 116–129.

Vreeland, V. 1974. Uptake of chlorobiphenyls by oysters. *Environmental Pollution,* **6,** 135–140.

Windsor, J. G., Jr., and R. A. Hites. 1979. Polycyclic aromatic hydrocarbons in Gulf of Maine sediments and Nova Scotia soils. *Geochimica et Cosmochimica Acta,* **43,** 27–33.

Young, D. R., T. C. Heesen, and D. J. McDermott. 1976. An offshore biomonitoring system for chlorinated hydrocarbons. *Marine Pollution Bulletin,* **7,** 157–159.

Youngblood, W. W., and M. Blumer. 1975. Polycyclic aromatic hydrocarbons in the environment: homologous series in soils and recent marine sediments. *Geochimica et Cosmochimica Acta,* **39,** 1303–1314.

CHAPTER 11

Comparative Accumulation of PAHs in Four Marine Invertebrates

Donald W. McLeese and Leslie E. Burridge

Biological Station
Fisheries and Environmental Sciences
Department of Fisheries and Oceans
St. Andrews, New Brunswick, Canada

ABSTRACT

The mean concentrations of five polycyclic aromatic hydrocarbons (PAHs) (phenanthrene, fluoranthene, pyrene, triphenylene, and perylene) in the clam, *Mya arenaria*, the mussel, *Mytilus edulis*, the shrimp, *Crangon septemspinosa*, and the polychaete worm, *Nereis virens*, were examined. After a 4-d exposure to the PAHs in water, the mean concentrations showed the following pattern: mussel > clam > shrimp hepatopancreas > whole shrimp ≥ polychaete. After exposure of the animals to contaminated sediment for 4 d, the general pattern changed to: polychaete ≥ mussel > clam ≥ whole shrimp ≥ shrimp hepatopancreas. Unlike mussels and clams, relative uptake of PAH by shrimp and polychaetes was greater from contaminated sediment than from water, indicating that the presence of sediment enhanced the uptake of PAH. During a 14-d postexposure period, concentrations of PAHs in the four species decreased by 25–100%. The uptake rate constant (K_1) for the five PAHs and four species ranged from 0.9 to 60.7 h^{-1}, and for each species, K_1 tended to increase with compounds of increasing molecular complexity or weight. The excretion rate constant (K_2) ranged from 0.0008 to 0.035 h^{-1} and tended to decrease with compounds of increasing molecular complexity or weight. The bioconcentration factor (BCF) (K_1/K_2) ranged from 140 to 14,000, was greatest for mussels and clams, and was usually smallest for shrimp. Large BCF values and small K_2 values for perylene in water and sediment tests with clams and mussels indicated that of the five PAHs, perylene had the highest potential for bioaccumulation problems. It is inferred that accumulation of PAHs by clams and mussels is mainly from PAHs in water, and that accumulation by shrimp and polychaetes is enhanced by the presence of contaminated sediment.

11.1. INTRODUCTION

Polycyclic aromatic hydrocarbons (PAHs) are relatively insoluble in water and are transported to sediments in association with particulate matter. Generally, PAH levels in sediment decrease with distance from urbanized areas. For example, total PAH concentrations in sediments from Boston, Massachusetts, to 64 km seaward range from 120,000 to 160 ng g^{-1} (dry wt), respectively. In the Gulf of Maine, levels of PAHs in the sediment range from 870 to 200 ng g^{-1}, and deep-ocean sediments range from 160 to 18 ng g^{-1} (Windsor and Hites, 1979).

Sources of PAHs in the marine environment include creosote oil, used to preserve wooden structures, industrial effluents, petroleum spillage, combustion of fossil fuels, and forest and brush fires (Neff, 1979).

It is well known that PAHs contaminate marine shellfish living near marinas and harbors (Cahnmann and Kuratsune, 1957; Andelman and Suess, 1970). Elevated levels of benz[*a*]pyrene, a carcinogenic PAH (U.S. National Academy of Sciences, 1972; Lo and Sandi, 1978), were found in mussels (*Mytilus edulis*) taken from creosoted pilings (Dunn and Stich, 1975); however, molluscs from an open-water area where there was no nearby source of creosote also contained PAHs. This indicates that PAH contamination of coastal areas could be quite widespread (Zitko, 1975). Harbor and channel dredging pose another type of PAH contamination problem. Disposal of the dredged material at sea redistributes sediments that may contain high levels of PAHs.

In our study, bioavailability of PAHs in water and in contaminated sediment was investigated by exposing clams (*Mya arenaria*), mussels (*Mytilus edulis*), shrimp (*Crangon septemspinosa*), and a polychaete (*Nereis virens*) to seawater containing a mixture of five PAHs and to sediment contaminated with wood-treatment plant wastes. Data for clams, reported previously (McLeese et al., 1985), are reviewed here as part of the information on the comparative accumulation of PAHs from water and from sediment by the four invertebrates.

11.2. EXPERIMENTAL

11.2.1. Invertebrate Exposure

For exposure to PAHs in water, groups of four invertebrates (*Mya arenaria, Mytilus edulis, Crangon septemspinosa, and Nereis virens*) were exposed simultaneously to seawater flowing at a rate of 1 liter min^{-1}. A solution containing five different PAHs (phenanthrene, fluoranthene, pyrene, triphenylene, and perylene) was added to the seawater at 0.6 ml min^{-1} from a Mariotte bottle. After a 96-h exposure, contaminant flow was terminated, and the animals were maintained in flowing seawater at 10°C for two weeks.

For the PAH sediment-exposure experiments, sediment was collected from a site in northeastern New Brunswick receiving wastes from a wood-treatment plant. The sediment consisted of 24% silt and clay and 76% sand, and the organic carbon content was 1%. Groups of the four invertebrates were maintained in a fiberglass tank (1 m x 1 m x 20 cm), containing a 4-cm layer of sediment. The temperature was maintained at 10°C and water flow was 1 liter min^{-1}. After a 96-h exposure, the animals were transferred to a similar setup with uncontaminated sediment.

Usually two animals of each species were sampled on days 0, 1, 2, 3, and 4 during exposure to water and on days 1, 2, and 4 during exposure to sediment. Two animals of each species were also sampled on days 1, 2, 4, 7, and 14 during the postexposure period. Water samples (100 ml) were taken on days 0, 1, 2, 3, and 4 during both exposure tests.

11.2.2. Chemical Analyses

A 15-g sample of contaminated sediment was taken on day 0 and extracted by the method described by McLeese et al. (1985), except that the samples were brought to final volume with acetonitrile.

Seawater samples (100 ml) were extracted by a method similar to that described by Wilkinson et al. (1979). The extract, 30 ml of methylene chloride, was dried over sodium sulfate and acetonitrile was added as a carrier. The mixture was then evaporated to 3 ml and brought to final volume under a gentle stream of nitrogen.

Whole tissues of each of the animals were homogenized by grinding with sodium sulfate, and the resulting mixture was extracted on a Soxhlet apparatus with hexane for 1 h. The cooled extract was brought to a volume of 100 ml, and an aliquot was cleaned with deactivated alumina, dried, and brought to final volume with acetonitrile. At each sampling, shrimp hepatopancreas was dissected from two or three shrimp, weighed, and homogenized with Na_2SO_4 at 6 times the weight of hepatopancreas tissue weight. The homogenate, layered on top of an alumina column, was eluted with 30 ml of hexane, and acetonitrile was added to the solution as a carrier. The mixture was evaporated to near dryness and brought to final volume with acetonitrile.

Samples were analyzed by high performance liquid chromatography (HPLC) using a Schoeffel SF 970 fluorometer (Schoeffel/McPherson, Acton, Massachusetts). The excretion wave length was 250 nm and the emission cut-off wave length was 379 nm. The solvent, 15% HPLC-grade water in HPLC-grade acetonitrile, was delivered at 1 ml min^{-1} by an Altex HPLC pump (Altex Scientific, Inc., Berkeley, California). The sample was passed over a Whatman/partisil RxS-10/25-ODS-2 reverse-phase column (Whatman, Inc., Clifton, New Jersey).

Extraction efficiencies (Table 11.1) were determined by spiking seawater and animal tissues with the five PAH com-

Table 11.1. Extraction Efficiencies (%) of PAHs from Water and Animals[a]

Compound	Seawater[b]	Clam[b]	Mussel[b]	Shrimp (whole)[b]	Shrimp (hepatopancreas)[c]	Polycheate[d]
Phenanthrene	100 ± 7	94 ± 4	86 ± 1	87 ± 4	95 ± 3	97 ± 4
Fluoranthene	74 ± 5	87 ± 6	94 ± 2	99 ± 3	92 ± 4	92 ± 7
Pyrene	89 ± 3	93 ± 8	110 ± 2	104 ± 7	98 ± 1	105 ± 16
Triphenylene	124 ± 9	82 ± 2	115 ± 7	111 ± 8	87 ± 0	88 ± 9
Perylene	77 ± 4	87 ± 8	118 ± 3	126 ± 2	108 ± 6	96 ± 4

[a] Mean and standard deviation are given.
[b] $n = 4$.
[c] $n = 2$.
[d] $n = 3$.

pounds, each at a level expected to occur in the tests. The extraction efficiencies were used to calculate PAH concen trations. The measured concentrations of each PAH in seawater and in contaminated sediment are listed in Table 11.2. The presence of these five PAHs in the sediment was con firmed by gas chromatography-mass spectrometry (GC-MS). Analysis of untreated laboratory seawater indicated the presence of phenanthrene (0.25 μg liter^{-1}) and fluoranthene (0.6 μg liter^{-1}) and the absence of pyrene, tripheylene, and perylene.

11.3. RESULTS

All data for mussels are listed in Table 11.3. The concentration of each PAH in mussels increased progressively during exposure. When placed in uncontaminated conditions for depuration, the levels of PAH declined progressively with time. The data points for fluoranthene and triphenylene are shown in Fig. 11.1.

The data for the exposure and postexposure phases for the five PAHs and the four invertebrates in water and sediment tests are summarized in Table 11.4. Based on the measured concentrations at a 4-d exposure, the uptake of phenanthrene, fluoranthene, triphenylene, and perylene from water follow the order: mussels > clams > shrimp ≥ polychaete. For pyrene in water, polychaetes ranked first. When exposed to contaminated sediment, the order of uptake for the five PAHs was polychaete ≥ mussels > clams > shrimp. The levels of PAH within the animals declined by 25–100% (Table 11.4) during the 14-d postexposure period.

Table 11.2. Concentrations of PAHs in Seawater, Surficial Water, and Sediment in the 96-h Animal Exposures[a]

Compound	Seawater (μg liter^{-1})[b]	Surficial Water (μg liter^{-1})[c]	Sediment (μg kg^{-1})[d]
Phenanthrene	4.3 ± 0.4	0.3	660
Fluoranthene	2.4 ± 0.2	1.0	710
Pyrene	1.7 ± 0.3	2.0	1180
Triphenylene	0.5 ± 0.1	0.4 ± 0.2	18
Perylene	0.4 ± 0.1	0.3 ± 0.2	8

[a] Mean and standard deviation are given.
[b] $n = 5$.
[c] $n = 4$.
[d] $n = 1$.

The ratios for the concentrations of PAHs in water to those in surficial water in the sediment tests ranged from 0.6 to 6 (Table 11.5). If PAH uptake was mainly from water, it would be expected that ratios of the levels of PAH in the animals from water and sediment tests at 4-d exposure would approximate these ratios. The ratios are close to those expected for clams, mussels, and shrimp hepatopancreas, indicating uptake primarily from water. In contrast, ratios that are less than expected would indicate that PAH uptake was enhanced by the presence of contaminated sediment. The ratios for polychaetes and whole shrimp were less than expected, indicating enhanced uptake.

11.4. DISCUSSION AND CONCLUSIONS

The full set of data from each animal and each PAH in water and sediment tests (Table 11.4) have been used to calculate uptake (K_1) and excretion (K_2) rate constants using the one-compartment model described by Zitko (1980); this model is also discussed by Mackay and Hughes (1984). The equation is as follows:

$$CF = K_1 \times CW \times [1 - \exp(-K_2 \times T)]/K_2 \qquad 1$$

where CF and CW are concentration in animals (mg kg^{-1}) and in water (mg liter^{-1}), respectively. Uptake and excretion rate constants (h^{-1}) are K_1 and K_2, respectively, and T is time in hours.

Table 11.3. Concentrations of PAHs in Mussels (mg kg^{-1}, wet wt)[a]

Sampling Day	Phenanthrene (4.3μg liter^{-1})[a]		Fluoranthene (2.4 μg liter^{-1})[b]		Pyrene (1.7 μg liter^{-1})[b]		Triphenylene (0.5 μg liter^{-1})[b]		Perylene (0.4 μg liter^{-1})[b]	
					Water Tests					
Exposure										
0	0.07	0.04	0.02	0.03	0	0	0	0	0	0
1	0.8	1.1	0.4	0.8	0.4	0.9	0.1	0.3	0.05	0.09
2	1.2	3.4	1.3	2.6	1.2	3.0	0.6	1.7	0.3	0.8
3	2.1	2.3	2.5	2.8	2.0	2.4	1.0	1.5	0.4	0.8
4	3.8	4.7	4.0	4.3	3.8	3.9	1.9	2.0	0.8	1.0
Postexposure										
1	2.0	2.4	2.2	3.1	1.9	2.8	0.8	1.6	0.4	0.9
2	1.5	2.3	3.0	3.6	2.3	2.7	1.7	2.0	0.6	0.9
4	1.2	1.2	1.4	2.5	1.1	1.4	0.5	1.8	0.3	1.2
7	0.4	1.0	1.3	1.8	1.0	1.3	0.7	0.9	0.4	0.4
14	0.2	1.4	0.7	2.5	0.6	1.3	0.5	1.1	0.4	0.6
					Sediment Tests					
Exposure										
0	0.04	0.04	0.02	0.03	0	0.01	0.01	0.01	0	0
1	0.04	0.4	1.2	1.6	1.0	1.5	0.3	0.5	0.04	0.05
2	1.0	1.1	3.2	3.8	3.2	4.0	1.1	1.4	0.1	0.2
4	0.9	1.2	4.1	5.7	4.5	6.2	1.4	2.2	0.1	0.3
Postexposure										
1	1.0	1.1	3.3	4.0	3.4	4.0	1.3	1.5	0.1	0.2
2	0.5	0.6	3.0	3.3	2.8	3.0	1.1	1.2	0.1	0.1
4	0.5	0.8	3.0	4.6	4.1	3.0	1.2	1.6	0.1	0.2
7	0.4	0.7	2.3	4.0	2.2	3.8	0.8	1.6	0.1	0.2
14	0.1	0.1	1.1	1.1	1.1	1.3	0.4	0.5	0.1	0.1

[a] Two organisms tested for each PAH.
[b] Initial exposure concentration.

From the excretion data, the K_2 values were estimated by using a computer program for the solution of non-linear regressions. Correlation coefficients in the regressions are listed in Table 11.6. By solving the equation by using an estimated K_2 value, *CF* at 4 d (Table 11.4) and a *T* of 96 h, K_1 values were obtained.

The K_1 and K_2 values are listed in Table 11.6 along with the bioconcentration factors (BCF), which in this case are estimated as K_1/K_2. The BCF represents the concentration factor to be expected under equilibrium conditions. Examples of the fitted results for fluoranthene and triphenylene are shown for mussels and polychaetes from water tests (Fig. 11.1a) and in sediment tests (Fig. 11.1b).

The PAH compounds, in order of increasing molecular complexity or molecular weight, are phenanthrene, fluoranthene, pyrene, triphenylene, and perylene. The uptake rate constants (K_1) ranged from 0.9 to 41.0 h^{-1} for water tests and from 1.0 to 60.7 h^{-1} for sediment tests (Table 11.6). For each species, the K_1 values tended to increase following the above-mentioned order of compounds. The major exception was perylene, because each of the 10 K_1 values was smaller than expected on the basis of the general trend.

The excretion rate constants (K_2) ranged from 0.0011 to 0.035 h^{-1} for water tests and from 0.0008 to 0.030 h^{-1} for sediment tests (Table 11.6). For each species, the K_2 values tended to decrease with increasing molecular complexity or weight.

Based on the estimated BCF factors and on estimated equilibrium concentrations, the uptake of PAHs from water and from sediment showed the following order: mussels > clams > polychaetes > shrimp. Based on the 4-d uptake data, it was seen that polychaetes ranked first rather than third in sediment tests; otherwise, the ranking in sediment and in water tests was similar.

Compounds with BCF values of more than 200–300 and K_2 values of less than 0.0008 h^{-1} may have high bioaccumulation potential (Zitko, 1980). The low excretion rate means

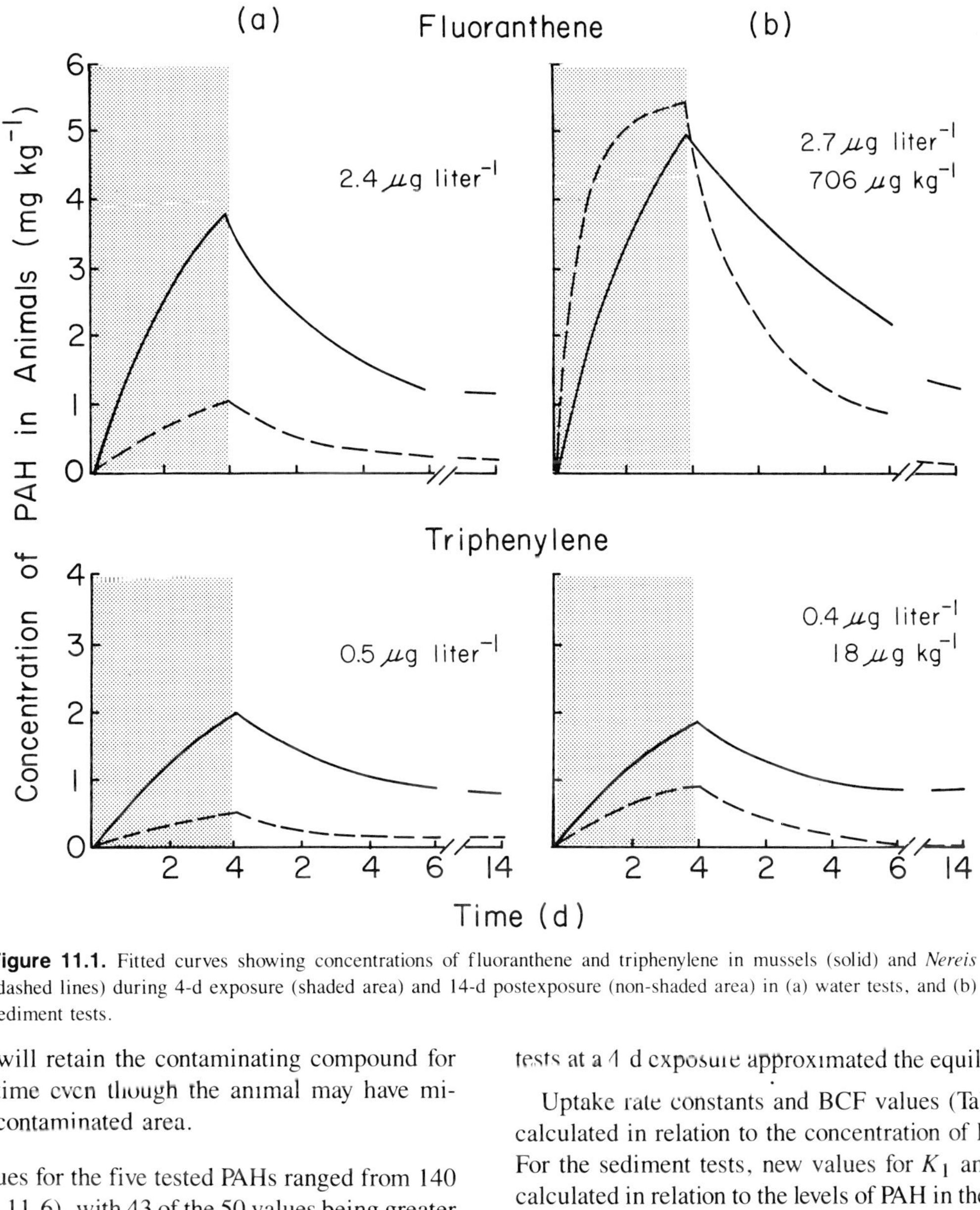

Figure 11.1. Fitted curves showing concentrations of fluoranthene and triphenylene in mussels (solid) and *Nereis* (dashed lines) during 4-d exposure (shaded area) and 14-d postexposure (non-shaded area) in (a) water tests, and (b) sediment tests.

that an animal will retain the contaminating compound for a considerable time even though the animal may have migrated to an uncontaminated area.

The BCF values for the five tested PAHs ranged from 140 to 14,000 (Table 11.6), with 43 of the 50 values being greater than 300. Only four of the 50 K_2 values are close to 0.0008 h^{-1}, however, and these are for perylene in clams and mussels. Therefore, it is concluded that perylene has the highest potential for bioaccumulation, at least in clams and mussels.

Comparison showed that the levels of PAHs in clams, mussels, and polychaetes at a 4-d exposure (Table 11.4) were less than the estimated equilibrium concentrations (Table 11.6). In contrast, the levels of PAHs in whole shrimp in water and sediment tests and shrimp hepatopancreas in water tests at a 4 d exposure approximated the equilibrium values.

Uptake rate constants and BCF values (Table 11.6) were calculated in relation to the concentration of PAHs in water. For the sediment tests, new values for K_1 and BCF can be calculated in relation to the levels of PAH in the contaminated sediment. It was assumed that K_2 values would not change. The new values were considerably smaller than the original ones. The magnitude of the difference is dependent on the ratio of the PAH level in surficial water to that in the sediment. Using phenanthrene as an example, the ratio was 0.7 µg liter^{-1}:664 µg kg^{-1}, or about 0.001. Consequently, the new K_1 and BCF values in relation to the sediment were about 0.001 times the original K_1 and BCF values. The factors for fluoranthene, pyrene, triphenylene, and perylene were about 0.004, 0.006, 0.02, and 0.04, respectively.

Table 11.4. Concentration of PAHs (mg kg^{-1}, wet wt) in Animals before (0-d) and during (4-d) Exposure to PAHs in Water or Sediment and after a Postexposure Period (14-d)

	Water Tests			Sediment Tests		
	PAH Concentration			PAH Concentration		
Compound and Animal	0-d	4-d	14-d	0-d	4-d	14-d
Phenanthrene						
Clam	0.01	1.7	0.2	0.03	0.2	0
	0.03	2.1	0.5	0.44	0.5	0.01
		2.3				
		2.7				
Mussel	0.04	3.7	0.2	0.04	0.9	0.1
	0.07	4.7	1.4	0.04	1.1	0.1
Shrimp (whole)	0.03	0.6	0.1	0.04	0.8	0
	0.03	1.6	0.5	0.12	0.8	0
Shrimp (hepatopancreas)	0	1.7	0	0	0.2	0
	0	1.7	0	0	0.7	0.1
		3.8				
Polychaete	0.04	1.1	0.1	0.11	3.7	0
	0.05	0.9	0.1	0.12	5.4	0
Fluoranthene						
Clam	0.01	1.6	0.7	0.02	1.3	0.1
	0.02	2.0	1.0	0.02	2.8	0.6
		2.2				
		3.0				
Mussel	0.02	4.0	0.7	0.02	4.1	1.1
	0.03	4.3	2.5	0.03	5.7	1.1
Shrimp (whole)	0.01	0.3	0	0.08	1.6	0
	0.02	0.9	0.1		2.2	0
Shrimp (hepatopancreas)	0	0.7	0	0	1.1	0
	0	1.1	0	0	1.2	0
		1.7				
Polychaete	0.01	0.6	0.1	0.05	3.7	0
	0.04	0.7	0.1	0.05	5.4	0.1
Pyrene						
Clam	0.01	1.3	0.9	0.01	1.2	0.2
	0.01	1.7	1.1	0.02	3.8	0.7
		3.0				
		3.4				
Mussel	0	3.8	0.6	0	4.5	1.0
	0	3.9	1.3	0.01	6.2	1.3
Pyrene (con't.)						
Shrimp (whole)	0	0.3	0	0.01	2.1	0
	0	0.5	0	0.01	2.3	0
Shrimp (hepatopancreas)	0	0.6	0	0	1.2	0
	0	0.7	0	0	2.1	0.4
		1.7				
Polychaete	0	4.6	0	0	4.6	0
	0	4.9	0.1	0	6.3	0
Triphenylene						
Clam	0	0.5	1.2	0	0.3	0.1
	0	0.8	1.1	0	0.8	0.2
		1.0				
		2.0				
Mussel	0	1.9	0.5	0.01	1.4	0.4
	0	2.0	1.1	0.01	2.2	0.5
Shrimp (whole)	0	0.1	0	0.03	0.3	0
	0	0.2	0	0.04	0.4	0
Shrimp (hepatopancreas)	0	0.2	0	0	0.2	0
	0	0.3	0	0	0.4	0.1
		0.4				
Polychaete	0	0.2	0.1	0	0.7	0
	0	0.2	0.1	0	0.5	0
Perylene						
Clam	0	0.1	0.2	0	0	0
	0	0.2	0.3	0	0.1	0.1
		0.3				
		0.6				
Mussel	0	0.8	0.4	0	0.2	0.1
	0	1.0	0.6	0	0.3	0.1
Shrimp (whole)	0	0.1	0	0.01	0	0
	0	0.1	0	0.01	0.1	0
Shrimp (hepatopancreas)	0	0.03	0	0	0.01	0.01
					0.02	0.01
Polychaete	0	0.01	0	0	0.1	0
	0	0.03	0.01	0	0.05	0

Table 11.5. Ratios of Concentrations of PAHs in Test Solutions and in Animals at a 4-d Exposure to PAHs in Water and Sediment[a]

Compound	Test Solutions Water: Surficial Water	Clam Water: Sediment	Mussel Water: Sediment	Shrimp (Whole) Water: Sediment	Shrimp (Hepatopancreas) Water: Sediment	Polychaete Water: Sediment
Phenanthrene	6	6 (9)	4 (2)	1.4 (0.8)	5 (1.9)	1.0 (0.9)
Fluoranthene	0.9	1.5 (1.9)	0.8 (0.9)	0.3 (0.2)	0.7 (0.3)	0.2 (0.3)
Pyrene	0.6	1.2 (1.9)	0.7 (0.5)	0.2 (0.2)	0.6 (0.2)	0.1 (0.2)
Triphenylene	1.2	3 (1.7)	1.1 (1.0)	0.4 (0.4)	0.8 (0.4)	0.4 (1.4)
Perylene	1.4	4 (5)	4 (1.5)	1.2 (0.6)	3 (2)	0.3 (0.4)

[a] Numbers in parentheses are ratios of concentrations of PAHs calculated using estimated equilibrium concentrations.

Table 11.6. Uptake (K_1) and Excretion (K_2) Rate Constants and Bioconcentration Factors (BCF) (K_1/K_2) for Four Species of Invertebrates Exposed to Five PAHs in Water and in Sediment

Animal or Compound	Water Tests: Rate Constants[a] K_1	Water Tests: Rate Constants[a] K_2	Water Tests: Correlation Coefficient	Water Tests: BCF (K_1/K_2)	Water Tests: Estimated Equilibrium Concentration (mg kg^{-1})	Sediment Tests: Rate Constants K_1	Sediment Tests: Rate Constants K_2	Sediment Tests: Correlation Coefficient	Sediment Tests: BCF (K_1/K_2)	Sediment Tests: Estimated Equilibrium Concentration (mg kg^{-1})
Phenanthrene										
Clam	8.2	0.0064	−0.88	1280	5.5	8.2	0.0092	−0.88	890	0.6
Mussel	16.1	0.013	−0.82	1240	5.3	20.9	0.006	−0.94	3480	2.4
Shrimp (whole)	6.8	0.032		210	0.9	31.4	0.020		1570	1.1
Shrimp (hepatopancreas)	12.7	0.032	−0.80	400	1.7	19.1	0.015	−0.76	1270	0.9
Polychaete	3.0	0.006	−0.95	500	2.2	11.0	0.030	−0.74	3670	2.5
Fluoranthene										
Clam	14.0	0.0034	−0.91	4120	9.9	9.9	0.005	−0.97	1980	5.3
Mussel	15.4	0.0026	−0.81	5920	14.2	22.6	0.0038	−0.93	5950	16.0
Shrimp (whole)	6.3	0.035		180	0.4	17.7	0.020		885	2.4
Shrimp (hepatopancreas)	10.0	0.032	−0.70	310	0.7	10.6	0.012	−0.91	880	2.4
Polychaete	3.6	0.005	−0.87	720	1.7	60.7	0.030	−0.90	2020	5.5
Pyrene										
Clam	18.0	0.0028	−0.89	6430	10.9	11.0	0.0052	−0.97	2120	5.7
Mussel	31.0	0.007	−0.71	4430	7.5	20.0	0.004	−0.93	5000	14.0
Shrimp (whole)	7.8	0.035		225	0.4	19.6	0.022		890	2.5
Shrimp (hepatopancreas)	13.9	0.028	−0.65	500	0.8	11.7	0.010	−0.89	1170	3.3
Polychaete	4.2	0.006	−0.92	700	1.2	53.6	0.025	−0.99	2150	6.0
Triphenylene										
Clam	36.0	0.0065	−0.30	5540	2.8	16.0	0.0041	−0.96	3900	1.6
Mussel	41.0	0.0036	−0.81	11390	5.7	53.0	0.0037	−0.93	14325	8.7
Shrimp (whole)	6.7	0.025		270	0.14	28.0	0.025		1120	0.4
Shrimp (hepatopancreas)	7.0	0.012	−0.90	590	0.3	24.2	0.014	−0.89	1730	0.7
Polychaete	5.1	0.002	−0.75	2560	1.3	59.0	0.026	−0.93	2270	0.9
Perylene										
Clam	11.0	0.0011	−0.57	10000	4.0	3.8	0.0014	−0.71	2170	0.8
Mussel	21.0	0.002	−0.65	10500	4.2	7.6	0.0008	−0.39	9500	2.8
Shrimp (whole)	4.4	0.025		175	0.07	7.7	0.020		385	0.11
Shrimp (hepatopancreas)	1.9	0.008	−0.90	235	0.09	1.0	0.007	−0.85	140	0.04
Polychaete	0.9	0.005	−0.90	180	0.07	3.8	0.007	−0.95	540	0.16

[a] Rate constants are K_1 h^{-1} and K_2 h^{-1}.

[b] Correlation coefficient for depuration data versus time.

Neff (1979) concluded that sediment-adsorbed PAHs are not readily assimilated by benthic invertebrates, although relatively few species have been studied. Roesijadi et al. (1978) exposed one group of deposit-feeding clams (*Macoma inguinata*) to sediment containing four PAHs and another group of clams (*Protothaca staminea*) to the surficial water in the same tank. At day 7, the BCFs calculated in relation to the level of PAHs in the sediment were 0.2 or less, and calculated in relation to PAH in surficial water were 10–1349. The authors concluded there was no biomagnification in relation to PAHs in the sediment. There was, however, accumulation from sediment, presumably from PAHs in the interstitial or surficial water.

Sediment appeared to have enhanced the uptake of PAHs by shrimp and polychaetes but not by clams, mussels, and shrimp hepatopancreas. Also, based on the ratios of estimated equilibrium concentrations (Table 11.5), uptake by shrimp and polychaetes appeared to be enhanced by the presence of sediment. Another indication of enhanced uptake by shrimp and polychaetes in the presence of sediment comes from a comparison of estimated BCF values from sediment and water tests. Most BCF values (8 of 10) for clams and mussels in sediment tests were less than or equal to those from water tests (Table 11.6). On the other hand, most BCF values (13 of 15) for shrimp and polychaetes in sediment tests were greater than those from water tests.

Particularly with polychaetes, PAHs in particulate matter (ingested sediment) within the digestive tract could increase the measured concentration of PAHs for the whole animal. Indirect evidence for enhancement from this source can be seen from a comparison of the magnitude of excretion rate constants (K_2) from the sediment and water tests. One would expect that excretion, and therefore K_2 values, for a particular PAH for an organism would be similar whether the exposures had been in water or sediment tests. This is the case for mussels, clams, and whole shrimp where the K_2 values from the water tests averaged 0.9 times (range from 0.4 to 1.8 times; $n = 15$) the K_2 values from the water tests (derived from Table 11.6). In contrast, the K_2 values for polychaetes from sediment tests averaged 6 times (range from 1.4 to 13 times; $n = 5$) the K_2 values from those in the water tests. Possibly, evacuation of contaminated sediment from the digestive tract during the depuration period of the worms resulted in the larger K_2 values.

For worms in the sediment tests, time was not allowed for the gut to empty before analysis of the whole polychaete. If time had been allowed, a decrease in PAH concentration could be expected both from the excretion process and from elimination of contaminated sediment from the gut. In other words, an accurate measurement of PAH concentration at the time of sampling could not have been attained.

Whole shrimp and polychaetes were analyzed; therefore, if surface sites on the integuments of these animals were saturated with PAHs from contact with sediment, the concentrations of PAHs calculated for the whole organism might appear to be elevated. This would give a false impression that sediment enhanced the uptake of PAH in the animals. However, for shrimp hepatopancreas, estimated equilibrium values for PAHs also indicated enhanced uptake in the presence of sediment. Because the hepatopancreas was removed before analysis, there could be no contribution of PAHs from the integument. Possibly, PAHs adhering to surface sites might be transported gradually across the integument to internal sites such as the hepatopancreas.

The 4-d exposure and the estimated equilibrium concentrations of PAHs for the molluscs (Tables 11.4 and 11.6) indicated no enhanced PAH uptake in the presence of contaminated sediment. Even though PAHs may have been attached to sites on the shells, it appears there was no transport across the shells to internal sites.

ACKNOWLEDGMENTS

Thanks are expressed to V. Zitko and S. Ray for their helpful reviews and to J. Hurley for typing the text.

REFERENCES

Andelman, J. B., and M. J. Suess. 1970. Polynuclear aromatic hydrocarbons in the water environment. *World Health Organization Bulletin*, **43**, 479–509.

Cahnmann, H. J., and M. Kuratsune. 1957. Determination of polycyclic aromatic hydrocarbons in oysters collected in polluted water. *Analytical Chemistry*, **29**, 1312–1317.

Dunn, B. P., and H. F. Stich. 1975. The use of mussels in estimating benzo[*a*]pyrene contamination of the marine environment. *Society for Experimental Biology and Medicine. Proceedings*, **150**, 49–51.

Lo, M. T., and E. Sandi. 1978. Polycyclic aromatic hydrocarbons (polynuclears) in foods. *Residue Review*, **69**, 35–86.

Mackay, D., and A. I. Hughes. 1984. Three-parameter equations describing the uptake of organic compounds by fish. *Environmental Science and Technology*, **18**, 439–444.

McLeese, D. W., S. Ray, and L. E. Burridge. 1985. Accumulation of polynuclear aromatic hydrocarbons from sea water and sediment by the clam *Mya arenaria*. *In:* Wastes in the Ocean, Vol. 6: Nearshore Waste Disposal, B. H. Ketchum, J. M.Capuzzo, W. V. Burt, I. W. Duedall, P. K. Park, and D. R. Kester (Eds.). Wiley-Interscience, New York, in press.

Neff, J. M. 1979. Polycyclic Aromatic Hydrocarbons in the Aquatic Environment. Applied Science, London, 262 pp.

Roesijadi, G., J. W. Anderson, and J. W. Blaylock. 1978. Uptake of hydrocarbons from marine sediments contaminated with Prudhoe Bay crude oil: Influence of feeding type of test species on availability of polycyclic aromatic hydrocarbons. *Journal of the Fisheries Research Board of Canada*, **35**, 608–614.

U. S. National Academy of Sciences. 1972. Particulate Polycyclic Organic Matter. National Academy of Sciences, Washington, D.C., 361 pp.

Wilkinson, S. E., P. E. Straup, and P. W. Jones. 1979. Quantitative analysis of selected PAH in aqueous effluent by high-performance liquid chromatography. *In*: Polynuclear Aromatic Hydrocarbons, P. W. Jones and R. P. Leber (Eds.). Ann Arbor Science Publishers, Ann Arbor, Michigan, pp. 217–229.

Windsor, J. G., and R. A. Hites. 1979. Polycyclic aromatic hydrocarbons in Gulf of Maine Sediments and Nova Scotia soils. *Geochimica et Cosmochimica Acta*, **43,** 27–33.

Zitko, V. 1975. Aromatic hydrocarbons in aquatic fauna. *Bulletin of Environmental Contamination and Toxicology*, **14**, 621–631.

Zitko, V. 1980. Relationships governing the behavior of pollutants in aquatic ecosystems and their use in risk assessment. *Canadian Technical Report of Fisheries and Aquatic Sciences,* **975**, 243–265.

Chapter 12

Pharmacokinetic Model for the Accumulation of PCBs in Marine Fish

Joseph M. O'Connor and John C. Pizza

New York University Medical Center
Laboratory for Environmental Studies
A. J. Lanza Laboratories
Tuxedo Park, New York

ABSTRACT

Pharmacokinetic studies were carried out with striped bass in order to determine assimilation and elimination rate constants for polychlorinated biphenyl (PCB) uptake from dietary sources. Efficient assimilation (85% of PCBs in a single dose) and a long elimination half-life (120 h) make it apparent that dietary sources of PCBs are important components of the overall body burden in striped bass. The pharmacokinetics of PCB uptake from food were incorporated into a model designed to predict body burdens. The model presented here takes into account several factors of significance in predicting PCB burdens in cold-blooded organisms with indeterminate growth. These are: (1) changes in diet associated with growth, (2) changes in elimination rate constant due to age-related decreases in metabolism, and (3) migratory movements that may cause changes in exposure to PCBs.

12.1. INTRODUCTION

Contaminants in the marine ecosystem adjacent to New York and New Jersey have been well documented (Stainken and Rollwagen, 1979; MacLeod et al., 1981; Anderson, 1982; Mueller et al., 1982). Among the more important are the polychlorinated biphenyls (PCBs), due primarily to their abundance in the Hudson–Raritan system (Bopp et al., 1981; O'Connor et al., 1982), their toxicity (U.S. National Academy of Sciences, 1979), and their potential to cause chronic effects in animal and human populations (Kuratsune et al., 1976; Mehrle et al., 1982).

The major source of PCB contamination in the New York Bight is ocean dumping of sewage sludge and dredged material (Table 12.1). When relative PCB contribution from direct discharge is considered and integrated according to typical flow patterns in the bight, expected PCB concentrations in the water should be greatest in the vicinity of the disposal sites in the bight. Various studies have shown that this is the case [Interstate Electronics Corporation (IEC), 1981; Pequegnat et al., 1980; MacLeod et al., 1981]. The increased PCB levels in the water column increase the potential for PCB uptake in all trophic levels of the ecosystem (Hamelink et al., 1971; Clayton et al., 1977; Wyman and O'Connors, 1980; Brown et al., 1982; Califano et al., 1982).

The majority of PCBs from dredged material placed in the bight system remains associated with deposited particulate matter. Coring studies at the dredged-material dumpsite in the New York Bight (New York University Medical Center, 1982) show that PCB levels vary with the depth of the core. Horzempa and Di Toro (1983) have shown that PCB from deposited sediments is slow: vertical migration is estimated to be on the order of millimeters per year. Thus, the contribution of dredged material to PCB levels in the New York Bight water column is associated primarily with desorption, or uncoupling (Boehm, 1983), which might occur during the dumping process. Tavolaro (1982) has estimated a dry-mass sediment loss of ~4% during dumping. Overall PCB losses during the dumping process may be ~15% due to selective loss of fine particles enriched in PCBs, as well as due to desorptive losses (Di Toro and Horzempa, 1982; Boehm, 1983).

In an attempt to determine how dredged-material dumping affects PCB body burdens in fish, several investigations have evaluated the rates and routes of PCB transport in marine ecosystems (O'Brien and Gere Engineers, 1979; Pavlou et al., 1982; Rubinstein et al., 1983). The results of such studies, when combined with appropriate pharmacokinetic data, allow predictive models to be formulated to estimate the extent to which PCBs in sediment may cause increased body burdens. If unacceptable PCB burdens in marine organisms can be related to dumping, and if the process of accumulation is well understood, managers and regulators can take steps to reduce the problem.

Since the early 1970s it has been thought that fish accumulated PCBs directly from water (Hamelink et al., 1971; Neely et al., 1974). Experimentally derived bioconcentration factors (BCFs) that predict steady-state burdens in fish have been published widely (Table 12.2), and uptake mechanisms based on octanol–water partition coefficients and lipid solubility of PCBs have been proposed (the equilibrium partition theory) (Neely et al., 1974; Branson et al., 1975; Mackay, 1982). Similar mechanisms have been proposed to explain PCB accumulation in zooplankton (Clayton et al., 1977; Pavlou and Dexter, 1979).

Table 12.1. Estimated PCB Input (kg y^{-1}) to the New York Bight Apex

Source	Maximum	Percentage of Total	Minimum	Percentage of Total
Atmospheric[a]	490	7	34	1
Municipal				
wastewater[b]	42	0.6	42	1
Dredged material[c]	3500	51	1800	61
Sewage sludge[d]	1300	19	750	26
Hudson plume[e]				
particulate	1037	15	62	2
dissolved	480	7	240	8
Total	6849	99.6	2928	99

[a] Assumes 1.14 m y^{-1}precipitation at 15 (minimum) and 215 (maximum) ng $liter^{-1}$ PCB.
[b] Assumes 3.75 x 10^5 m^3 direct wastewater flow; all secondary at 0.3 μg $liter^{-1}$ PCB.
[c] Minimum from Bopp et al. (1981); maximum from O'Connor et al. (1982).
[d] Based on estimates from West and Hatcher (1980), Bopp et al. (1981), and O'Connor et al. (1982).
[e] Plume flow assumed to be 6.6 x 10^{10} liters d^{-1}, carrying 3 (minimum) and 50 (maximum) mg $liter^{-1}$ solids at 0.86 μg $liter^{-1}$ PCBs for particulate load, and 10 ng $liter^{-1}$ (minimum) and 20 ng $liter^{-1}$ (maximum) for dissolved load. (Data from Mueller et al., 1982.)

Three factors militate against the equilibrium partition theory as the full explanation for PCB burdens in marine fishes. First, the concept was developed using pure, dissolved compounds in relatively particle-free water. Under natural circumstances seawater contains many particles to which PCBs are likely to adsorb (Hiraizumi et al., 1979; Nau-Ritter et al., 1982). For striped bass (*Morone saxatilis*), the presence of particles in bioassay water decreases the quantity of PCBs available for uptake, and body burden is directly related to available rather than total PCBs (Califano et al., 1982). Second, published bioconcentration factor (BCF) data are generally derived from experiments of long duration ($\geq$ 5 d). Organisms must be fed during the experiment, and the proportion of PCBs accumulated by way of contaminated food is not taken into account. Many studies have shown that food material, both living and dead, accumulates PCBs rapidly (Wyman and O'Connors, 1980; Peters and O'Connor, 1982), providing a secondary route of PCB uptake in the test chamber. For example, the common striped bass food organisms *Gammarus tigrinus* and *Neomysis americana* accumulated up to 2 μg g^{-1} PCBs from water after < 10 h of exposure to a PCB concentration of 1 μg $liter^{-1}$ (Peters and O'Connor, 1982). Inanimate food may accumulate PCBs just as rapidly (Wyman and O'Connors, 1980; Nau-Ritter et al., 1982). Thus, BCF values calculated from long-term exposures include a dietary uptake component not accounted for in application of the data to equilibrium partitioning theory.

Third, and perhaps most important, equilibrium partition calculations for PCB bioconcentrations generally yield estimates which are low, relative to field observations (Table 12.3). Given the potential importance of PCBs as a toxicant in natural systems, it is unwise to rely heavily on such estimates. Because all parts of the marine food web contain PCBs (MacLeod et al., 1981; O'Connor et al., 1982), and cross-gut assimilation of PCBs in fish approaches 90% (Pizza and O'Connor, 1983), we suggest that a large fraction of the PCBs in fish is derived from their food. Recently published models for contaminant transport in fish and plankton demonstrate that dietary sources may be the prime determinant of PCB body burdens (Thomann, 1981; Brown et al., 1982).

Studies of PCB uptake from diet have been carried out in our laboratory with a variety of fishes and crustaceans (Califano et al., 1980, 1982; Peters and O'Connor, 1982; Pinkney, 1982; Pizza and O'Connor, 1983). This chapter provides the results of our food chain transport studies of PCBs in marine fishes, principally striped bass. The concepts presented are based on pharmacokinetic relationships (Goldstein et al., 1974). Pharmacokinetics has been applied to the dynamics of pesticide uptake in fish by Krzeminski et al. (1977). Similar principles underlie the recent efforts of Thomann (1981) to model PCB accumulation in fish. We propose a seasonally variable pharmacokinetic model to predict dietary accumulation of PCBs in fish.

12.2. METHODS

The techniques used in these studies were developed by Califano et al. (1982) and by Pizza (1983). Complete details of the pharmacokinetic approach to PCB studies are available in reports by Pizza (1983) and Pizza and O'Connor (1983). To summarize, an evaluation of the uptake and excretion of a compound by an organism requires empirical data regarding the rate of assimilation from the absorption site (k_a), and the rate of loss of the compound (k_e) from some physiological pool. The pool we used was the entire body mass of the subject organism (a single-compartment model). Use of a single-compartment model for PCBs is justified for at least two reasons. First, several investigations with PCBs and other persistent halogenated organic compounds have shown that accumulation and elimination data are best described by single-exponent equations (Norstrom et al., 1976; Mayer et al., 1977; Morales et al., 1979; vanVeld, 1980; Thomann, 1981). Second, calculated elimination rate constants of PCBs from individual tissues of fish have been found to be similar to one another and to the whole-body mass (Guiney and Peterson, 1980; Pizza, 1983). Tissues are considered synonymous with compartments (Moriarity, 1975), so the single-compartment model is justified. The mathematics for a single-compartment model are straightforward and based on the exponential expression for a decay curve:

$$M = M_0 e^{-k_a t} \qquad 1$$

where M is the quantity remaining at time t, M_0 is the quantity of PCBs placed at the absorption site, and k_a is the absorption rate constant.

Table 12.2. Bioconcentration of Aroclor in Fish

Organism	Commercial Aroclor Mixture	Exposure Concentration (μg liter^{-1})	Exposure Time (d)	BCF[a] (10^4)	Reference
Channel catfish	1248	5.8	77	5.6	Mayer et al. (1977)
(*Ictalurus punctatus*)	1254	2.4	77	6.1	Mayer et al. (1977)
Bluegill sunfish	1248	2–10	chronic	2.6 7.1	Stallings and Mayer (1972)
(*Lepomis macrochirus*)	1254	2–10	chronic		
Brook trout (fry)	1254	6.2	118	4.6	Mauck et al. (1978)
(*Salvelinus fontinalis*)					
Spot	1254	1	56	3.7	Hansen et al. (1971)
(*Leiostomus xanthurus*)					
Pinfish	1016	1	56	1.7	Hansen et al. (1974)
(*Lagodon rhomboides*)					
Rainbow trout	2,2′,4,4′-	1.6	5	2.9	Branson et al. (1975)
(*Salmo gairdneri*)	tetrachlorobiphenyl	9.0			
Fathead minnow	1248	3.0	250	12.0	DeFoe et al. (1978)
(*Pimephales promelas*)	1260	2.1	250	27.0	

[a] Bioconcentration factor determined from the concentration of the Aroclor mixture in the fish divided by the concentration in the exposure water.

Elimination rate constants (k_e) are derived from the same expression with different notation:

$$X = X_{0}e^{-k_e t} \quad 2$$

where X is the whole-body PCB burden at time t, X_0 is the whole-body PCB level at time zero, and k_e is the elimination rate constant. Note that k_a and k_e are estimated based on actual mass of contaminant and, therefore, represent rate constants rather than exponential constants (Moriarty, 1975).

Values for PCB assimilation were determined by feeding striped bass known quantities of ^{14}C-labeled PCBs [Aroclor 1254 (Monsanto Co., St. Louis, Missouri)] in their normal food organisms and sampling at fixed intervals to determine the quantity of PCBs in the gut, whole body, and fecal material. Sampling of fish continued for 120 h, during which time they were not fed. Tissues were analyzed for total ^{14}C-PCB by liquid scintillation counting (Pizza, 1983). Experiments were conducted to determine k_a and k_e from single and multiple feedings. Manipulation of the empirical data was that suggested by Goldstein et al. (1974), and the equations are documented in Table 12.4.

Table 12.3. Expected and Observed PCB Body Burdens in Fish from the New York Bight

Parameter	Minimum	Maximum
Water column PCB concentration (ng liter^{-1})[a]	10	40
Particulate:dissolved PCB ratio[b]	0.67	0.67
Dissolved PCB (ng liter^{-1})	6.7	27
Bioconcentration factor[c]	1 x 10^4	1 x 10^4
Expected PCB concentration (μg g^{-1})[d]	0.07	0.27
Observed concentrations (μg g^{-1})[e]		
Striped bass (*Morone saxatilis*)	0.6	3.8
Winter flounder (*Pseudopleuronectes americanus*)	0.1	—
Atlantic mackerel (*Scomber scombrus*)	0.5	0.7
Bluefish (*Pomatomus saltatrix*)	0.7	3.6
American eel (*Anguilla rostrata*)	0.5	0.8
Tautog (*Tautaga onitis*)	0.6	—

[a] Concentrations derived from Lee and Jones (1978), Interstate Electronics Corporation (1981), Pequegnat et al. (1980), and MacLeod et al. (1981).

[b] Various authors suggest particulate:dissolved PCB ratios ranging from zero to about one. Based on suggested values from Brown et al. (1982), Nau-Ritter (1980), and Pavlou and Dexter (1979), we arrived at a value in which two-thirds of the total water column PCB may be in the dissolved state.

[c] Based on chronic bioassay data in which values range from 1.7 x 10^4 to 6.1 x 10^4 (Table 12.2) for various species. Assuming some fraction is associated with feeding, we suggest 1 x 10^4 to be a reasonable and conservative BCF approximation.

[d] Calculated from equilibrium partitioning using New York Bight data as μg^{-1} PCB in water times BCF; the water value equals ng (g x 10^{-9}) ÷ 1030, the weight of one liter of seawater at 30‰ salinity.

[e] Data from O'Connor et al. (1982), New York State Department of Environmental Conservation (1981), and Belton et al. (1983).

12.3. RESULTS

By direct application of the PCB dose to striped bass we have imposed an apparent zero-order (instantaneous) uptake process (Fig. 12.1). Assimilation of PCBs from the gut to the rest of the fish is describable by a first-order process in which elimination from the gut during the first 24 h of the experiment represents assimilation into the whole-body compartment (Fig. 12.1a). During the first 48 h of exposure, while the PCBs in the gut declined from 100 to 4% of the administered dose, the proportion of the dose in the whole fish remained at or above 80%. No more than 10% of the administered dose appeared in fecal matter. Later in the experiment, elimination rate constants for the whole body (0–120 h) and for the gut (48–120 h) were similar, suggesting that after the assimilation of PCBs from the gut was complete, the rate of PCB removal from the gut and from the rest of the fish were similar.

For a single dose of PCBs, the body burden follows a time course reflective of the ratio k_e:k_a, as shown in Fig. 12.2. When striped bass were given single doses of PCBs there was an initial and rapid assimilation from the alimentary tract to other tissues. Fish in contaminated environments, however, do not accumulate PCBs as single, isolated dietary doses; rather, they take up PCBs from water and from multiple doses of PCBs in food. Our multiple-dose study showed the gradual approach to plateau PCB levels expected for young striped bass exposed to sequential doses of PCBs in food given at 48-h intervals (Fig. 12.3). This interval was chosen in order to observe the approach to steady-state. Mathematically, the feeding interval is unimportant; what is essential is that the kinetics of the system show that a plateau is reached at dose n = 17 (Table 12.4). In our experiments, a plateau was reached in 34 d. For a fish feeding two times daily, a plateau would be reached in 8.5 d.

12.4. APPLICATION TO MODELLING

Thomann (1981) documented the complexity of models describing contaminant accumulation in fishes, and he noted the following: (1) the lack of data relevant to the function referred to as the "food-chain multiplier," and (2) the need to account for age-specific changes in respiration, feeding, and growth as determinants of predicted PCB body burdens in striped bass. In the preliminary model presented here we provide the food-chain multiplier as a function of fish growth, changes in metabolism, changes in food ration, and changes in dietary PCB levels.

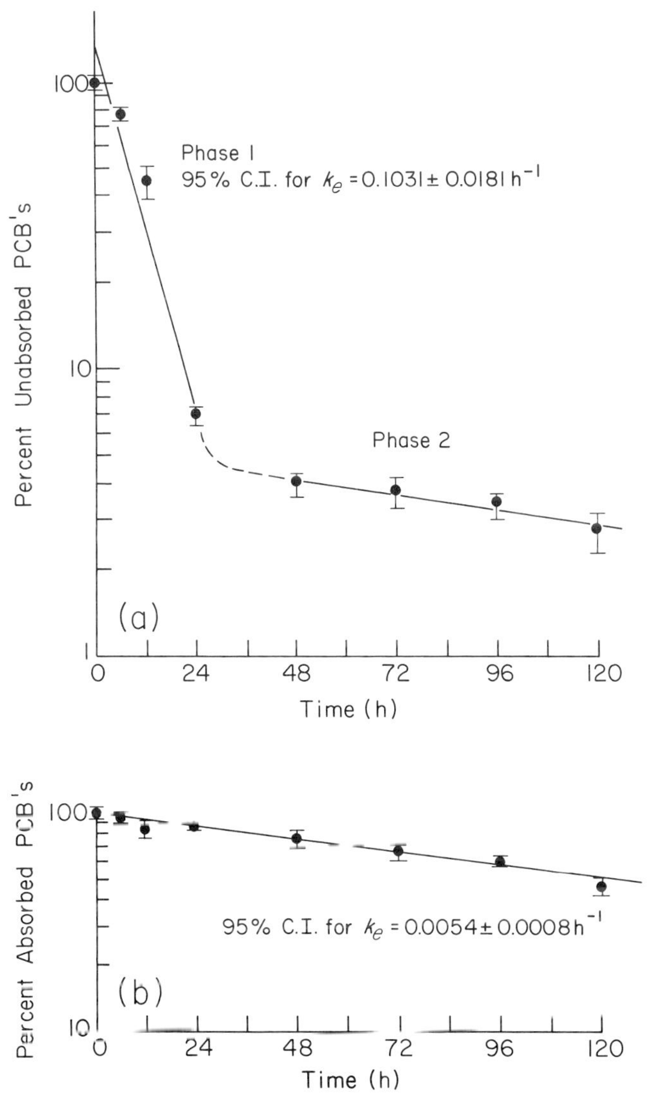

Figure 12.1. Experimental determination of PCB concentrations in striped bass after forced feeding (C.I. is confidence interval): (a) loss of PCBs from the alimentary tract showing a two-phase elimination, phase 1 describes elimination in parallel with other tissues; and (b) whole-body elimination of PCBs. Note the similarity of slope with phase 2 of alimentary tract elimination.

Table 12.4. Pharmacokinetic Expressions Applied to PCB Dietary Uptake Studies[a]

Parameter	Expression
Absorption rate (k_a)	$M = M_{0e}^{-k_a t}$
Elimination rate (k_e)	$X = X_{0e}^{-k_e t}$
Absorption half-time ($t_{1/2}$)	$t_{1/2} = - \frac{-\ln 0.5}{k_a}$
Fractional absorption single dose	$X/M_0 = [e^{-k_e t/(k_e/k_a)} - e^{-k_e t}]/[k_e/k_a) - 1]$
Time to maximum absorption single dose ($k_a \neq k_e$)	$t_{max} = [2.3 \log (k_a/k_e)]/(k_a - k_e)$
Maximum fraction absorbed single dose	$X_{max}/M_0 = (k_a/k_e)^{k_a/(k_e - k_a)}$
Body burden at end of dosing interval multiple doses	$X = X_{0e}^{-k_e t*} + X_0 (e^{-k_e t*})^2 + X_0 + (e^{-k_e t*})^n$
Body burden after dosing multiple doses	$X_n = X_0 (1 - e^{-k_e k*n})/(1 - e^{-k_e t*})$
Plateau (steady state burden multiple dose)	$X_\infty = X_0/1 - e^{-k_e t*}$
Fraction of plateau at n doses	$f = 1 - e^{-k_e t*n}$

[a] Equations from Goldstein et al. (1974).

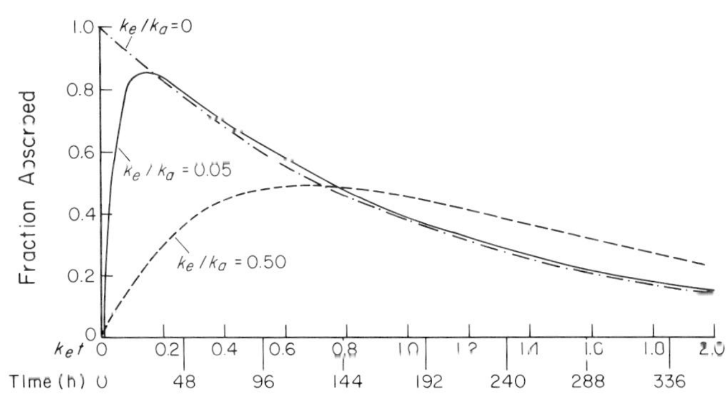

Figure 12.2. The fractional absorption of a single dose of first-order absorption and elimination. This study is represented by the solid curve, where $k_a = 0.103$ h^{-1}, $k_e = 0.005$ h^{-1}, $k_a/k_e = 0.05$, and the maximum absorbance level of 85% is reached by $t_{max} = 30$ h. The other curves are presented for comparison at the same k_e (Goldstein et al., 1974). The $k_e/k_a = 0$ ($k_a = \infty$) is attained by a constant input rate. For $k_e/k_a = 0.50$, for a k_a smaller than that of the solid curve, a lower maximum would be attained at a later time.

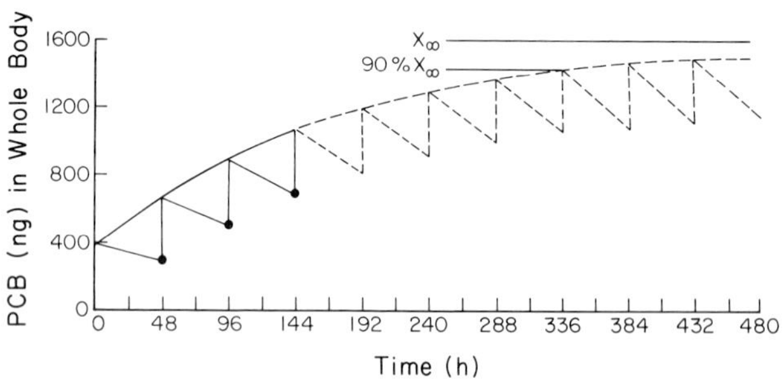

Figure 12.3. Curve of the cumulative retention of PCBs from multiple doses administered at 48-h intervals, starting at time zero. Solid lines represent actual levels attained during the experiment, and dashed lines are the calculated extension of the data. The plateau burden (X_∞) is the steady-state level attained from peak values after sufficient dosing.

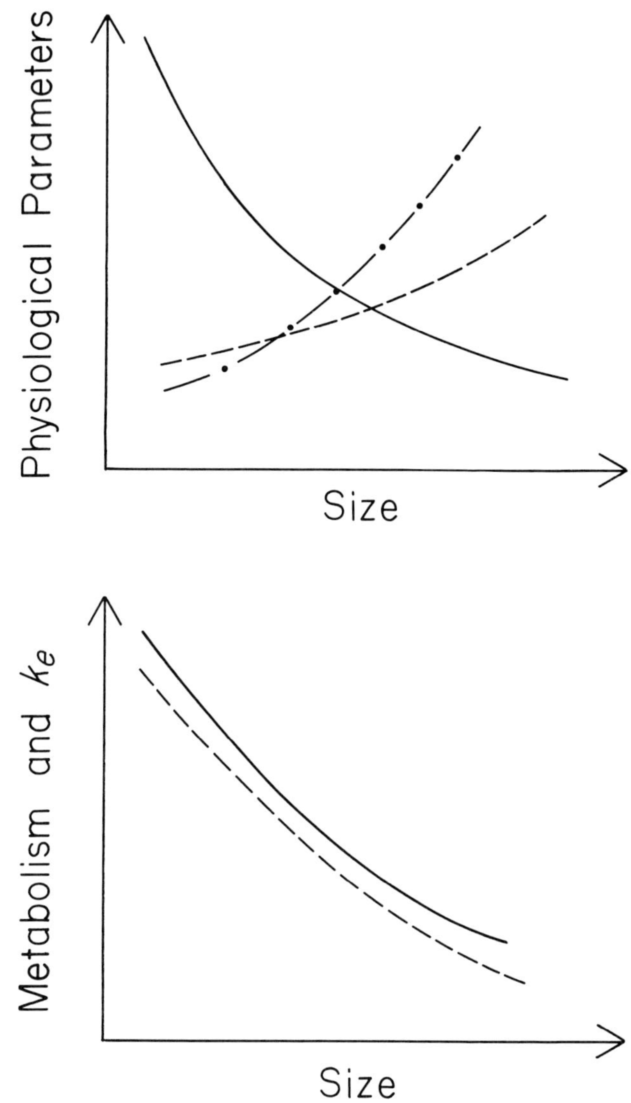

Figure 12.4. Physiological parameter responses to increased size of the organism: (a) assimilation (———), food ration (—·—·), and respiration (- - - -); (b) metabolism (- - - -) and elimination constant, k_e (———).

Analysis of PCB accumulation by striped bass from their diet reveals that changes in the plateau PCB burden depend on two factors: the k_e and the rate at which PCBs are taken in via food (Pizza, 1983). The relationship between physiological growth and required ration size, as well as weight-specific metabolism, is depicted schematically in Fig. 12.4. The expected effects that changes in these factors may have on food-assimilation efficiency and the k_e for PCBs are also shown in Fig. 12.4 (Califano et al., 1982; Pizza and O'Connor, 1983). The physiological data suggest that for several reasons there should be no steady-state PCB level in growing fish. First, the efficiency with which food is assimilated declines with increasing age; this means there is an increase in the food ration required for growth (Fry, 1971). Because the volume of food eaten increases, the PCB intake with food also increases; therefore, the PCB burden should increase continuously. Second, the quantity of water required for oxygen exchange must increase with size, because oxygen consumption increases with size; therefore, the direct uptake of PCBs from water increases with fish size (McKim and Heath, 1983). The importance of the latter in contributing to the body burden may be questioned, however, because respiratory requirements may be extremely variable and depend on ambient levels of dissolved oxygen, temperature, time of day, time in the feeding cycle, and other environmental factors (Neumann et al., 1982). Third, the overall metabolic rate of a fish declines with age and size (Neumann et al., 1982), resulting in a slowing of physiological processes. Pizza (1983) analyzed data for rainbow trout of different ages and found a decline in k_e associated with age. The reduction in k_e conformed to a power curve function:

$$E_t = E_0 x^b \qquad 3$$

where E_t is k_e at time t, E_0 is k_e at time zero, x is fish weight, and b is a constant estimated at -0.03 (Thomann and Connolly, 1982; Pizza, 1983). The little data available on the k_e of striped bass of different ages appear to conform to a similar curve (Califano, 1981; Califano et al., 1982; Pizza, 1983).

The pharmacokinetic model incorporates these mathematical relationships under conditions representative of striped bass biology in the Hudson Estuary: (1) active feeding during the first growing season on zooplankton containing 1 μg g^{-1} PCB (O'Connor et al., 1982), (2) an instantaneous growth rate of 0.02 d^{-1} (Dey, 1981), (3) decreasing k_e at a rate proportional to weight: $W^{-0.3}$ (Brett, 1979; Pizza, 1983), (4) increasing ration size to maintain a daily intake of 10% body weight, (5) a feeding interval of 12 h, with one-half of the daily ration (and PCB dose) consumed during each feeding period, and (6) additional feeding on zooplankton containing either 0.5 or 0.1 μg g^{-1} PCB (dry weight). The model was run for two seasons of growth (total time, 330 d). This exercise illustrates the temporal relationship of PCB input and output rate constant within the boundaries of a realistic time scale (Fig. 12.5). Since the concentration in the diet was 1.0 μg g^{-1}, the solid curve shows the computed bioaccumulation factor (BAF). The initial rise in concentration (0–30 d) was the predicted shift in plateau. Once a plateau was attained, the typical asymptote did not occur; instead, the BAF skewed upward, as determined by the rate of decline in k_e.

After one season of growth, the k_e declined to 0.0044 h^{-1} and the computed BAF was 0.76. This value is much lower than a BAF of 0.95 that would be calculated by dividing the feeding rate by the k_e determined after 5 months of growth (0.0044 h^{-1}). This discrepancy is explained by the fact that a BAF of 0.95 would only be attained with sufficient time (~44 d) and cessation of the decline in k_e allowing for the establishment of a new steady state. Such conditions are unlikely, and we stress that bioaccmumulation in fish is rarely characterized by a steady state; rather, it is a process of uninterrupted plateau shifts operating throughout the life of the fish. This point is illustrated by the dashed curves in Fig. 12.5. For a diet containing 0.5 μg g^{-1} PCBs, the time required for the plateau shift was the same as the 1.0 μg

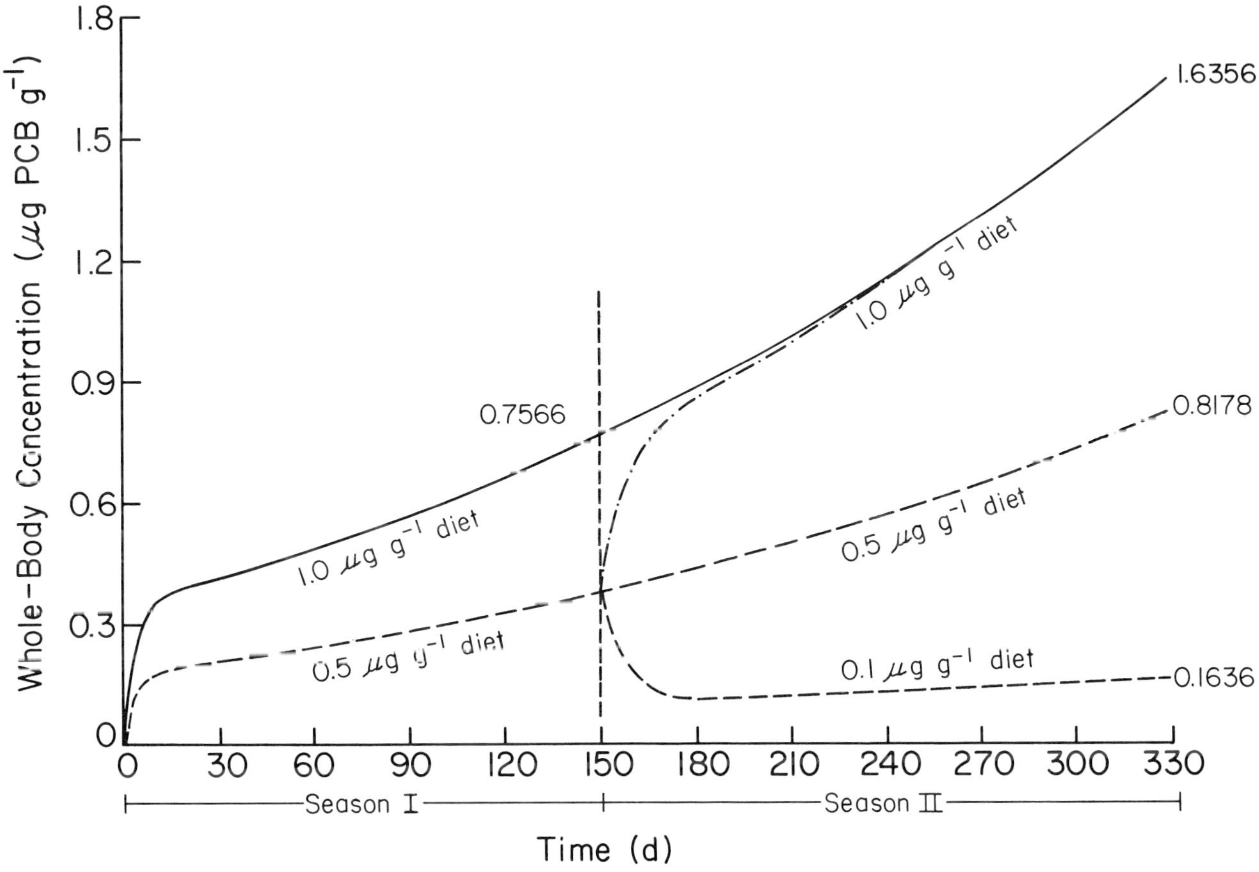

Figure 12.5. Outcome of the pharmacokinetic model for dietary PCB accumulation in striped bass.

g^{-1} situation, and the resultant whole-body concentration was one-half that attained at the higher dosage. Dividing the whole-body concentrations at any point in the curves by the PCB concentration in the respective diet showed that the BAF was the same for the two conditions.

If the PCB concentration in the diet is decreased, a plateau shift will occur at a rate determined by the declining k_e and the whole-body concentration will decrease. Since the k_e continues to decline with growth, the plateau will eventually skew upward, maintaining the integrity of the BAF. This is demonstrated by dividing the whole-body PCB concentration by the PCB concentration of the diet.

If the concentration of PCBs in the diet is increased, then the whole-body burden should reach a higher level. The model predicts that an increase of 1.0 μg g^{-1} dietary PCB would cause a k_e-governed plateau shift in the whole-body PCB concentration to a concentration that would be attained from a continuous 1.0-μg g^{-1} dietary exposure (Fig. 12.5).

The growth rate in the model was determined by an assumed instantaneous rate constant of 0.02 d^{-1}. This value is consistent with the growth rate of juvenile Hudson River striped bass, and corresponds to growth observed in the field. The model was run for consecutive growth intervals (seasons I and II). Fluctuations in growth due to photoperiod, temperature, and other factors (Brett, 1979) were not considered. Such perturbations should elicit a response consistent with the k_e- and input-governed plateau shift.

A period of overwintering was not included in the calculations because relatively little is known of the physiological conditions of Hudson River striped bass at this time of year. A reduction in feeding rate (Brett, 1979) and a decline in metabolic rate (Neumann et al., 1981) would be expected. The decline in body burden caused by a decreased PCB input should more than offset the increased burden owed to depressed k_e. It is important to realize that any reduction in k_e would also cause an increase in the time required for the plateau shift to occur. Overall, a slow decline in body burden would be expected from overwintering, unless utilization of stored lipids accelerates the reduction. Whatever level the body burden attains during periods of reduced metabolism, prediction based on growth would not change; once the plateau shift is complete, the PCB concentration in the fish should reflect the operant PCB input rate (dose) and k_e.

When interpreted over a specific and realistic time span during life stages, the growth-related pharmacokinetic model suggests that the long-term PCB history of the fish may be unimportant in estimating body burdens. Field data for PCBs in fish are probably only indicative of the most recent exposure regime experienced by the organisms. Such data do not necessarily provide any information on life-history exposure to PCBs and certainly do not reflect PCB dynamics in any part of the ecosystem other than the organisms analyzed. On the one hand, if fish with very low PCB concentrations occur in samples, it is possible that they may have recently completed elimination of a substantial PCB body burden obtained in a contaminated environment. Fish with high PCB burdens may have accumulated that material over very few days, starting from a burden at or near zero. For these reasons, predictions of PCB burdens should incorporate as much available ecological data as possible, including PCB levels in food organisms, metabolic rates, probable migratory behavior of the organism, and the relative importance of PCB uptake from food and from water.

Data are available for young-of-the-year striped bass taken in 1978 from the Indian Point portion of the Hudson River (Califano et al., 1982; Mehrle et al., 1982). Concentrations of PCBs in these two samples were 1.59 and 2.62 μg g^{-1} (wet weight), or about 6.4 and 10.4 μg g^{-1} (dry weight), respectively. O'Connor's (1982) data for striped bass food organisms (*Gammarus* spp.) from the same portion of the Hudson averaged ~ 7 μg g^{-1} (dry weight) PCBs. Given a BAF of 0.76 and a food source with 7 μg g^{-1} PCBs, a fish might contain 5.3 μg g^{-1} PCBs (dry weight) due to diet alone.

This calculation suggests that 51 to 83% of the PCBs in striped bass might be due to dietary uptake alone; however, direct uptake from water is an important source of PCBs to fish and cannot be ignored. We have stressed the dietary route of PCB uptake because it has been largely ignored in the literature and because use of food as a dosing mechanism is an effective means for obtaining pharmacokinetic data. It would appear that PCBs accumulated directly from water by equilibrium partitioning are handled by fish in a manner similar to PCBs obtained from food (McKim and Heath, 1983; Pizza, 1983). When adequate pharmacokinetic data from PCB uptake from water become available, the pharmacokinetic model may be expanded to include doses associated with the water route.

Feeding rate is a critical factor in determining how much of the PCB burden is acquired via the diet. The model operates under two assumptions concerning the input rate that could affect the outcome of the predicted body burden. The first assumption is that the dietary PCB absorption efficiency is 100%. The present study, using Aroclor 1254, showed that this level is not far from the efficiency observed and should suffice because it also serves to simplify the model. For the limited data available, 51–83% is a reasonable estimate of the PCB burden obtained via the diet. The second assumption involves the daily ration used in the model. A daily requirement of 10% body weight was chosen because this level of feeding is reported by many studies (Chesney and Estevez, 1976; Phillips and Buhler, 1978). This level is abritrary, however, because insufficient work has been performed on the relation of physiological condition and growth rate to ration size. This problem is compounded by the fact that species, temperature, and body size affect the growth–ration relation. The few available studies usually deal with salmonoids at temperatures generally <20°C (Brett, 1979). The submaximum growth rations reported in these studies are ≤8%. In the present study, in which striped bass received an amount of food that filled the stomach without extension, a daily ration of 5% would have been feasible if given two times per day. Laboratory striped bass weighing ~1 g (dry) might consume two times this quantity of live *Gammarus tigrinus* when fed passively at 12-h intervals. A 10% daily ration was used in the model. Once again, before verification of the 51–83% dietary contribution is possible, an accurate estimate of the daily ration is needed. It is also important to note that the BAF values (Fig. 12.5) would be directly affected by a reduction or an increase in the ration.

Dietary accumulation is most strongly influenced by feeding and metabolic rates. The changes in these rates as growth occurs will determine the ultimate dietary contribution to the body burden in mature fish. Certain environmental factors such as reduced temperature could serve to decrease k_e and increase BAF due to a depressive effect on metabolic rates in striped bass and other fish (Neumann et al., 1981). The impact of the reduced k_e under low temperature conditions, however, might be offset to some extent by simultaneous reduction in feeding during the winter months or by metabolic compensation to slowly changing environmental conditions (Fry, 1971; Vetter, 1982).

The concept of a growth-related plateau shift has application to estimating body burdens in fish exposed to varying conditions of PCB input during different life stages. We know, for example, that when the Hudson River striped bass migrate from riverine nursery areas to the lower estuary and marine waters, they show a significant reduction in PCB body burden (Lawler, Matusky and Skelly Engineers, 1980; MacLeod et al., 1981; O'Connor et al., 1982). This reduction could be due to reduced PCB input because the PCB concentration in water and food in coastal regions is less than in the estuary (Pierce et al., 1981; O'Connor et al., 1982). With knowledge of operant k_e and PCB levels in food and water, the body burden and time for reaching the new and lower level should be possible to calculate.

The data from the present study, combined with observations from the field, suggest strongly that diet is an important source of PCBs to striped bass. The combined effect of efficient cross-gut assimilation of PCBs and low k_e dictates that PCB uptake from the food web should be given additional attention in toxicological studies.

Our empirical studies and models have direct application to ongoing ocean dumping studies in the following ways:

1. Our hypothesis that the PCB burden in fish is strongly related to diet renders the mass-loading approach to oceanic PCB pollution ineffective in estimating contaminant levels in fish. Both dredged materials and sewage sludge must be evaluated more carefully to assess real quantities of PCBs being released into the food chain and the water column, rather than merely estimating mass loads placed in the oceanic environment.
2. The pharmacokinetics of PCBs in fish suggests that any attempt at isolation of contaminated materials from entry into the food chain will have the effect of lowering PCB body burdens in the large, predatory species that are often an important part of the human diet. Thus, maintaining a surficial layer of sediments with low or non-available PCBs will, in all likelihood, result in a trend toward lower body burdens at all trophic levels.
3. The overall outcome of the model suggests that PCB levels in fish will not increase if input rates and input sources remain similar to those of recent years.

ACKNOWLEDGMENTS

This work was supported by a Cooperative Research Agreement (CR 808006) between New York University and the U.S. Environmental Protection Agency, Gulf Breeze Environmental Research Laboratory, and by the National Institute of Environmental Health Sciences, Grant No. ES 00260.

REFERENCES

Anderson, J. W. 1982. The transport of petroleum hydrocarbons from sediments to benthos and the potential effects. *In:* Ecological Stress and the New York Bight: Science and Management, G. F. Mayer (Ed.). Estuarine Research Federation, Columbia, South Carolina, pp. 156–179.

Belton, T. J., B. Ruppel, K. Lockwood, and M. Boriek. 1983. PCBs in Selected Finfish Caught within New Jersey Water 1981–1982. Technical Report, Office of Science Research, U.S. Environmental Protection Agency, Trenton, New Jersey, 43 pp.

Boehm, P. D. 1983. Coupling of particulate organic pollutants between the estuary and continental shelf and the sediments and water column in the New York Bight region. *Canadian Journal of Fisheries and Aquatic Sciences*, **40** (Supplement 2), 262–276.

Bopp, R. F., H. J. Simpson, C. R. Olsen, and N. Kostyk. 1981. Polychlorinated biphenyls in sediments of the tidal Hudson River, New York. *Environmental Science and Technology*, **15**, 210–216.

Branson, D. R., G. E. Blau, H. C. Alexander, and W. B. Neely. 1975. Bioconcentration of 2,2′,4,4′-tetrachlorobiphenyl in rainbow trout as measured by an accelerated test. *Transactions of the American Fisheries Society*, **104**, 785–792.

Brett, J. R. 1979. Environmental factors and growth. *In:* Fish Physiology, Vol. 8: Bioenergetics and Growth, W. S. Hoar, D. J. Randall, and J. R. Brett (Eds.). Academic Press, New York, pp. 599–675.

Brown, M. P., J. J. A. McLaughlin, J. M. O'Connor, and K. Wyman. 1982. A mathematical model of PCB bioaccumulation in plankton. *Ecological Modelling*, **15**, 29–47.

Califano, R. J. 1981. Accumulation and Tissue Distribution of Polychlorinated Biphenyls in Early Life Stages of the Striped Bass, *Morone saxatilis*. Ph.D. Thesis, New York University, New York, 165 pp.

Califano, R. J., J. M. O'Connor, and L. S. Peters. 1980. Uptake, retention and elimination of PCB (Aroclor 1254) larval striped bass (*Morone saxatilis*). *Bulletin of Environmental Contamination and Toxicology*, **24**, 467–472.

Califano, R. J., J. M. O'Connor, and J. A. Hernandez. 1982. PCB dynamics in Hudson River striped bass. 1: Accumulation in early life history stages. *Aquatic Toxicology*, **2**, 187–204.

Chesney, E. J., and J. I. Estevez. 1976. Energetics of winter flounder (*Pseudopleuronectes americanus*) fed the polychaete, *Nereis virens*, under experimental conditions. *Transactions of the American Fisheries Society*, **105**, 592–595.

Clayton, J. R., S. P. Pavlou, and N. F. Breitner. 1977. Polychlorinated biphenyls in coastal marine zooplankton: bioaccumulation by equilibrium partitioning. *Environmental Science and Technology*, **11**, 676–682.

DeFoe, D. L., G. D. Veith, and R. W. Carlson. 1978. Effects of Aroclor 1248 and 1260 on the fathead minnow (*Pimephales promelas*). *Journal of the Fisheries Research Board of Canada*, **35**, 997–1002.

Dey, W. P. 1981. Mortality and growth of young-of-the-year striped bass in the Hudson River estuary. *Transactions of the American Fisheries Society*, **110**, 151–157.

Di Toro, D. M., and L. M. Horzempa. 1982. Reversible and resistant components of PCB adsorption–desorption: isotherms. *Environmental Science and Technology*, **16**, 594–602.

Fry, F. E. J. 1971. The effect of environmental factors on the physiology of fish. *In:* Fish Physiology, Vol. 6: Environmental Relations and Behavior, W. S. Hoar and D. J. Randall (Eds.). Academic Press, New York, pp. 1–99.

Goldstein, A., L. Aronow, and S. M. Kalman. 1974. Principles of Drug Action: The Basis of Pharmacology. Wiley-Interscience, New York, 854 pp.

Guiney, P. D., and R. E. Peterson. 1980. Distribution and elimination of a polychlorinated biphenyl after acute dietary exposure in yellow perch and rainbow trout. *Archives of Environmental Contamination and Toxicology*, **9**, 667–674.

Hamelink, J. L., R. C. Waybrant, and R. C. Ball. 1971. A proposal: Exchange equilibria control the degree chlorinated hydrocarbons are biologically magnified in lentic environments. *Transactions of the American Fisheries Society*, **100**, 207–214.

Hansen, D. J., P. R. Parrish, J. I. Lowe, A. J. Wilson, Jr., and P. D. Wilson. 1971. Chronic toxicity, uptake and retention of Aroclor 1254 in two estuarine fishes. *Bulletin of Environmental Contamination and Toxicology*, **6**, 113–119.

Hansen, D., P. R. Parrish, and J. Forester. 1974. Aroclor 1016: toxicity to and uptake by estuarine animals. *Environmental Research*, **7**, 363–373.

Hiraizumi, Y., M. Takahashi, and H. Nishimura. 1979. Adsorption of polychlorinated biphenyl onto sea bed sediment, marine plankton, and other adsorbing agents. *Environmental Science and Technology*, **13**, 380–584.

Horzempa, L. M., and D. M. Di Toro. 1983. The extent of reversibility of polychlorinated biphenyl adsorption. *Water Research*, **17**, 851–859.

Interstate Electronics Corporation. 1981. Environmental Impact Statement for New York Dredged Material Ocean Disposal Site Designation. Report to the U.S. Environmental Protection Agency, Interstate Electronics Corporation, Anaheim, California, 148 pp.

Krzeminski, S. F., J. T. Gilbert, and J. A. Ritts. 1977. A pharmacokinetic model for predicting pesticide residues in fish. *Archives of Environmental Contamination and Toxicology*, **5**, 157–166.

Kuratsune, M., Y. Masuda, and J. Nagayama. 1976. Some recent findings concerning Yusho. *In:* Proceedings of the National Conference on Polychlorinated Biphenyls. EPA-560/6-75-004, U.S. Environmental Protection Agency, Washington, D.C., pp. 14–30.

Lawler, Matusky and Skelly Engineers. 1980. Biological and Water Quality Data Collected in the Hudson River near the Proposed Westway Project during 1979, Vol. II. Report to the New York State Department of Transportation. LMS Engineers, Pearl River, New York, 4 chapters (paginated separately) + appendices.

Lee, G. F., and R. A. Jones. 1977. An Assessment of the Environmental Significance and Chemical Contaminants Present in Dredged Sediments Dumped in the New York Bight. Report to the U.S. Army Corps of Engineers, New York District, 62 pp.

Mackay, D. 1982. Correlation of bioconcentration factors. *Environmental Science and Technology*, **16**, 274–278.

MacLeod, W. D., L. S. Ramos, A. J. Friedman, D. G. Burrows, P. G. Prohaska, D. L. Fisher, and D. W. Brown. 1981. Analysis of Residual Chlorinated Hydrocarbons, Aromatic Hydrocarbons and Related Compounds in Selected Sources, Sinks and Biota of the New York Bight. Technical Memorandum OMPA-6, National Oceanic and Atmospheric Administration, Boulder, Colorado, 128 pp.

Mauck, W. L., P. M. Mehrle, and F. L. Mayer. 1978. Effect of the polychlorinated biphenyl Aroclor 1254 on growth, survival, and bone development in brook trout (*Salvelinus fontinalis*). *Journal of the Fisheries Research Board of Canada*, **35**, 1084–1088.

Mayer, R. L., P. M. Mehrle, and H. O. Sanders. 1977. Residue dynamics and biological effects of polychlorinated biphenyls in aquatic organisms. *Archives of Environmental Contamination and Toxicology*, **5**, 501–511.

McKim, J. M., and E. M. Heath. 1983. Dose determinations from waterborne 2,5,2′,5′ (^{14}C) tetrachlorobiphenyl and related pharmacokinetics in two species of trout (*Salmo gairdneri* and *Salvelinus forntinalus*): a mass balance approach. *Toxicology and Applied Pharmacology*, **108**, 177–187.

Mehrle, P. M., T. A. Haines, S. Hamilton, J. L. Ludke, R. L. Mayer, and M. A. Ribick. 1982. Relationship between body contaminants and bone development in east-coast striped bass. *Transactions of the American Fisheries Society*, **111**, 231–241.

Morales, N. M., D. B. Tuhey, W. A. Colburn, and H. B. Mathews. 1979. Pharmacokinetics of multiple oral doses of selected polychlorinated biphenyls in mice. *Toxicology and Applied Pharmacology*, **48**, 397–407.

Moriarty, F. 1975. Exposures and residues. *In:* Organochlorine Insecticides: Persistant Organic Pollutants, F. Moriarty (Ed.). Academic Press, New York, pp. 29–72.

Mueller, J. A., T. A. Gerrish, and M. C. Casey. 1982. Contaminant Inputs to the Hudson Raritan Estuary. OMPA-21, U.S. National Oceanic and Atmospheric Administration, Boulder, Colorado, 135 pp.

Nau-Ritter, G. M. 1980. The Dynamics of PCB Transfers among Marine Phytoplankton, Clay Particles and Water. M.S. Thesis, Marine Sciences Research Center, State University of New York, Stony Brook, 177 pp.

Nau-Ritter, G. M., C. F. Wurster, and R. G. Rowland. 1982. Partitioning of [^{14}C] PCB between water and particulates with various organic contents. *Water Research,* **16,** 1615–1618.

Neely, W. B., D. R. Branson, and G. E. Blau. 1974. Partition coefficients to measure bioconcentration potential of organic chemicals in fish. *Environmental Science and Technology,* **8,** 1113–1115.

Neumann, D. A., J. M. O'Connor, and J. A. Sherk, Jr. 1981. Oxygen consumption of white perch (*Morone americanus*), striped bass (*M. saxatilis*), and spot (*Leiostomus xanthurus*). *Comparative Biochemistry and Physiology,* **69**(A), 467–478.

Neumann, D. A., J. M. O'Connor, J. A. Sherk, Jr., and K. V. Wood. 1982. Respiratory response of striped bass (*Morone saxatilis*) to suspended solids. *Estuaries,* **5,** 28–39.

New York State Department of Environmental Conservation. 1981. Toxic Substances in Fish and Wildlife: 1979 and 1980 Annual Reports, Vol. 4, No. 1. Technical Report 81-1, Division of Fish and Wildlife, New York State Department of Environmental Conservation, Albany, 138 pp.

New York University Medical Center. 1982. Identifying Chemical Signatures for Disposed Dredged Materials. Report to the U.S. Army Corps of Engineers, New York District, New York University Medical Center, 79 pp.

Norstrom, R. J., A. E. McKinnon, and A. S. W. deFreitas. 1976. A bioenergetics-based model for pollutant accumulation by fish. Simulation of PCB and methylmercury residue levels in Ottowa River yellow perch (*Perca flavescens*). *Journal of the Fisheries Research Board of Canada,* **33,** 248–267.

O'Brien and Gere Engineers. 1979. Bioaccumulation studies on *Homarus americanus.* Report to U.S. Army Corps of Engineers, New York District, O'Brien and Gere Engineers, New York, 52 pp.

O'Connor, J. M. 1982. Biological Monitoring of PCBs in Hudson River Plankton. Report to New York State Department of Environmental Conservation, Albany, 202 pp.

O'Connor, J. M., J. B. Klotz, and T. J. Kneip. 1982. Sources, sinks and distribution of organic contaminants in the New York Bight ecosystem. *In:* Ecological Stress and the New York Bight: Science and Management, G. F. Mayer (Ed.). Estaurine Research Federation, Columbia, South Carolina, pp. 631–653.

Pavlou, S. P., and R. N. Dexter. 1979. Distribution of polychlorinated biphenyls (PCB) in estuarine ecosystems. Testing the concept of equilibrium partitioning in the marine environment. *Environmental Science and Technology,* **13,** 65–71.

Pavlou, S. P., W. Horn, R. N. Dexter, D. E. Anderson, and E. A. Quinlan. 1982. Release, distribution, and impacts of polychlorinated biphenyls (PCBs) induced by dredged material disposal activities at a deep-water estuarine site. *Environment International,* **7,** 99–117.

Pequegnat, W. E., B. M. James, E. A. Kennedy, A. D. Fredericks, R. R. Fay, and F. Hubbard. 1980. Application of the Biotal Ocean Monitor System to a Study of the Impacts of Ocean Dumping of Dredged Material in the New York Bight. TerEco Technical Report to the U.S. Army Corps of Engineers, New York District, 61 pp.

Peters, L. S., and J. M. O'Connor. 1982. Factors affecting short-term PCB and DDT accumulation in zooplankton and fish from the Hudson Estuary. *In:* Ecological Stress and the New York Bight: Science and Management, G. F. Mayer (Ed.). Estuarine Research Federation, Columbia, South Carolina, pp. 451–465.

Phillips, G. R., and D. R. Buhler. 1978. The relative contributions of methylmercury from food or water to rainbow trout (*Salmo gairdneri*) in a controlled laboratory environment. *Transactions of the American Fisheries Society,* **107,** 853–861.

Pierce, R. J., T. D. Wright, and J. M. O'Connor. 1981. Biomagnification: relationship to dredged material disposal in the coastal marine environment (Abstract). *Estuaries,* **4,** 258–259.

Pinkney, A. E. 1982. Uptake and Retention of ^{14}C-Aroclor 1254 in *Gammarus tigrinus* after Dietary Exposure with the Fungus *Fusarium oxysporum.* M.S. Thesis, New York University, New York, 70 pp.

Pizza, J. C. 1983. Pharmacokinetics and Distribution of Dietary Polychlorinated Biphenyls (PCBs) in Hudson River Striped Bass, *Morone saxatilis.* Ph.D. Thesis, New York University, New York, 109 pp.

Pizza, J. C., and J. M. O'Connor. 1983. PCB dynamics in Hudson River striped bass. II. Accumulation from dietary sources. *Aquatic Toxicology,* **3,** 313–327.

Rubinstein, N. I., E. Lores, and N. Gregory. 1983. Accumulation of PCBs, mercury and cadmium by *Nereis virens, Mercenaria mercenaria* and *Palaemonetes pugio* from contaminated harbor sediments. *Aquatic Toxicology,* **3,** 249–260.

Stainken, D., and J. Rollwagen. 1979. PCB residues in bivalves and sediments of Raritan Bay. *Bulletin of Environmental Contamination and Toxicology,* **23,** 690–697.

Stallings, D., and F. L. Mayer. 1972. Toxicity of PCBs to fish and environmental residues. *Environmental Health Perspectives,* **1,** 159–164.

Tavolaro, J. 1982. Sediment Budget Study for Dredged Material in the New York Region. U.S. Army Corps of Engineers, New York District, New York, 27 pp.

Thomann, R. V. 1981. Equilibrium model of fate of microcontaminants in diverse aquatic food chains. *Canadian Journal of Fisheries and Aquatic Sciences,* **38,** 280–296.

Thomann, R. V., and J. P. Connolly. 1982. An Age Dependent Model of PCB in a Lake Michigan Food Chain. ERL-Duluth CR805916010, Report to the U.S. Environmental Protection Agency, Duluth, Minnesota, 110 pp.

U.S. National Academy of Sciences. 1979. Polychlorinated Biphenyls. National Academy of Sciences Press, Washington, D.C., 182 pp.

vanVeld, P. 1980. Uptake, Distribution, Metabolism and Clearance of Kepone by Channel Catfish (*Ictalurus punctatus*). M.S. Thesis, School of Marine Science, College of William and Mary, Williamsburg, Virginia, 73 pp.

Vetter, R. D. 1982. Seasonal metabolic compensation in sympatric seatrout: adaptation to the estuary. *Transactions of the American Fisheries Society,* **111,** 193–198.

West, R. H., and P. G. Hatcher. 1980. Polychlorinated biphenyls in sewage sludge and sediments of the New York Bight. *Marine Pollution Bulletin,* **11,** 126–129.

Wyman, K. D., and H. B. O'Connors, Jr. 1980. Implication of short term PCB uptake by small estuarine copepods (*Genus Acartia*) from PCB-contaminated water, inorganic sediments and phytoplankton. *Estuarine and Coastal Marine Science,* **11,** 121–131.

PART IV: RADIONUCLIDES IN THE MARINE ENVIRONMENT

Chapter 13

Tritium and Technetium Uptake and Distribution in Three Marine Algae

Gérasime Arapis, Silvano Bonotto, Albert Bossus, Gustaaf Nuyts, Georg B. Gerber, and René Kirchmann

Department of Radiobiology
Centre d'Etude de l'Energie Nucléaire
Studiecentrum voor Kernenergie
Mol, Belgium

Jacques Colard and Pierre Mathot

Department of Physics
Centre d'Etude de l'Energie Nucléaire
Studiecentrum voor Kernenergie
Mol, Belgium

Michel Cogneau

Laboratoire de Chimie Inorganique et Nucléaire
Louvain-la-Neuve, Belgium

Dr. Arapis' present address: Laboratoire de Biologie Cellulaire Végétale, Université de Paris VII, 2, Place Jussieu, F-75221 Paris, Cédex 05, France.

ABSTRACT

The uptake and distribution of ^3H and technetium were studied in three marine algae: *Acetabularia acetabulum, Boergesenia forbesii,* and *Ascophyllum nodosum*. Tritium-containing water (HTO) and ^{95m}Tc-potassium pertechnetate were used in the experiments. In *Acetabularia,* uptake and release of HTO was rapid and occurred according to an apical–basal gradient. Moreover, uptake of HTO was optimal at a salinity of 37‰. Tritium was readily incorporated but did not accumulate in the organic matter of growing *Acetabularia* cells. Organically bound ^{3}H was unevenly distributed among cellular constituents, a large part (70%) of which was in small organic molecules. *Acetabularia* accumulated ^{95m}Tc and distributed it in the cell according to an apical–basal gradient. *Boergesenia* took up ^{95m}Tc but did not accumulate it. The puncturing technique showed that ^{95m}Tc crossed the cell wall and entered the cytoplasmic layer and vacuole. *Ascophyllum* accumulated ^{95m}Tc in all organs, stalk, lateral branches, and aerocysts. However, the tips of the stalk and of the lateral branches, which have growing tissues, accumulated more ^{95m}Tc than did other parts of the plant. The results underline the usefulness of marine algae in studying the interaction of artificial radioactivity with living organisms. Because under natural and artificial conditions the algae are grazed by marine animals, the algae may be successfully used to study the transfer of radionuclides in the food web.

13.1. INTRODUCTION

The contamination of the marine environment by synthetic radioactivity has received considerable attention because of its potentially harmful effects on living organisms [International Atomic Energy Agency (IAEA), 1975, 1979, 1981; Nuclear Energy Agency–Organisation for Economic Cooperation and Development, 1980; Commission of the European Communities (CEC), 1984a]. In spite of these efforts, the behavior of radioactive material in the sea deserves further investigation. More biological data are needed on radionuclides of importance in radioprotection and especially on long-lived radionuclides (CEC, 1984a). The data are needed for a deeper comprehension of the radionuclides' global geochemical cycle, and for a more accurate assessment of the risk that they pose to humans and other living organisms through possible transfer of radioactivity along the marine food web.

Part of the organic matter produced by unicellular and multicellular algae is released into the external medium and used by microorganisms (Allen, 1956; Sepers, 1977; Larsson and Hagström, 1979; Watanabe, 1980; Nienhuis, 1981; Rice and Tenore, 1981; Wolter, 1982; Jensen, 1983). Another part of the organic matter is transferred to higher trophic levels by grazing animals. Radionuclides taken up by marine algae from seawater might arrive to humans via seafoods. For this reason, marine and freshwater algae are particularly useful for studying transport pathways of radioactive-waste material in aquatic systems (Nakahara et al., 1975; Bonotto et al., 1978a, 1981, 1983a; Holm and Persson, 1980).

The marine green algae *Acetabularia, Boergesenia,* and *Valonia,* due to their large cell size, have been utilized to determine the precise cellular location of radionuclides taken up by the algae. It has been possible through the puncturing technique (Bonotto et al., 1983a) to demonstrate that radionuclides not only contaminate the outside of the algae, but they also penetrate the cells and are inhomogeneously distributed among the main compartments of the cells. Several other species of marine algae (green, red, and brown) have been used to study the incorporation of radionuclides for the purpose of learning about radioprotection (Gutknecht, 1963; Tremblay and Mehran, 1964; Mehran and Tremblay, 1965; Nakahara et al., 1975; Bonotto et al., 1978a, 1981, 1982a, 1982b; Masson et al., 1981).

This chapter deals with uptake and distribution of ^{3}H and technetium in the marine algae *Acetabularia, Boergesenia,* and *Ascophyllum*. A considerable amount of literature has been devoted to ^{3}H (Moghissi and Carter, 1973; IAEA, 1979), an isotope that is naturally produced by cosmic rays or ejected by stars, but originates mainly from fission and activation of several elements in nuclear weapons and nuclear reactors (Bonka, 1979). The importance of ^{3}H may increase if fusion technology is successfully developed (Coyle, 1979). Because of this, studies concerning the fate of ^{3}H in aquatic and terrestrial ecosystems are being pursued in many laboratories (Assises Internationales de Radioécologie, 1982; National Institute of Radiological Sciences, 1982).

It is known that ^{3}H is released from nuclear facilities as HTO, the predominant chemical form, and as hydrogen gas (Bonka, 1979). Moreover, organic ^{3}H has been found in the liquid effluents from nuclear power plants (Kirchmann et al., 1977); however, the chemical form of the released organic ^{3}H and its origin remain unknown. Our experiments were done with HTO, a form of ^{3}H that is rapidly exchanged by algal cells (Bonotto et al., 1982a). Once in the cell, HTO is distributed to the various cellular compartments and metabolized. Of particular interest is the splitting of the HTO molecule, which occurs in the chloroplasts through complex chemical reactions (Metzler, 1977) that lead to the incorporation of ^{3}H first into photosynthetic organic compounds and later into the various cell constituents.

Technetium, a radioelement without a stable isotope, is produced in nature by spontaneous fission of uranium or by cosmic-ray activation of molybdenum, niobium, and zirconium (Saas et al., 1980; Till et al., 1979). However, the major source of technetium in the environment originates from fission of uranium, thorium, and plutonium from nuclear weapons and from nuclear power plants (Wildung et al., 1979). Medical use of radiation delivers a relatively small amount of technetium to the environment (Birks, 1975). The most important isotope of technetium is ^{99}Tc, which is produced with a relatively high yield (6.2%) from ^{235}U and has a long half-life of 2.1 × 10^5 y (Wildung et al., 1979).

Despite recent estimates showing that the concentration of technetium in the environment may reach 170,000 kg by the year 2000 (Burkholder et al., 1975), information on the fate of technetium remains insufficient for use in assessing its impact on the terrestrial and marine environments and, indirectly, on humans (Birks, 1975; Gearing et al., 1975; Spies, 1975; Wildung et al., 1979; Fowler et al., 1981, 1983; Masson et al., 1981; Pentreath, 1981; Beasley et al., 1982; Fisher, 1982). For this reason, the CEC has recently organized a meeting dedicated to technetium (CEC, 1984b). The results reported in this chapter contribute to a better understanding of the interaction of ^{3}H and technetium with marine algae.

13.2. MATERIALS AND METHODS

13.2.1. Harvest and Culture

Ascophyllum nodosum is a large brown alga that is abundant along the North Sea coast. The plants were collected in the southern part of The Netherlands, near Borsele, cleaned mechanically of epiphytes and small animals, washed, and kept at 21°C in seawater with 10 μg ml^{-1} of chloramphenicol added to inhibit bacterial growth. *Acetabularia acetabulum (mediterranea)* was cultivated according to the method described by Lateur and Bonotto (1973) in natural seawater enriched with $NaNO_3$ (0.1 g liter^{-1}), Na_2HPO_4 (0.01 g liter^{-1}), and an earth extract (Bonotto, 1976). Under the culture conditions, *Acetabularia* cells completed their biological cycle in about 150 d. The method described by Enomoto and Hirose (1972) was used to cultivate *Boergesenia forbesii* from artificially induced aplanospores in enriched seawater.

Both *Acetabularia* and *Boergesenia* are marine green algae. Cells of *Acetabularia* have only one nucleus, but they contain millions of chloroplasts and mitochondria. Cells of *Boergesenia*, on the other hand, possess many nuclei and organelles.

13.2.2. Labeling

Groups of *Acetabularia* and *Boergesenia* cells were selected for their uniform size and kept in a 30- to 100-ml culture medium of natural seawater containing (0.3–1.2) × 10^5 Bq ml^{-1} of ^{3}H as HTO, or (0.6–18) × 10^2 Bq ml^{-1} of ^{95m}Tc-potassium pertechnetate, a chemical that is relatively stable in seawater (Spies, 1975; Fowler et al., 1981).

The macroalga *Ascophyllum* was kept in 5 liters of filtered natural seawater with ≤ 1.5 × 10^2 Bq ml^{-1} of ^{95m}Tc. At the end of the labeling periods, which ranged from minutes to days, depending on the type of experiment, the algae were rapidly washed (15–60 s) in seawater, blotted on filter paper, and submitted to cell fractionation. The cells were then lyophilized or fixed with 45% formalin in seawater for γ-ray spectrometry. Tritium was obtained from the Institut National des Radioéléments (Fleurus, Belgium). Technetium-95m was produced at the University of Louvain-la-Neuve, Belgium, from molybdenum irradiated in the cyclotron. Short experiments (in terms of minutes to hours) were done during the day, under light (approximately 3 W m^{-2}), at 21°C. Long experiments (in terms of days) were done under a regime of 12 h of light and 12 h of dark at 21°C. Control experiments with HTO and *Acetabularia* were conducted in the dark.

13.2.3. Puncturing

Algae with large cells, such as *Acetabularia*, *Boergesenia*, and *Valonia*, offer the opportunity to show if radionuclides can cross the cell wall. This is important because, once inside the cell, the radioactive material would undergo metabolic transformation. The vacuolar sap of *Acetabularia* may be withdrawn only with a micromanipulator. In contrast, the vacuoles of *Boergesenia* and *Valonia* are so large (up to 0.5 cm) that their sap can be easily obtained by puncturing the cell with a thin needle mounted on a 1-ml plastic syringe. Withdrawal of the vacuolar sap leaves the cells flat, with the cytoplasmic layer sticking to the cell walls (Bonotto et al., 1983a).

Large *Boergesenia* cells (1.5 cm × 0.5 cm) were selected for the experiments. After labeling, the cells were washed with seawater, blotted on filter paper, and put on a Parafilm "M" sheet (American Can Company, Dixie/Marathon, Greenwich, Connecticut) to avoid water loss by capillarity. The vacuolar saps of several cells were combined and measured by liquid scintillation for ^{3}H in a Packard Tri-Carb spectrometer (Packard Instrument Corp., Downers Grove, Illinois) or by γ-spectrometry for ^{95m}Tc in a Packard Auto-Gamma Spectrometer (for small samples) or in a lithium-drifted germanium [Ge(Li)] detector connected to a multichannel analyzer and interfaced to a minicomputer for spectrum analyses (Bonotto et al., 1983a). The uncertainty in the counting statistics (1-σ) for ^{95m}Tc was <1%.

13.2.4. Fractionation of *Acetabularia*

Groups of 10–20 cells of *Acetabularia* were labeled, rapidly washed, blotted on filter paper, frozen on solid CO_2, and cut into basal, medial, and apical fragments. The fragments were analyzed for ^{3}H by liquid scintillation in a Packard Tri-Carb spectrometer or for ^{95m}Tc by γ-spectrometry in a Packard Auto-Gamma spectrometer. Chlorophyll analysis was done according to the method described by Arnon (1949).

13.2.5. Protein and Organic Matter

The distribution of ^{3}H in the organic matter of *Acetabularia* was studied after a 6- to 7-d exposure to HTO (2.46×10^6 Bq ml^{-1}). The cells were lyophilized and reduced to powder in a mortar. The powder was directly measured for ^{3}H content by a microincineration technique (Kirchmann et al., 1977) or successively extracted with ice-cold 5.5% trichloroacetic acid (TCA), warm methanol (60°C for 1 h), hot TCA (90°C for 30 min), and hot 1 *N* NaOH (90°C for 30 min). The first extract contained small organic molecules; the second contained lipids, chlorophylls, and carotenoids; the third contained nucleic acids and some proteins; and the fourth contained mostly proteins. After the extractions, a residue remained that contained various structural materials, mainly consisting of polysaccharides from the cell wall (Bonotto et al., 1982a, 1982b). The extracts and the residue were analyzed for radioactivity by liquid scintillation.

In other experiments, total proteins of HTO-labeled *Acetabularia* cells were extracted for 3 min with hot electrophoresis buffer (100°C) containing 0.06 *M* Tris (hydroxymethyl) aminomethane, 0.005 *M* EDTA•2Na•$2H_2O$, 0.64 *M* β-mercaptoethanol, and 2% sodiumdodecyl sulfate at a pH of 6.8. After extraction, the proteins were analyzed by gel electrophoresis according to a procedure described by Bonotto et al. (1978b). The gels were stained with Coomassie brilliant blue to reveal protein bands and sliced into 1-mm fragments. The slices were counted for radioactivity by using a 1:10 mixture of Soluene-350 and Insta-Fluor TM (Packard Instrument Corp.).

13.3. RESULTS

13.3.1. Tritiated Water

Experiments with HTO were done with *Acetabularia* and *Boergesenia*. Uptake and release of HTO in *Acetabularia* were rapid (Fig. 13.1). Computer analysis of the HTO release revealed the presence of three metabolic components that represented ~40%, 50%, and 1.6% of the total radioactivity, respectively, and had half-lives of a few seconds, ~1 min, and 5 h, respectively. The second component probably represented water in the vacuole. In *Boergesenia*, HTO uptake and release was slower than it was in *Acetabularia*. The puncturing technique revealed that HTO uptake by the vacuole of *Boergesenia* was characterized by essentially one metabolic component with a half-life of 12.4 min.

Because *Acetabularia* cells possess circadian rhythms (Vanden Driessche, 1975), it is of interest to know if HTO is taken up at different rates at different times of the day. The results obtained in two experiments (Fig. 13.2) did not reveal important differences in the water-exchange capability of *Acetabularia* during the day. However, at the beginning of the day, which coincides with the dark–light transition, the cells seem to take up less HTO, probably because their physiological conditions were not yet optimal after the dark period (Pâques et al., 1979).

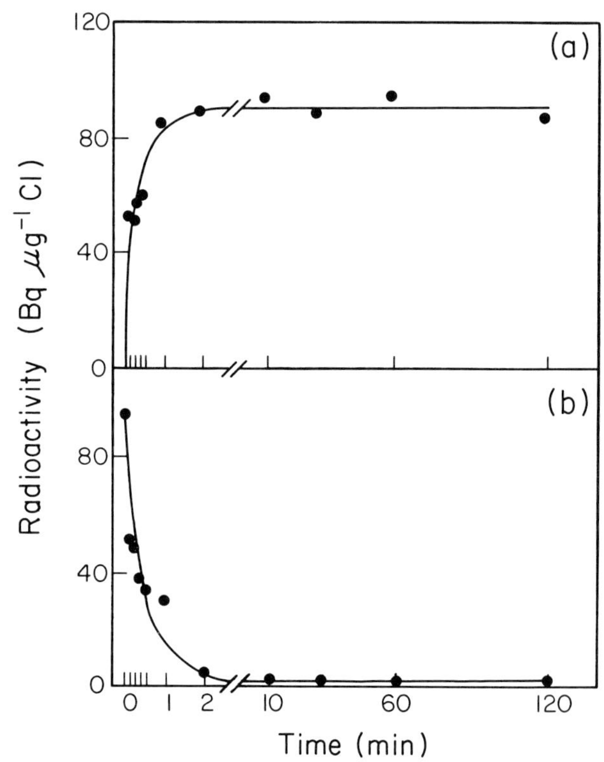

Figure 13.1. Uptake (a) and release (b) kinetics of tritiated water (HTO) in vegetative cells (stage 4 according to Bonotto and Kirchmann, 1970) of *Acetabularia acetabulum (mediterranea)*.

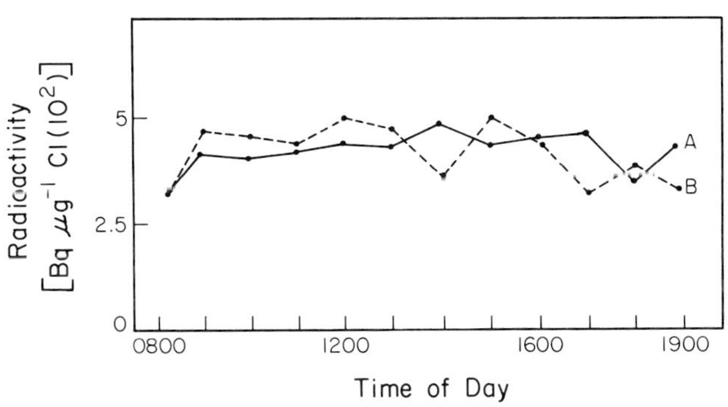

Figure 13.2. Uptake of tritiated water (HTO) in vegetative cells of *Acetabularia acetabulum (mediterranea)* versus time of day in two separate trials in light (1000 lux).

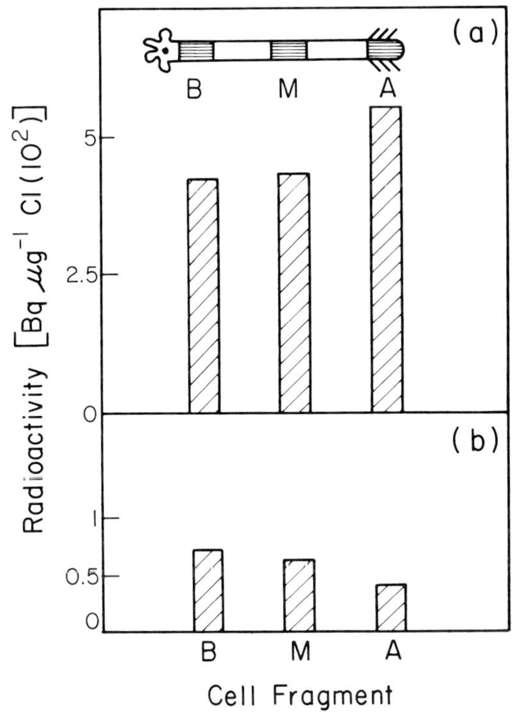

Figure 13.3. Apical-basal gradient of uptake (a) and release (b) of tritiated water (HTO) in vegetative cells of *Acetabularia acetabulum (mediterranea)* in light (1000 lux). Specific cell regions are illustrated and defined as: B, basal; M, medial; and A, apical.

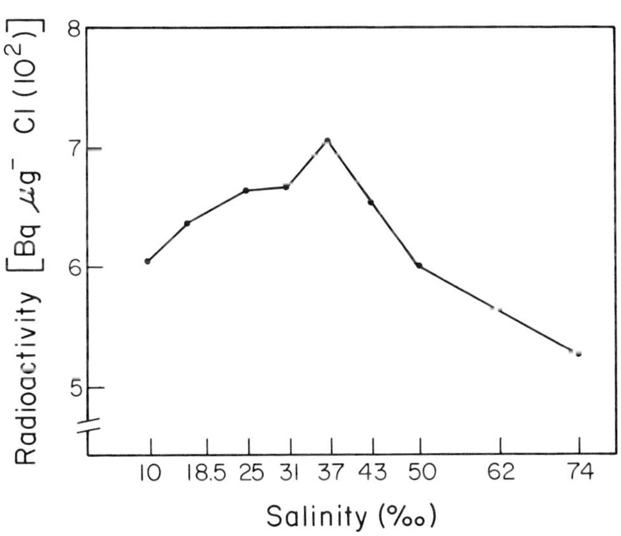

Figure 13.4. Effect of salinity on the uptake of tritiated water (HTO) in vegetative cells of *Acetabularia acetabulum (mediterranea)*.

Because vegetative cells of *Acetabularia* display morphological and physiological gradients (Puiseux-Dao et al., 1972; Bonotto et al., 1976), we investigated whether the exchange of HTO follows a gradient within the cell. Figure 13.3 shows clearly that uptake and release of HTO during 1 min occurs according to an apical–basal gradient, with the apical region showing the highest exchange rate.

Uptake of HTO in *Acetabularia* may be affected by physical and chemical factors (Arapis et al., 1986a). Figure 13.4 shows that maximal uptake of HTO occurs at a salinity of 37‰, which represents the optimal salt concentration for cell development, and decreases toward the lower and higher salt concentrations. It is important to note that the highest salinity (74‰) induced partial cytolysis and should be considered toxic to *Acetabularia*.

13.3.2. Incorporation of ^3H in Total Proteins and Organic Matter

At first, *Acetabularia* kept in HTO showed a phase of rapid ^3H incorporation into the organic matter of the cells. The specific activity of the organic matter was 6% of that of HTO. Later, incorporation was rather slow and increased almost linearly with time, probably because the growth rate decreased. Extraction procedures showed that after a 6-d period of incubation in HTO, the algae contained 70% of the organically bound ^3H in the form of small acid-soluble molecules. Presumably, these molecules represented material dissolved in the vacuole. The remaining radioactivity was distributed among lipids and pigments (7%), nucleic acids (7%), proteins (14%), and residual structural material (1%). Similar results were obtained from *Boergesenia* (Bonotto et al., 1982a). Control experiments, done in the dark with *Acetabularia,* suggested that ^3H incorporation occurred mainly through photosynthesis, which caused the tritiated water molecule to split. Under dark conditions, the incorporation of ^3H into the total organic matter was small or negligible, depending on the incubation length.

Analysis by electrophoresis with polyacrylamide gels revealed the presence of at least 50 polypeptides in extracts of the total proteins in whole cells of *Acetabularia*. Tritium from HTO was incorporated into several main polypeptides, whose molecular weights ranged from 100 to 14 kilodaltons (Fig. 13.5).

13.3.3. Uptake of ^{95m}Tc

Both vegetative and reproductive *Acetabularia* cells extracted ^{95m}Tc from the surrounding medium (Bonotto and Kirchmann, 1970). Concentration factors (defined as the ratio of the radioactivity per gram of fresh plant material to the radioactivity per milliliter of medium) were 25,46, and 59 after 8 h, 4 d, and 11 d of incubation, respectively (Fig. 13.6). After 30 or 120 min of incubation, vegetative cells fixed more ^{95m}Tc in their apical region than in their basal region (Fig. 13.7). Thus, ^{95m}Tc seems to be distributed in the cell according to an apical–basal gradient.

Vegetative *Boergesenia* cells did not concentrate ^{95m}Tc. Even after a relatively long incubation (10 d), the concentration factors remained below 1 (Fig. 13.8). However, ^{95m}Tc must have crossed the cell wall and the cytoplasmic layer, because after a 120-min incubation a measurable amount (96 Bq ml^{-1}) was found in the vacuolar sap. It is not yet known if technetium changes its oxidation state when it penetrates the vacuole.

The brown alga *Ascophyllum* accumulated ~40% of the ^{95m}Tc present in the external medium within a few hours

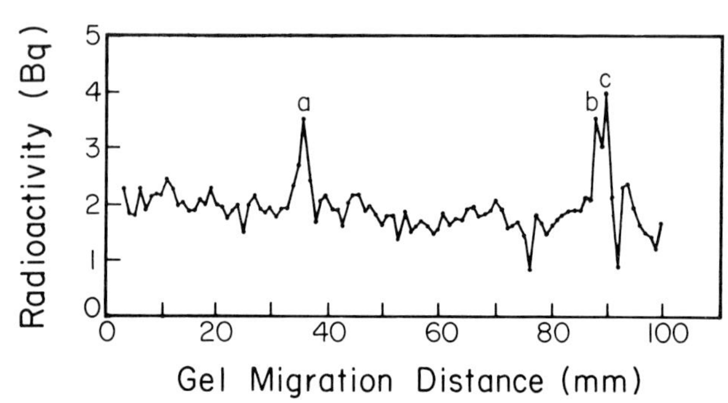

Figure 13.5. Distribution of ^3H in the total proteins of *Acetabularia acetabulum (mediterranea)*, electrophoresed on a polyacrylamide gel. Labeled polypetides are represented by a, b, and c and have molecular weights of 53, 19, and 17 kilodaltons, respectively.

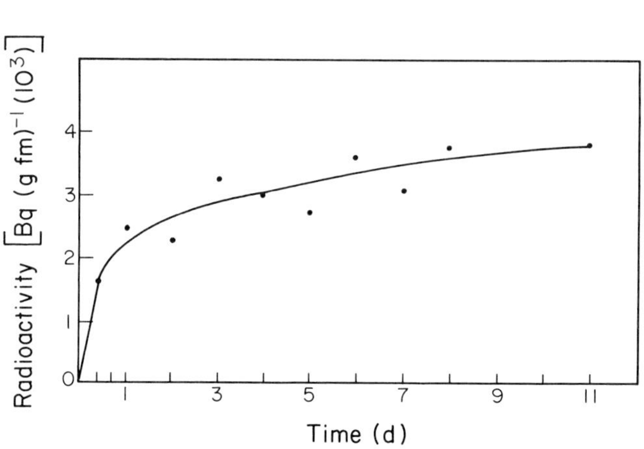

Figure 13.6. Uptake kinetics of ^{95m}Tc, as potassium pertechnetate, in vegetative cells of *Acetabularia acetabulum (mediterranea)*. The culture medium has not been renewed and ^{95m}Tc is probably exhausted at the end of incubation. The notation "(g fm)$^{-1}$" denotes "per gram of fresh plant matter."

(Fig. 13.9). Concentration factors of 50 were found after a 4-d incubation. All of the organs of the plant (stalk, lateral branches, and aerocysts) took up ^{95m}Tc. On the other hand, lateral branches, which contained growing young tissues, were found to accumulate twice as much ^{95m}Tc as did the stalk and four times more than did the aerocysts, which only had a concentration factor of 4. In order to determine if the uptake of ^{95m}Tc was different along the stalk, 1-cm fragments were cut from different sections (~10 cm each), starting from the base. Figure 13.10 shows that the radioactivity taken up by the fragments varied between 700 and 1700 Bq g^{-1} of fresh plant matter and that it was not distributed according to an apical–basal gradient. The distribution of ^{95m}Tc along the stalk deserves further investigation, because the apical region of some plants were labeled more strongly than were other regions of those plants.

13.4. DISCUSSION

Studies of the uptake and distribution of radionuclides in marine organisms are of general interest for two main reasons: (1) to assess the impact of radionuclide releases on life

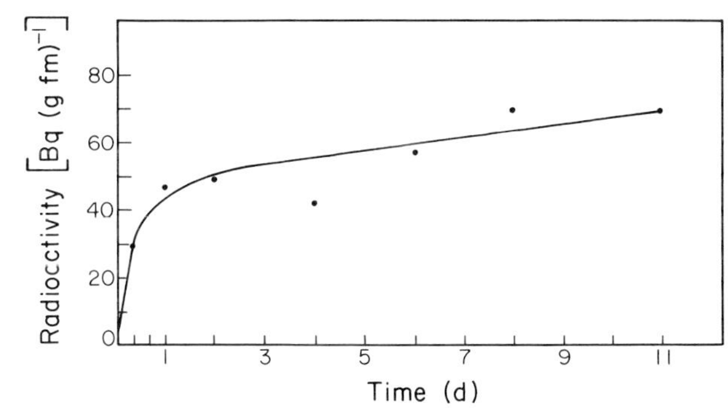

Figure 13.8. Uptake kinetics of ^{95m}Tc, as potassium pertechnetate, in vegetative cells of *Boergesenia forbesii*. The notation "(g fm)$^{-1}$" denotes "per gram of fresh plant matter."

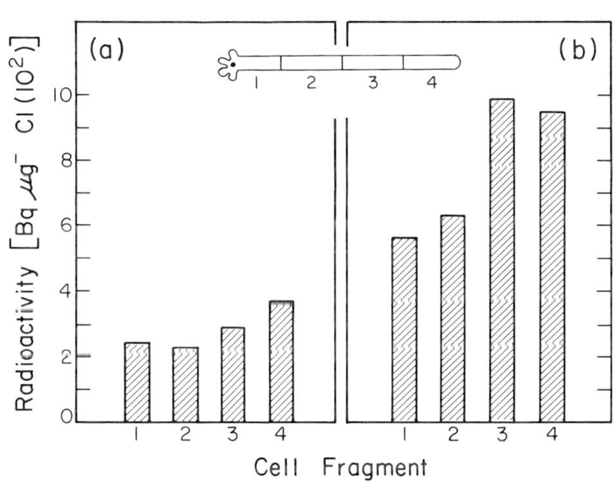

Figure 13.7. Apico-basal gradient of ^{95m}Tc, uptake in vegetative *Acetabularia acetabulum* (*mediterranea*) cells incubated for 30 min (a) or 120 min (b) in light (1000 lux) in normal culture medium with ^{95m}Tc-potassium pertechnetate.

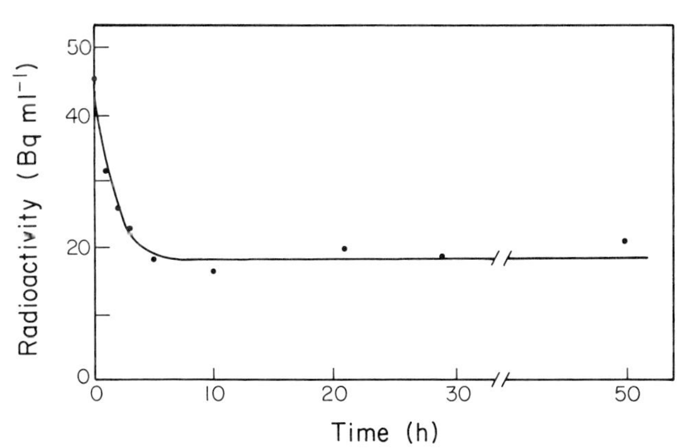

Figure 13.9. Extraction kinetics of ^{95m}Tc, as potassium pertechnetate, from seawater by a whole plant of *Ascophyllum nodosum*.

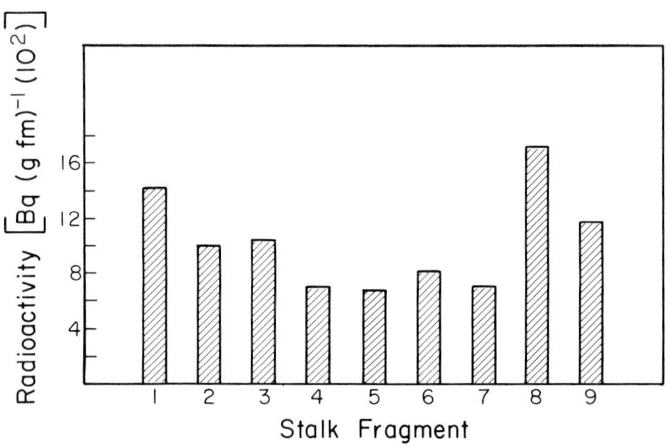

Figure 13.10. Distribution of ^{95m}Tc along with the stalk of *Ascophyllum nodosum*. The numbers represent stalk fragments from the base (1) to the apex (9). The notation "(g fm)$^{-1}$" denotes "per gram of fresh plant matter."

in the marine environment, and (2) to estimate the contamination of marine food webs that lead to humans.

Marine algae are important links to marine food webs because they are eaten by many different animals (Bonotto et al., 1983a). In addition, several species of marine algae are commercially harvested for food for livestock and humans (Bonotto, 1979). It is known that unicellular and multicellular marine algae are capable of accumulating certain radionuclides (Rice, 1956; Rice and Willis, 1959; Gutknecht, 1961, 1963; Pillai et al., 1964; Tremblay and Mehran, 1964; Mehran and Tremblay, 1965; Czapke, 1966; Hodge et al., 1974; Nakahara et al., 1975; Bonotto et al., 1978a; Holm and Persson, 1980; Noshkin, 1980). In spite of the considerable amount of literature, the information on the interaction of radionuclides with marine organisms remains insufficient to make valid assessments of the impact of artificial radionuclides on the marine environment. Our work on ^{3}H and technetium, two radionuclides important in radioprotection (Feinendegen, 1967; Moghissi and Carter, 1973; Carsten, 1979; IAEA, 1979; CEC, 1984a, 1984b), will contribute to a better understanding of the behavior of artificial radioactivity on the biogeochemical cycle.

Because 99% of environmental ^{3}H is contained in water, it is important to study the uptake of HTO by marine algae and the subsequent incorporation of ^{3}H into the cell constituents. The apical–basal gradient of HTO uptake and release found in *Acetabularia* is not yet understood. The possible relation of the apical–basal relationship to other cellular physiological gradients (Puiseux-Dao et al., 1972; Bonotto et al., 1976) and to the gradient of photosynthetic capacity of the chloroplasts (Issinger et al., 1971; Arapis et al., 1984b) remains to be investigated. Uptake of HTO by *Acetabularia* may not change significantly during the day, but it is affected by the salinity of the water. In a similar way, the salinity influences the uptake and the incorporation of tritiated leucine and adenine (Bonotto et al., 1984).

Most of the ^{3}H from HTO appears to be incorporated into the algae by photosynthesis; however, some ^{3}H may be incorporated into organic matter by metabolic pathways not directly related to photosynthesis (Inomata and Higuchi, 1976). Additional experiments with longer periods of darkness (on the order of several days) are needed in order to elucidate this point.

It has been shown that *Acetabularia* discriminates against incorporation of ^{3}H in organic matter (Bonotto et al., 1982a). Once organically bound, ^{3}H will have the same turnover as do the compounds in which it is incorporated, provided that it is present in a non-exchangeable form (Guenot et al., 1982). Though animals grazing on marine algae consume the whole plant (Cinelli et al., 1984), the fate of the ingested radioactivity may depend mostly on its chemical form. This assumption is supported by experimental work done in Japan on ^{3}H accumulation in aquatic organisms through the food web (Mukade and Higuchi, 1974; Higuchi et al., 1980; Komatsu et al., 1981). These authors have observed a different behavior of ^{3}H in brine shrimp fed ^{3}H-containing diatoms (having ^{3}H-labeled organic molecules) and HTO, as opposed to those brine shrimp fed unlabeled diatoms and HTO. Thus, it is reasonable to expect that the efficiency of radioactivity transfer along marine food webs depends on the chemical form of the radionuclide. This expectation should also apply to the technetium isotopes released into the marine environment.

In contrast to other marine green algae (*Ulva lactuca, Cladophora rupestris,* and *Codium tomentosum*) (Masson et al., 1981), *Acetabularia* is capable of extracting ^{95m}Tc from the external medium, and as is the case with HTO, it is taken up according to an apical–basal gradient. This observation raises the question of which mechanism is involved in ^{95m}Tc entering the cell. It has been suggested by Fisher (1982) that an anion such as TcO_4^- would not be passively taken up by algal cells that carry a negative surface charge. Uptake of ^{95m}Tc by *Acetabularia* cells may resemble that observed in higher terrestrial plants, which are known to remove technetium from the soils or nutrient solutions (Wildung et al., 1977, 1979; Routson and Cataldo, 1978; Mousny et al., 1979; Myttenaere et al., 1980; Vandecasteele et al., 1982). Studies of the membrane potential (resting potential of − 170 mV between the cytoplasm and the external medium) and of ionic regulation by *Acetabularia* indicate that Cl^- and SO_4^{-2} are probably actively transported into the cell and that active transport of Na^+ is outwards (Gradmann, 1970; Saddler, 1970a, 1970b, 1970c). Since anions are actively transported into *Acetabularia,* it is possible that TcO_4^- is also taken up by active transport.

In spite of the lack of accumulation, as evidenced by concentration factors <1 even after a 10-d incubation, experiments on *Boergesenia* cells demonstrated that ^{95m}Tc is capable of crossing the cell wall–cytoplasmic layer complex, because the radionuclide was found in the vacuole. Thus, this marine green alga is valuable for studies of the cellular distribution of radionuclides.

Results obtained for ^{95m}Tc in *Ascophyllum* confirm the capability of this brown alga to accumulate technetium (Masson et al., 1981). Our observations did not reveal an apical–basal uptake gradient of ^{95m}Tc in *Ascophyllum,* as was found for ^{60}Co in the brown alga *Eisenia bicyclis* (Nakahara et al., 1975). It should be noted, however, that the tips of the stalks, and lateral branches, where growth occurs, accumulated twice as much ^{95m}Tc as did other parts of the stalk, and that the tips accumulated four times as much as did the aerocysts in short-term experiments (24 or 50 h).

This observation is of ecological importance, because young soft tissues are preferred by some grazing animals. Results reported in the literature and in this chapter show that several brown algae (*Ascophyllum nodosum, Cystoseira compressa, Ectocarpus confervoides, Fucus serratus, Fucus vesiculosus,* and *Pelvetia canaliculata*) can markedly accumulate technetium (Pentreath, 1981; Fisher, 1982; Bonotto et al., 1983b). This unique behavior of brown algae is not yet understood. Fisher (1982) suggests that accumulation of technetium by brown algae may occur through a mechanism similar to that used by terrestrial vascular plants, but no evidence is presently available to support this hypothesis. It is known that some brown algae (*Ascophyllum nodosum, Fucus vesiculosus, Halidrys siliquosa,* and *Himanthalia lorea*) strongly accumulate ^{226}Ra (Kirchmann et al., 1974). Because these brown algae possess alginates, which are known to fix ^{226}Ra (Bonotto, 1976), it is possible that, in a native chemical form, these polysaccharides also play a role in ^{95m}Tc fixation. This hypothesis may be proven experimentally by *in vivo* and *in vitro* experiments with algae, crude extracts, and purified alginates. Surface adsorption may also be an important mechanism in radionuclide fixation by marine algae, as suggested by Rice (1956), Gutknecht (1961, 1963), and Nakahara et al. (1975). Additional effects of materials such as inorganic and organic substances adhering to the plant surfaces, microorganisms (Nakahara et al., 1975), or living bryozoans (Hodge et al., 1974) should also be taken into consideration.

Finally, the results reported in this chapter deal only with uptake and distribution of tritium and technetium in marine algae. Work is in progress in order to investigate the retention capabilities of *Acetabularia, Boergesenia,* and *Ascophyllum,* for these two important radionuclides.

ACKNOWLEDGMENTS

We thank C. Huysmans–Vanbaelen and M. C. Hemelaers–Gielen for typewriting the text, and we acknowledge the financial support of the European Communities (Contracts CCE No. BIO-B-431-B and CCE No. BIO-B-485-82-B) and of the International Atomic Energy Agency of Vienna (Contract No. 2579/R1/RB). Gérasime Arapis is a scientific and technical fellow of the Commission of the European Communities. This is publication No. 1975 of the Commission of European Communities.

REFERENCES

Allen, M. B. 1956. Excretion of organic compounds by *Chlamydomonas*. *Archiv für Mikrobiologie,* **24,** 163–168.

Arapis, G., S. Puiseux-Dao, and S. Bonotto. 1984a. Factors affecting uptake and distribution of tritium of the giant unicellular marine alga *Acetabularia. In:* Radiation Protection, Proceedings of the European Seminar on the Risks from Tritium Exposure (Mol, Belgium, 22–24 November 1982), G. Gerber and C. Myttenaere (Eds.). EUR 9065 EN, Commission of the European Communities, Luxembourg, pp. 167–169.

Arapis, G., S. Bonotto, and S. Puiseux-Dao. 1984b. Incorporation of ^{3}H and ^{14}C into *Acetabularia acetabulum.* Sixth International Symposium on *Acetabularia* (Pisa, Italy, 1–3 October 1984), Book of Abstracts. Centre D'Etude de l'Energie Nucléaire–Studiecentrum voor Kernergie, Mol, Belgium, pp. 27–28.

Arnon, D. J. 1949. Copper enzymes in isolated chloroplasts. Polyphenoloxidase in *Beta vulgaris. Plant Physiology,* **24,** 1–15.

Assises Internationales de Radioécologie. 1982. Assises Internationales de Radioécologie (Wageningen, The Netherlands, 28–30 June 1982). *Annales de l'Association Belge de Radioprotection,* **7,** 127–373.

Beasley, T. M., H. V. Lorz, and J. J. Gonor. 1982. Biokinetic behaviour of technetium in the red abalone *Haliotis rufescens:* a re-assessment. *Health Physics,* **43,** 501–507.

Birks, J. L. 1975. Medical radionuclides in marine environment. *Nature,* **255,** 621–622.

Bonka, H. 1979. Production and emission of tritium from nuclear facilities, and the resulting problems. *In:* Behaviour of Tritium in the Environment. International Atomic Energy Agency, Vienna, pp. 105–123.

Bonotto, S. 1976. Cultivation of plants. Multicellular plants. *In:* Marine Ecology, O. Kinne (Ed.). Wiley-Interscience, New York, pp. 467–501.

Bonotto, S. 1979. List of multicellular algae of commercial use. *In:* Marine Algae in Pharmaceutical Science. H. A. Hoppe, T. Levring, and Y. Tanaka (Eds.). Walter de Gruyter, Berlin, pp. 121–137.

Bonotto, S., and R. Kirchmann. 1970. Sur les processus morphogénétiques d'*Acetabularia mediterranea. Bulletin de la Société Royale de Botanique de Belgique,* **103,** 255–272.

Bonotto, S., P. Lurquin, and A. Mazza. 1976. Recent advances in research on the marine alga *Acetabularia. In:* Advances in Marine Biology, Vol. 14, F. S. Russell and M. Yonge (Eds.). Academic Press, London, pp. 123–250.

Bonotto, S., A. Bossus, G. Nuyts, R. Kirchmann, G. Cantillon, and R. Declerck. 1978a. Contamination d'organismes marins par le ^{3}H, le ^{134}Cs et le ^{60}Co. *Revue Internationale d'Océanographie Médicale,* **49,** 127–133.

Bonotto, S., G. Nuyts, L. Baugnet-Mahieu, J. Gilles, E. Dujardin, and C. Sironval. 1978b. Evidence for heterogeneity of chloroplasts in *Acetabularia:* fluorescence emission and protein synthesis. *In:* Chloroplast Development, G. Akoyunoglou and J. H. Argyroudi-Akoyunoglou (Eds.). Elsevier/North-Holland Biomedical Press, Amsterdam, pp. 333–344.

Bonotto, S., G. Carraro, S. Strack, A. Lüttke, J. Colard, G. Koch, and R. Kirchmann. 1981. Ten years of investigation on radioactive contamination of the marine environment. Incorporation, by marine algae and animals, of hydrogen-3 and other radionuclides present in effluents of nuclear or industrial origin. *In:* Impacts of Radionuclide Releases into the Marine Environment. International Atomic Energy Agency, Vienna, pp. 649–660.

Bonotto, S., G. B. Gerber, G. Arapis, and R. Kirchmann. 1982a. Modelization of tritium transfer into the organic compartments of algae. *Annales de l'Association Belge de Radioprotection,* **7,** 283–292.

Bonotto, S., G. Nuyts, A. Bossus, and R. Kirchmann. 1982b. Apport de la biochimie à l'étude de la pollution radioactive du milieu marin: Distribution du ^{3}H dans les differents constituants cellulaires des algues marines *Acetabularia, Boergesenia* et *Dunaliella. In:* Indices Biochimiques et Milieux Marins, Journées du GABIM (Brest, 18–20 November, 1981. Publications Centre National pour l'Exploitation des Océans (Actes Colloque) No. 14, pp. 399–410.

Bonotto, S., A. Bossus, G. Nuyts, R. Kirchmann, P. Mathot, J. Colard, and F. Cinelli. 1983a. Experimental uptake study of ^{60}Co, ^{137}Cs, ^{125}Sb and ^{65}Zn in four marine algae. *In:* Wastes in the Ocean, Vol. 3: Radioactive Wastes and the Ocean, P. K. Park, D. R. Kester, I. W. Duedall, and B. H. Ketchum (Eds.). Wiley-Interscience, New York, pp. 287–300.

Bonotto, S., G. B. Gerber, M. Cogneau, and R. Kirchmann. 1983b. Behavior of technetium in several marine algae. *Annales de l'Association Belge de Radioprotection,* **8,** 237–243.

Bonotto, S., G. Arapis, J. Mellet, D. Hoursiangou-Neubrun, F. Cinelli, P. Belcari, G. B. Gerber, and R. Kirchmann. 1984. Behaviour of tritium in the marine environment: uptake and distribution of tritiated water and of tritiated organic compounds in unicellular and multicellular algae. *In:* Proceedings of the Eleventh International Seaweed Symposium (Qingdao, People's Republic of China, 19–25 June 1983), C. J. Bird and M. A. Ragan (Eds.). Dr. W. Junk Publishers, Dordrecht, The Netherlands, pp. 187–189.

Burkholder, H. C., M. O. Cloninger, D. A. Baker, and G. Jansen. 1975. Incentives for Partitioning High-Level Waste. U.S. Energy Research and Development Agency Report BNWL-1927, National Technical Information Service, Springfield, Illinois, 233 pp.

Carsten, A. L. 1979. Tritium in the environment. *In:* Advances in Radiation Biology, Vol. 8, J. T. Lett and H. Adler (Eds.). Academic Press, New York, pp. 419–458.

Cinelli, F., P. Belcari, S. De Ranieri, G. Arapis, S. Bonotto, G. B. Gerber, and R. Kirchmann. 1984. Behaviour of tritium in the marine environment: studies with the macroalgae *Laurencia obtusa* and *Cystoseira compressa. In:* Radiation Protection, Proceedings of the International Symposium on the Behavior of Long-lived Radionuclides in the Marine Environment (La Spezia, Italy, 28–30 September 1983), A. Cigna and C. Myttenaere (Eds.). EUR 9214 EN, Commission of the European Communities, Luxembourg, pp. 271–282.

Commission of the European Communities (CEC). 1984a. Radiation Protection, Proceedings of the International Symposium on the Behaviour of Long-Lived Radionuclides in the Marine Environment (La Spezia, Italy, 28–30 September 1984), A. Cigna and C. Myttenaere (Eds.). EUR 9214 EN, Commission of the European Communities, Luxembourg, 462 pp.

CEC. 1984b. Scientific Seminar on the Behaviour of Technetium in the Environment (Cadarache, France, 23–26 October 1984). Commission of the European Communities, Radiation Protection Programme, in press.

Coyle, P. E. 1979. Laser fusion. Status, future and tritium control. *In:* Behaviour of Tritium in the Environment. International Atomic Energy Agency, Vienna, pp. 139–153.

Czapke, K. 1966. Seaweeds as radioactivity indicators of marine environment. Proceedings of the Fifth International Seaweed Symposium (Halifax, Canada), E. G. Young and J. L. McLachlan (Eds.). Pergamon Press, Oxford, pp. 371–373.

Enomoto, S., and H. Hirose. 1972. Culture studies on artificially induced aplanospores and their development in the marine algae *Boergesenia forbesii* (Harvey) Feldmann (Chlorophyceae, Syphonocladales). *Phycologia,* **11,** 119–122.

Feinendegen, L. E. 1967. Tritium-Labeled Molecules in Biology and Medicine. Academic Press, New York, 430 pp.

Fisher, N. S. 1982. Bioaccumulation of technetium by marine phytoplankton. *Environmental Science and Technology,* **16,** 579–581.

Fowler, S. W., G. Benayoun, P. Parsi, M. W. A. Essa, and E. H. Schulte. 1981. Experimental studies on the bioavailability of technetium in selected marine organisms. *In:* Impact of Radionuclide Releases into the Marine Environment. International Atomic Energy Agency, Vienna, pp. 319–339.

Fowler, S. W., S. R. Aston, G. Benayoun, and P. Parsi. 1983. Bioavailability of technetium from artifically labelled north-east Atlantic deep-sea sediments. *Marine Environmental Research,* **8,** 87–100.

Gearing, P., C. Van Baalen, and P. L. Parker. 1975. Biochemical effects of technetium-99-pertechnetate on microorganisms. *Plant Physiology,* **55,** 240–246.

Gradmann, D. 1970. Einfluss von Licht, Temperatur und Aussenmedium auf das elektrische Verhalten von *Acetabularia crenulata. Planta,* **93,** 323–353.

Guenot, J., C. Caput, and Y. Belot. 1982. Rôle des réactions de photosynthèse et d'échange dans l'incorporation du tritium dans la matière organique des végétaux. *Annales de l'Association Belge de Radioprotection,* **7,** 293–305.

Gutknecht, J. 1961. Mechanism of radioactive zinc uptake by *Ulva lactuca*. *Limnology and Oceanography,* **6,** 426–431.

Gutknecht, J. 1963. ^{65}Zn uptake by benthic marine algae. *Limnology and Oceanography,* **8,** 31–38.

Higuchi, M., K. Komatsu, and A. Mukade. 1980. Assessment of tritium effects on brine shrimp, *Artemia salina,* reared in model ecosystem contaminated with tritiated water. *In:* Radiation Effects on Aquatic Organisms, N. Egami (Ed.). Japan Science Societies Press, Tokyo, pp. 13–25.

Hodge, V. F., F. L. Hoffman, and T. R. Folsom. 1974. Rapid accumulation of plutonium and polonium on giant brown algae. *Health Physics,* **27,** 29–35.

Holm, E., and B. Persson. 1980. Behavior of natural (Th, U) and artifical (Pu, Am) actinides in coastal waters. *In:* Third NEA Seminar on Marine Radioecology (Tokyo, 1–5 October 1979). Organisation for Economic Co-operation and Development, Paris, pp. 237–243.

Inomata, T., and M. Higuchi. 1976. Incorporation of tritium into cell materials of *Rhodopseudomonas* from tritiated water in the medium under aerobic conditions. *Journal of Biochemistry,* **80,** 569–578.

International Atomic Energy Agency (IAEA). 1975. Impacts of Nuclear Releases into the Aquatic Environment. International Atomic Energy Agency, Vienna, 524 pp.

IAEA. 1979. Behaviour of Tritium in the Environment. International Atomic Energy Agency, Vienna, 711 pp.

IAEA. 1981. Impacts of Radionuclide Releases into the Marine Environment. International Atomic Energy Agency, Vienna, 748 pp.

Issinger, O., I. Maas, and H. Clauss. 1971. Photosynthese Intensität der Stielregionen von *Acetabularia mediterranea*. *Planta,* **101,** 360–364.

Jensen, L. M. 1983. Phytoplankton release of extracellular organic carbon, molecular weight composition and bacterial assimilation. *Marine Ecology Progress Series,* **11,** 39–48.

Kirchmann, R., S. Bonotto, and G. Cantillon. 1974. Contamination des organismes vivants d'un estuaire par le radium-226 provenant d'effluents liquides d'origine industrielle. *In:* Environmental Surveillance around Nuclear Installations, Vol. 2. International Atomic Energy Agency, Vienna, pp. 167–183.

Kirchmann, R., C. Piron, M. Meurice, E. Fagniart, and C. Sironval. 1977. Le tritium des effluents liquides d'installations nucléaries. *Annales de l'Association Belge de Radioprotection,* **2,** 163–182.

Komatsu, K., M. Higuchi, and M. Sakka. 1981. Accumulation of tritium in aquatic organisms through a food-chain with three trophic levels. *Journal of Radiation Research,* **22,** 226–241.

Larsson, U., and A. Hagström. 1979. Phytoplankton exudate release as an energy source for the growth of pelagic bacteria. *Marine Biology,* **52,** 199–206.

Lateur, L., and S. Bonotto. 1973. Culture of *Acetabularia mediterranea* in the laboratory. *Bulletin de la Société Royale de Botanique de Belgique,* **106,** 17–38.

Masson, M., G. Aprosi, A. Lanièce, P. Gueguenіat, and Y. Belot. 1981. Approches experimentales de l'étude des transferts du technétium à des sédiments et à des espèces marines benthiques. *In:* Impacts of Radionuclide Releases into the Marine Environment. International Atomic Energy Agency, Vienna, pp. 341–359.

Mehran, A. R., and J. L. Tremblay. 1965. Absorption de zinc par certaines fucacées du niveau moyen de la zones des marées. *Revue Canadienne de Biologie,* **24,** 29–34.

Metzler, D. E. 1977. Biochemistry. The Chemical Reactions of Living Cells, International Edition. Academic Press, New York, 1129 pp.

Mousny, J. M., P. Roucoux, and C. Myttenaere. 1979. Absorption and translocation of technetium in pea plant. *Environmental and Experimental Botany,* **19,** 263–268.

Moghissi, A. A., and M. W. Carter. 1973. Tritium. Messenger Graphics, Publishers, Phoenix, Arizona, 807 pp.

Mukade, A., and M. Higuchi. 1974. Tritium incorporation into the cellular materials and its biological effects to *Artemia salina* reared in tritiated seawater. *Bulletin of Plankton Society of Japan,* **21,** 32–40.

Myttenaere, C., D. Daoust, and P. Roucoux. 1980. Leaching of technetium from foliage by simluated rain. *Environmental and Experimental Botany,* **20,** 415–419.

Nakahara, M., T. Koyanagi, and M. Saiki. 1975. Concentration of radioactive cobalt by seaweeds in the food chain. *In:* Impacts of Nuclear Releases into the Aquatic Environment. International Atomic Energy Agency, Vienna, pp. 301–312.

National Institute of Radiological Sciences. 1982. Proceedings of the Workshop on Tritium Radiobiology and Health Physics, H. Matzudaira, T. Yamaguchi, T. Nakazawa, and C. Sito (Eds.). National Institute of Radiological Sciences, Chiba, Japan, 318 pp.

Nienhuis, P. H. 1981. Distribution of organic matter in living marine organisms. *In:* Marine Organic Chemistry, E. K. Duursma and R. Dawson (Eds.). Elsevier Scientific Publishing Company, Amsterdam, pp 31–69.

Noshkin, V. E. 1980. Transuranium radionuclides in components of the benthic environment of Enewetak atoll. *In:* Transuranic Elements in the Environment, W. C. Hanson (Ed.). U.S. Department of Energy, Technical Information Center, Washington, D.C., pp. 578–601.

Nuclear Energy Agency–Organisation for Economic Co-operation and Development. 1980. Proceedings of the Third NEA Seminar on Marine Radioecology (Tokyo, Japan, 1–5 October 1979). Organisation for Economic Co-operation and Development, Paris, 409 pp.

Pâques, M., C. Sironval, and S. Bonotto. 1979. On chloroplast movement in the stalk of *Acetabularia mediterranea*. *In:* Developmental Biology of *Acetabularia,* S. Bonotto, V. Kefeli, and S. Puiseux-Dao (Eds.). Elsevier/North-Holland Biomedical Press, Amsterdam, pp. 179–182.

Pentreath, R. J. 1981. The biological availability to marine organisms of transuranium and other long-lived nuclides. *In:* Impacts of Radionuclide Releases into the Marine Environment. International Atomic Energy Agency, Vienna, pp. 241–272.

Pillai, K. C., R. C. Smith, and T. R. Folsom. 1964. Plutonium in the marine environment. *Nature,* **203,** 568–571.

Puiseux-Dao, S., A. C. Dazy, D. Hoursiangou-Neubrun, and H. Borghi. 1972. Observations on the apicobasal gradient of plastidal population in *Acetabularia*. *Protoplasma,* **75,** 484–485.

Rice, D. L., and K. R. Tenore. 1981. Dynamics of carbon and nitrogen during the decomposition of detritus derived from estuarine macrophytes. *Estuarine, Coastal and Shelf Science,* **13,** 681–690.

Rice, T. R. 1956. The accumulation and exchange of strontium by marine planktonic algae. *Limnology and Oceanography,* **1,** 123–138.

Rice, T. R., and V. M. Willis. 1959. Uptake, accumulation and loss of radioactive cerium-144 by marine planktonic algae. *Limnology and Oceanography,* **4,** 277–290.

Routson, R. C., and D. A. Cataldo. 1978. Accumulation of ^{99}Tc by tumbleweed and cheatgrass grown on acid soils. *Health Physics,* **34,** 685–690.

Saas, A., J. L. Denardi, C. Colle, and J. M. Quinault. 1980. Cycle du molybdène et du technétium dans l'environnement. Evolution physicochimique et mobilité dans les sols et les vegetaux. *In:* Deuxième Symposium International de Radioécologie (Cadarache, France, 19–22 June 1979). Commissariat à l'Energie Atomique, Electrictié de France, Centre d'Etudes Nucléaires, Cadarache, France, pp. 443–489.

Saddler, H. D. W. 1970a. The ionic relations of *Acetabularia mediterranea*. *Journal of Experimental Botany,* **21,** 345–359.

Saddler, H. D. W. 1970b. Fluxes of sodium and potassium in *Acetabularia mediterranea*. *Journal of Experimental Botany*, **21**, 605–616.

Saddler, H. D. W. 1970c. The membrane potential of *Acetabularia mediterranea*. *Journal of General Physiology*, **55**, 802–821.

Sepers, A. B. J. 1977. The utilization of dissolved organic compounds in aquatic environment. *Hydrobiologia*, **52**, 39–54.

Spies, R. B. 1975. Uptake of technetium from seawater by red abalone *Haliotis rufescens*. *Health Physics*, **29**, 695.

Till, J. E., F. O. Hoffman, and D. E. Dunning. 1979. A new look at ^{99}Tc releases to the atmosphere. *Health Physics*, **36**, 21–30.

Tremblay, J. L., and A. R. Mehran. 1964. L'absorption du zinc par certaines algues marines de la zone des marées. *Revue Canadienne de Biologie*, **23**, 117–122.

Vandecasteele, C. M., A. Delmotte, P. Roucoux, and C. Van Hove. 1982. Technetium et organismes diazotrophes: Toxicité, localisation, facteurs de transfert. *In:* Environmental Migration of Long-lived Radionuclides. International Atomic Energy Agency, Vienna, pp. 275–286.

Vanden Driessche, T. 1975. Circadian rhythms and molecular biology. *Biosystems*, **6**, 188–201.

Watanabe, Y. 1980. A study of the excretion and extracellular products of natural phytoplankton in Lake Nakanuma, Japan. *Internationale Revue der gesamten Hydrobiologie*, **65**, 809–834.

Wildung, R. E., T. R. Garland, and D. A. Cataldo. 1977. Accumulation of technetium by plants. *Health Physics*, **32**, 314–317.

Wildung, R. E., K. M. McFadden, and T. R. Garland. 1979. Technetium sources and behaviour in the environment. *Journal of Environmental Quality*, **8**, 156–161.

Wolter, K. 1982. Bacterial incorporation of organic substances released by natural phytoplankton populations. *Marine Ecology Progress Series*, **7**, 287–295.

Chapter 14

Effects of Radiation on Frequency of Chromosomal Aberrations and Sister Chromatid Exchange in *Neanthes arenaceodentata*

Florence L. Harrison and David W. Rice, Jr.

Environmental Sciences Division
Lawrence Livermore National Laboratory
Livermore, California

Dan H. Moore

Biomedical Sciences Division
Lawrence Livermore National Laboratory
Livermore, California

Marilyn Varela

Office of Radiation Programs
U.S. Environmental Protection Agency
Washington, D.C.

ABSTRACT

Traditional bioassays are unsuitable for assessing sublethal effects of low levels of radioactivity because mortality and phenotypic responses are not anticipated. We compared the usefulness of chromosomal aberration (CA) and sister chromatid exchange (SCE) induction as measures of low-level radiation effects in the larvae of the sediment-dwelling marine worm, *Neanthes arenaceodentata*. Results from X-ray irradiation indicated that dose-related increases occur in chromosome and chromatid deletions, but an X-ray dose ⩾2 Gy was required to observe a significant increase. Worm larvae receiving ^{60}Co irradiation showed elevated SCE frequencies; a significant increase in SCE frequency was observed at 0.6 Gy. Both SCEs and CAs may be useful for measuring effects on genetic material induced by low levels of radiation.

14.1. INTRODUCTION

Since the onset of the nuclear age, radioactive wastes have been disposed of on land and in the ocean. In addition, the testing of nuclear weapons has contributed measurable quantities of radionuclides to the ocean. In the United States, low-level, solid radioactive wastes were disposed of in the coastal areas both of the Atlantic and Pacific Oceans (Joseph et al., 1971; Hagen, 1983). Although these practices were discontinued by 1970, little effort was made until recently to determine the subsequent fate and distribution of the radionuclides in these wastes. Information now available indicates that some synthetic radionuclides, from ocean disposal, are present in bottom sediments, but that there is little or no accumulation of these radionuclides by organisms in the human food chain (Dyer, 1976; Noshkin et al., 1978).

Increased use of ocean waters for radioactive waste disposal is being considered because land disposal has become difficult and costly. Some opposition to ocean disposal of nuclear waste has been based on fear of irreparable consequences and on the lack of data documenting the effects of waste disposal into the ocean. This lack of data can be attributed in part to the absence of appropriate bioassays. Traditional bioassays that use mortality and phenotypic responses as end points are unsuitable for assessing the sublethal effects that may be expected from low-level radioactive wastes.

The U.S. Marine Protection, Research and Sanctuaries Act requires the U.S. Environmental Protection Agency, in reviewing requests for permits, to (1) determine that ocean "dumping will not unreasonably degrade or endanger human health, welfare, or amenities, or the marine environment, ecological systems, or economic potentialities" and (2) establish regulations and criteria to implement a permit program (U.S. Congress, 1972). One possible monitoring criterion utilizes a bioassay technique that requires a methodology for detecting the response of marine organisms to low levels of chronic radiation. With this methodology, post-disposal monitoring could verify the assumptions regarding doses to marine organisms and ultimately evaluate the impact of radiation on the organisms.

Deleterious effects of radiation on organisms are well documented (Templeton et al., 1976; U.S. National Academy of Sciences, 1980). Increased cell death and mutations have been related to increased radiation doses. Changes in genetic material include base damage, single-strand breaks, double-strand breaks, hydrogen-bond rupture, and cross-linking between DNA and proteins (Yu, 1976). Some lesions can be detected by examining cells in metaphase for chromosomal aberrations (CAs) and sister chromatid exchanges (SCEs). The induction of chromosomal aberrations in cells exposed to radiation is well documented (Blaylock and Trabelka, 1978) compared to information available for SCEs (Kato, 1979).

The CAs are caused by the breakage of chromosomes followed by either the subsequent rejoining of the broken ends to form new combinations or the failure of broken ends to rejoin. They are identified by changes in chromosomal structure that include deletions, translocations, and rings. It is well established that a substantial part of the changes in DNA induced by ionizing radiation consists of single- and double-strand breaks in the phospho-diester backbone of the DNA molecule. According to Evans (1977), however, all CAs do not result from one, two, or three specific lesions, but are caused by a variety of changes in the DNA that lead to helix disruption, helix distortion, or interference with the normal replication process of the cell.

The SCEs represent the interchange of DNA replication products at apparently homologous loci (Latt et al., 1981). This exchange, which does not alter the chromosome morphology, was demonstrated first by autoradiographic techniques using tritiated thymidine (Taylor, 1958). Currently, these exchanges are distinguished by exposing cells to 5-bromodeoxyuridine (BrdUrd) for two rounds of replication and by a combined staining with fluorochrome plus Giemsa (Perry and Wolff, 1974). Data on SCE frequency in cells exposed to various physical and chemical agents indicate that in mammalian cell systems, SCEs are a sensitive indicator of DNA alterations caused by environmental mutagens and carcinogens.

The effects of ionizing radiation on the frequencies of SCEs have been studied in a number of mammalian cell systems exposed to either β-rays (Gibson and Prescott, 1972; Kato, 1974); β-plus X-rays (Marin and Prescott, 1964; Gatti et al., 1974); X-rays (Perry and Evans, 1975; Yu, 1976; Galloway, 1977; Livingston and Dethlefsen, 1979; Morgan and Crossen, 1980; Geard et al., 1981; Nagasawa and Little, 1981; Popescu et al., 1981; Renault et al., 1982; Morgan et al., 1983); or γ-rays (Solomon and Bobrow, 1975; Abramovsky et al., 1978; Littlefield et al., 1979). However, very little data from these experiments are applicable to whole-animal, *in vivo* irradiation. Increases in SCEs were found following *in vivo* radiation of mice with X-rays (Nakanishi and Schneider, 1979). In this study, as in the majority of those using cell systems, dose rates much greater than those expected at oceanic disposal sites were used.

Studies are required to characterize the incidence of CAs and SCEs in organisms that have been irradiated with doses expected at radioactive waste-disposal sites. Nereid worms are indigenous to marine disposal sites used by the United States in the past, and it is expected that they will be present in future designated areas as well. Because they are benthic and do not migrate, they are well suited to studies of radionuclides and other contaminants that become associated with sediments.

Pesch and Pesch (1980a) proposed that the marine polychaete *Neanthes arenaceodentata* be used as an *in vivo* cytogenetic model for marine genetic toxicology. *Neanthes arenaceodentata* is suitable for cytogenetic studies because it has 18 large chromosomes. This is in contrast to many invertebrates and fishes that have large numbers of small chromosomes. The effects of ionizing radiation on this species have been assessed by quantifying the number of CAs induced by ^{60}Co radiation; at a dose rate of 7.5×10^{-2} Gy h^{-1} and a total dose of 1.8 Gy an increase in CAs was found (Pesch et al., 1981). Also, a preliminary study of *N. arenaceodentata* has been performed to determine the usefulness of SCE induction as a measure of low-level radiation effects (Harrison and Rice, 1981). Larvae exposed to ^{60}Co radiation at intermediate doses (0.1–0.6 Gy) had SCE frequencies about two times that of the control larvae, but those exposed to high doses (1.6–3.0 Gy) had SCE frequencies that were approximately those of the control larvae.

In the present study, our objective was to assess the feasibility of a cytogenetic approach to detect alterations caused by radiation in the genetic material of a marine organism. We evaluated the responses to irradiation by using the classic cytogenetic approach of quantifying the frequency of CAs and by using the more recently developed technique of quantifying the frequency of SCEs.

14.2. EXPERIMENTAL

14.2.1. Worm Culturing and Handling

Neanthes arenaceodentata were cultured following methods recommended by Reish (1974). Their life cycle is well known and is completed in 3–4 months at 20–22°C (Reish, 1957). Mated pairs of adult worms were shipped by Reish in inflated plastic bags containing ~100 ml of seawater. The worms were shipped in the tubes they had constructed from algae they were fed; only one death occurred during shipping.

On arrival at the laboratory, each mated pair of adult worms was placed in a 4-liter glass beaker. The adult worms that produced larvae were cultured using semistatic conditions; the water was aerated continuously and three-fourths of the volume in the beakers was exchanged weekly. Thereafter, adult worms were cultured in 2-liter beakers with flow-through conditions; flow rate through the beakers was 100 ml min^{-1}. Adult worms were maintained at a mean culture temperature of 19.4 ± 1.4°C for 20–30 d in our laboratory before larvae were harvested. The adult worms were fed frozen *Enteromorpha* sp. ad libitum, and uneaten food was removed weekly.

14.2.2. Irradiation

X-ray irradiation was conducted in plastic 100- × 20-mm petri dishes containing 10 ml of seawater. Three thermoluminescent dosimeters (TLDs) were placed in the water with worms to determine the X-ray dose delivered. Larvae harvested 1–3 d after hatching (3- to 5-setiger stage) were irradiated with X-rays delivered at a rate of 0.7 Gy min^{-1} from a 40-keV X-ray machine. The larvae were then examined for CA induction. The doses ranged from 0.08 to 3.8 Gy, and a 28-h BrdUrd exposure time was used to obtain the high proportion of first division cells required for CA scoring.

Exposures to ^{60}Co were for 24 h at rates that varied from 4.8×10^{-5} to 1.2×10^{-1} Gy h^{-1}; total doses received varied from 0.01 to 3 Gy. All irradiations were conducted in our low-level radiation facility equipped with a 4.4×10^{10} Bq ^{60}Co source. For each exposure, 50–75 worm larvae harvested from 1–3 broods were placed in a cylindrical plastic chamber (2.5 cm in diameter) containing 30 ml of filtered seawater. A plexiglass sheet (5 × 7 × 0.6 cm) was placed in front of each exposure chamber to ensure electron equilibrium. Different dose rates and total doses were obtained by varying the distance between the chamber and the source. Delivered dose was determined from three TLDs placed behind each exposure chamber. A 54-h BrdUrd exposure time was used to obtain the high proportion of second division cells required for SCE scoring.

Two groups of control worms were tested during each experiment. Neither group was irradiated, but one was treated with 5×10^{-7} *M* mitomycin C (MMC), a drug known to increase the frequencies of SCEs; this group served as a positive control. Both controls were maintained in the exposure facility during the irradiation of the other groups of worms.

14.2.3. Cytogenetic Preparation and Scoring of Worm Chromosomes

Immediately following irradiation for experiments 1–3, each treatment group of larvae was transferred under amber light with a large-bore plastic pipette to 100- × 20-mm plastic culture dishes. Each dish contained 30 ml of 3×10^{-5} *M* BrdUrd in filtered seawater. The same procedure was followed in experiments 4–30, except 50 ml of seawater in 100-ml glass beakers was used. Nonirradiated control groups were also transferred to the same concentration of BrdUrd or to BrdUrd plus 5×10^{-7} *M* MMC. The BrdUrd and MMC exposures were carried out in the dark, and colchicine (final concentration of 0.4 mg ml^{-1}) was added to the seawater 4 h before the termination of the BrdUrd exposure. Colchicine is a microtubule disruptor that results in the accumulation of cells in metaphase.

The method of larvae harvest and larval tissue preparation developed by Pesch and Pesch (1980b) was followed. Larvae were transferred to 15-ml conical plastic tubes, the seawater decanted, and 10 ml of 0.075 *M* potassium chloride added. After 12 min, this solution was decanted and the larvae were fixed in three changes of methanol plus acetic acid (3:1). The first fixative change was performed after 5 min and the remaining changes after 15 min. Fixed larvae (50–75) were placed in the depression of a ceramic spot dish and mashed twice with broad-tipped forceps. Next, 1 ml of 60% acetic acid was added, and the mixture was mashed continuously for 1 min. Two drops of the worm tissues suspended in acetic acid were deposited on the end of a clean microscope slide held at 45°C. Using a disposable plastic pipet, 10–15 successive transfers of the original drops of tissue suspension to clean areas of the slide were made. This process resulted in the deposition of cells in a series of rings along the length of the slide. The slides were dried at 45°C before staining. Generally, 4 slides could be made from the macerated tissues of 50–75 larvae. The slides were prepared within 1 h of the start of fixation to ensure spreading of chromosomes.

Differential staining of the sister chromatids was accomplished according to the procedure described by Minkler et al. (1978). Preparations were first stained for 10 min in 5μg ml^{-1} Hoechst 33258 solution (Aldrich Chemical Company, Inc., Milwaukee, Wisconsin) in 0.9% sodium chloride (pH 6). Hoechst-stained slides were rinsed for 5 min in distilled water and air dried for at least 20 min. They were then placed in a shallow, clear plasic tray and covered with 0.067 *M* phosphate buffer (pH 6.8) to a depth of 5 mm. Slides were exposed to ultraviolet light in an M-99 printer (400-W General Electric Mercury Lamp, Colight, Inc., Minneapolis, Minnesota) for 45 min. They then were transferred to 10% Giemsa stain in 0.067 *M* phosphate buffer for 6–10 min, air dried, and mounted with coverslips using Permount (Fischer Scientific Company, Fairlawn, New Jersey). Worm tissue fixation, slide preparation, and staining were all carried out under amber light.

Slides were scored by scanning the entire slide with a Zeiss Universal microscope (Carl Zeiss, Inc., Oberkochen, West Germany) equipped with a 10X objective, 63X objective, 1.25X optovar, and 12.5X oculars. All slides were scored blind, and four people performed the scoring.

The proportions of metaphases identified as first, second, and third divisions after the beginning of BrdUrd exposure were recorded. First-division metaphases were examined for the number of CAs and second-division metaphases for the number of SCEs per metaphase. The number of chromosomes scored was recorded for all metaphases examined. Data for CAs were recorded only for cells that had 17 or 18 chromosomes that could be scored; data for SCEs were recorded for those that had 15–18 chromosomes that could be scored.

14.3. RESULTS

14.3.1. Chromosomal Aberrations

Almost all CAs induced by X-rays in cells of *N. arenaceodentata* larvae were chormosomal gaps and chromosome and chromatid deletions (CDs); however, only data on CDs are included because the scoring of gaps is subjective.

The frequency of CDs in cells that received no irradiation was low (Table 14.1). The mean number of CDs per cell for the 14 control experiments was 0.06, and the individual means from the experiments ranged from 0.00 to 0.22. Differences in the frequencies of CDs reported by different scorers and for different slides were tested for homogeneity of binomial proportions using Cochran's test (Snedecor and Cochran, 1967). There was excellent agreement between scorers when the same slides were scored independently; all differences between scorers were easily accounted for by binomial sampling variability. There was also good agreement in frequencies reported for different slides within the same experiment; differences did not exceed those expected from binomial sampling. In contrast to these results, there was significant heterogeneity among the CD rates from different experiments [$p = 0.0016$ based on Cochran's test for homogeneous binomial proportions (Snedecor and Cochran, 1967)].

Table 14.1. **Chromosome and Chromatid Deletions (CDs) in Cells from *N. arenaceodentata*[a]**

Experiment Number	Number of Scorers[b]	Cells Scored[c]	Number of CDs[c]	CDs per Cell
1	3	78	0	0.00
2	3	33	0	0.00
3	3	62	2	0.032
4	3	28	6	0.21
5	1	30	1	0.033
6	3	50	5	0.10
7	3	50	3	0.060
8	2	51	11	0.22
9	1	9	0	0.00
10	3	11	2	0.18
11	2	41	2	0.049
12	1	89	5	0.056
20	1	16	1	0.062
21	3	95	1	0.010

[a] Nonirradiated larvae exposed to BrdUrd for 28 h.
[b] Four scorers were available for the experiments.
[c] For slides scored by more than one person for an individual experiment, the number of cells scored and CDs reported were averaged.

Table 14.2. **Chromosome and Chromatid Deletions (CDs) in Cells from *N. arenaceodentata*[a]**

Dose (Gy)	Experiment Number	Number of Scorers[b]	Cells Scored[c]	Number of CDs[c]	CDs per Cell
0.08	2	1	41	2	0.049
0.18	5	1	30	2	0.067
0.19	4	1	54	2	0.037
0.22	10	1	29	3	0.10
0.24	2	1	40	8	0.20
0.37	5	1	109	4	0.037
0.45	4	2	22	1	0.045
0.47	2	2	28	2	0.071
0.48	1	2	112	12	0.11
0.88	6	2	66	5	0.076
0.88	12	2	95	7	0.074
1.0	1	2	57	10	0.18
1.6	8	3	12	2	0.17
1.7	12	3	52	7	0.13
2.0	7	3	21	12	0.57
2.2	8	3	61	10	0.26
2.5	3	3	48	36	0.75
2.6	12	2	51	11	0.22
3.6	11	3	34	12	0.35
3.8	12	2	56	14	0.25

[a] Larvae exposed to X-rays and then BrdUrd for 28 h.
[b] Four scorers were available for the experiments.
[c] For slides scored by more than one person for an individual experiment, the number of cells scored and CDs reported were averaged.

The frequency of CDs was determined also in cells exposed to X-rays (Table 14.2). Weighted (by the number of cells scored) least-squares linear regression was used to determine if there was a linear relationship between dose and CD rate. The regression was highly significant ($F = 76.6$ with 1 and 14 degrees of freedom and $p < 0.01$), and the estimated slope was 0.094 CDs $cell^{-1}$ Gy^{-1} of radiation (standard error = 0.017). However, a prediction based on the least-squares best-fit line is not very reliable because of the extreme heterogeneity of the reponses at each dose level. Nevertheless, it appears that doses above 2.0 Gy lead to increased frequencies of CDs.

In the experimental cells scored, the number of CDs per individual cell ranged from one to greater than four. The percentage of the total CDs that occurred singly (one per cell) and multiply (more than one per cell) was compared to the total dose delivered (Table 14.3). The fraction of the total CDs scored that occurred as more than one CD per cell did not appear to be dose-related.

14.3.2. Sister Chromatid Exchange

The frequency of SCEs in cells from worm larvae exposed to BrdUrd for 54 h, but not to radiation, was determined in 18 different experiments. The SCE frequency per chromosome for the individual experiments ranged from 0.096 to 0.38 (Table 14.4). These experiments were performed over a 17-month period. Occasionally, larvae were harvested after varying BrdUrd exposure times (including 54 h). Using data from larvae harvested at times ranging from 28 to 66 h, we determined that the cell cycle time was about 28 h and had little effect on SCE frequency (Harrison et al., 1984).

An analysis of individual results revealed that at zero dose, the experiment to-experiment variability was large compared with the variability between scorers and slides. The standard deviation from scorer to scorer was 0.033, from slide to slide 0.037, and from experiment to experiment 0.056.

The frequency distribution of cellular SCEs was skewed, with medians consistently lower than mean SCEs because of the presence of variable numbers of high-frequency cells (HFCs) (Fig. 14.1). The HFC was defined by pooling all SCEs from the controls (1059 cells from 18 experiments) and finding the 90th percentile of this pooled distribution. In order to have 95% confidence that the estimated percentile would contain 90% of the cells, we used the nonparametric procedure described by Walsh (1962) and found the *k*th-

Table 14.3. Chromosome and Chromatid Deletions (CDs) Occurring as One, Two, Three, or Four CDs per Cell in *N. arenaceodentata*[a]

Dose (Gy)	Cells Scored	Number of CDs	Percent of CDs (per cell) 1 CD	2 CD	3 CD	≥4 CD
Control	643	39	63	21	16	0
0.08	41	2	100	0	0	0
0.18–0.24	153	15	60	40	0	0
0.37	109	4	100	0	0	0
0.45–0.48	164	15	60	13	0	27
0.88–1.0	228	22	60	24	0	16
1.6–1.7	62	9	57	0	0	43
2.0–2.2	82	28	63	15	22	0
2.5–2.6	99	47	42	20	38	0
3.6–3.8	90	26	56	25	0	19

[a] Larvae exposed to X-rays and to BrdUrd for 28 h.

Table 14.4. Sister Chromatid Exchanges (SCEs) per Chromosome in Cells from *N. arenaceodentata*[a]

Experiment Number	Mean SCEs per Chromosome	Standard Deviation	Median SCEs per Chromosome	Cells Scored	High Frequency Cells (%)[b]
13	0.096	0.08	0.09	24	0
14	0.20	0.22	0.14	18	16.7
15[c]	0.27	0.22	0.20	21	14.3
16	0.13	0.10	0.13	33	0
17	0.22	0.15	0.18	65	4.6
18	0.30	0.20	0.25	52	26.9
19	0.18	0.19	0.13	40	7.5
20	0.19	0.15	0.18	9	11.1
21	0.13	0.09	0.11	12	0
22	0.17	0.14	0.12	32	3.1
23	0.12	0.10	0.089	52	0
24	0.22	0.25	0.17	21	9.5
25	0.26	0.24	0.17	24	20.8
26	0.15	0.12	0.11	67	1.5
27	0.18	0.13	0.17	73	4.1
28	0.38	0.51	0.17	20	20
29	0.18	0.17	0.17	22	9.1
30	0.17	0.32	0.063	18	11.1

[a] Nonirradiated (control) larvae exposed to BrdUrd for 54 h.
[b] High-frequency cells are cells with more than 0.44 SCEs per chromosome.
[c] Larvae harvested after a 48-h exposure to BrdUrd.

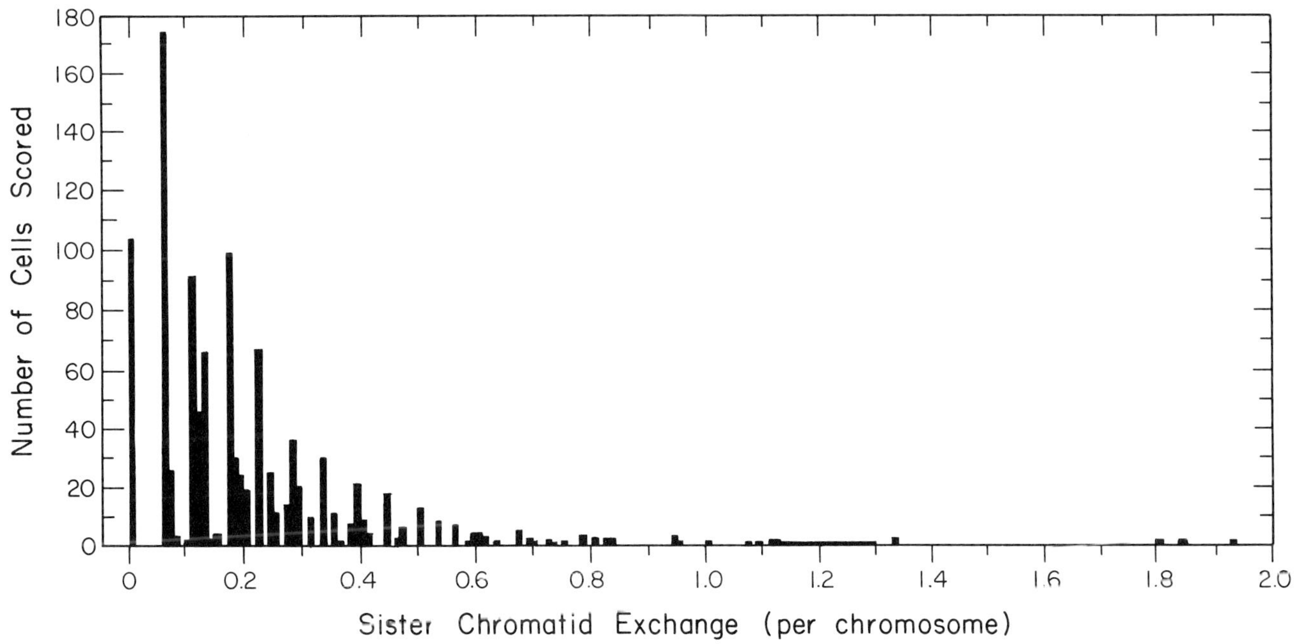

Figure 14.1. The sister chromatid exchange frequency distribution observed in cells of *N. arenaceodentata* larvae that received no irradiation (1057 cells from 18 experiments).

largest SCE value from the pooled sample, where k is given by

$$k = 1059(1.0-0.90) + 0.5 - 1.645\sqrt{1059(0.90)(0.10)} \quad (1)$$

where 1059 is our sample size and 0.90 is the percentile expressed as a fraction. The 1.645 comes from the 95th percentile of the standard normal distribution. The 90th-largest SCE frequency in our sample was 0.44 SCE per chromosome. Thus, we define an HFC as a cell with more than 0.44 SCE per chromosome.

There is clearly variability in the numbers of HFCs in our control samples. Fortunately, the mean SCEs are normally distributed with an overall mean (weighted by number of cells scored) of 0.19 SCEs per chromosome and a weighted standard deviation of the means of 0.056. Normality of the means was tested using Filliben's order statistic correlation test (Filliben, 1975); a value of 0.988 was obtained, which is well above the 5% critical value of 0.938.

Worm larvae exposed to different doses of ^{60}Co and then examined for SCE induction after a 54-h exposure to BrdUrd had mean frequencies of SCE that varied with the total dose (Table 14.5). Analysis of results from individual experiments conducted at comparable doses revealed again that there was significant variability from experiment to experiment. This was caused mainly by significantly higher proportions of HFCs in a few of the experiments.

The frequencies of SCEs in cells from larvae receiving 0.6 Gy of ^{60}Co radiation were clearly different from those not receiving radiation (Fig. 14.2). The frequency distribution is similiar to that of larvae exposed to MMC; both groups are characterized by having increased incidences of HFCs.

A least-squares linear regression of mean SCE on radiation dose was performed. When a weighted (by number of cells scored) linear regression was performed for all doses, the slope of the best-fit line was not significantly different from zero. This was caused by the low responses at the two highest doses. When these data pairs were omitted, a significant ($p = 0.025$) slope resulted (0.241 ± 0.008 increase in SCE per chromosome per Gy). Again, the best-fit line was not very useful for predicting responses because of the large experiment-to-experiment variability; however, the significant regression indicates that over this range of radiation there is a general rise in SCE frequency with increased radiation dose. A significant increase in SCE frequency occurred at a dose of 0.6 Gy but not at 0.3 Gy (Bonferroni t-test adjusted

Table 14.5. Sister Chromatid Exchanges (SCEs) per Chromosome in Cells from *N. arenaceodentata*[a]

Dose (Gy)	Experiment Number	Mean SCEs per Chromosome	Standard Deviation	Median SCEs per Chromosome	Cells Scored	High Frequency Cells (%)[b]
0.001	22	0.22	0.22	0.17	81	12.3
0.001	23	0.17	0.13	0.13	42	2.4
0.001	24	0.16	0.15	0.11	29	6.3
0.001	25	0.17	0.14	0.12	10	0
0.01	26	0.17	0.14	0.17	152	5.3
0.1	15	0.42	0.34	0.44	25	32
0.1	27	0.17	0.19	0.12	140	5.7
0.3	13	0.55	0.49	0.47	24	50
0.3	14	0.16	0.18	0.11	15	13.3
0.3	15	0.34	0.31	0.19	25	32
0.3	16	0.12	0.11	0.11	49	2
0.3	17	0.23	0.23	0.12	19	10.5
0.3	18	0.30	0.27	0.26	12	25
0.3	19	0.17	0.22	0.13	32	3.1
0.3	21	0.20	0.22	0.12	22	13.6
0.6	13	0.41	0.54	0.13	25	24
0.6	14	0.20	0.28	0.06	22	18.2
0.6	15	0.47	0.39	0.33	32	40.6
0.6	17	0.37	0.22	0.39	11	36.4
0.6	18	0.32	0.36	0.22	17	23.5
0.6	19	0.15	0.09	0.12	23	0
0.6	20	0.46	0.60	0.24	15	33.3
0.6	21	0.25	0.22	0.20	6	16.7
1.7	13	0.24	0.21	0.16	10	10
1.7	14	0.15	0.12	0.13	13	0
3.0	13	0.24	0.40	0.17	23	4.3
3.0	14	0.18	0.27	0.17	31	6.5

[a] Larvae were irradiated with ^{60}Co and exposed to BrdUrd for 54 h.
[b] High-frequency cells are cells with more than 0.44 SCEs per chromosome.

for seven multiple comparisons; $p = 0.0003$) (Miller, 1966). The absence of a significant difference from control at 0.3 Gy may be a false negative. A difference may have been detected if the sample size were larger.

14.4. DISCUSSION

The induction of CAs by irradiation has been demonstrated both *in vivo* and *in vitro* in many mammals and mammalian cell systems (Gebhart, 1981). Although the response to radiation is dose related in mammalian systems, the rate of induction differs with the test system and the cell cycle at the time of irradiation. In the worm larvae, we found that the response was dose-related also. The CAs that we quantified, CDs, are one-break aberrations that are considered to be induced linearly with dose. Our studies on the background levels of CA induction indicate that in worm larvae, as in mammalian systems, the incidence of CAs is low. Neither the incidence of true point (intragenic) mutations, which are base-pair changes in DNA, nor gross chromosomal (polygenic) mutations brought about by the breakage of chromosomes is easily quantified at low radiation doses. This is because very few of these events are found, and it is difficult to get a significant increase over background values unless large numbers of cells are scored.

A reliable method that uses peripheral lymphocytes for determining the exposure to ionizing radiation in humans has been developed. Because these cells are long-lived and nondividing, aberrations accumulate. Because few lymphocytes cycle *in vivo*, aberrations can persist rather than be converted to lethal events as a result of genetic imbalance in cell division. These cells have been shown to act as an integrating dosimeter for ionizing radiation. We suggest that a comparable integrating dosimeter for CAs should be looked for in the worm. Blood cells, reproductive cells, or cells that accumulate at a site of injury or excision in preparation for regeneration may be useful for this purpose.

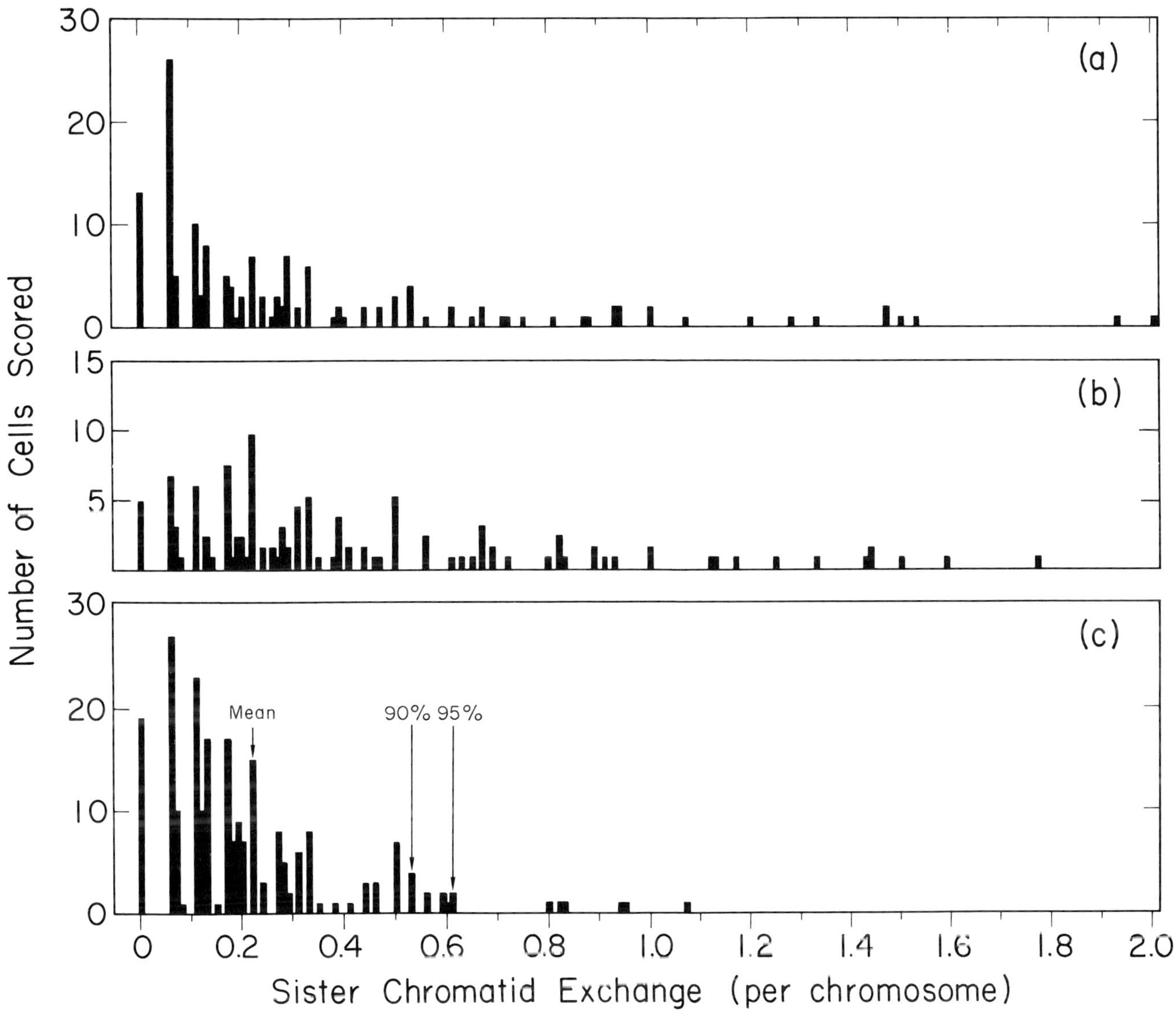

Figure 14.2. The sister chromatid exchange frequency distributions observed in cells of *N. arenaceodentata* larvae that (a) received 0.6 Gy of ^{60}Co radiation, (b) were treated with 5 × 10^{-7} *M* mitomycin C (positive control), and (c) received no irradiation (control). Larvae were harvested after a 54-h exposure to BrdUrd. Mitomycin C was present throughout these 54 h; the irradiation preceeded BrdUrd exposure. The total number of cells scored for irradiated, positive-control, and control larvae was 152, 143, and 241, respectively, and the mean SCE per chromosome was 0.34, 0.40, and 0.22, respectively.

The induction of SCEs is a sensitive indicator of changes due to mutagenic chemicals. Data available from mammalian cells indicate that the response is more sensitive to chemicals than to ionizing radiation (Gebhart, 1981). In contrast, our results for worm larvae show that a response significantly different from those of controls was obtained at a lower radiation dose using the SCE rather than the CA frequency as an end point. However, the SCE response to doses ≥1.7 Gy was not related to dose; the frequency of SCEs appeared to plateau or decline. A plateau in the SCE induction rate in mammals was found for β-radiation dose rates between 0.01 and 0.014 Gy h^{-1} (Gibson and Prescott, 1972; Kato, 1979). More recently, however, Nagasawa and Little (1981) reported a dose-related response in SCE frequencies in the density-inhibited plateau phase of cultures of mouse 10T-½ cells irradiated with X-rays. At high doses, the existence of a plateau or a decline in SCE induction may limit the usefulness of *N. arenaceodentata* as an indicator of *in vivo* environmental exposure to radiation.

The usefulness of SCE frequency in tissues as an integrating dosimeter for chronic exposure of mammalian cells has not yet been established. For those chemical mutagens that are S (the phase of DNA replication) dependent, DNA repair can potentially remove adducts before the cells enter S phase, resulting in increased variability of response. An S dependency means that the substance must be present during replicative DNA synthesis, or that the lesions it produces in the chromatin require DNA synthesis to be translated into a structured change. Because the induction of SCEs by radiation is S dependent, such increased variability can be expected. We still, however, suggest considering the use of SCE induction in cells as a dosimeter while searching for an integrating system for CAs.

At the Windscale (now called Sellafield) radioactive waste disposal site in the Irish Sea, the maximum possible dose rate is estimated to be about 4.5×10^{-4} Gy h^{-1} (Woodhead, 1980). The lowest dose rate that we used in our study that gave a response statistically greater than that of the control was the induction of SCEs by 2.5×10^{-2} Gy h^{-1} of ^{60}Co (0.6-Gy total dose). Although a small increase in mean SCE frequency was observed at 4.8×10^{-4} Gy h^{-1} of ^{60}Co (0.01-Gy total dose), no significant increase in the number of HFCs was found. The maximum time over which the doses were delivered in our system was 24 h; in natural systems, doses delivered may be over the life span of the organism. Increased sensitivity of our bioassay could be obtained by increasing the duration of the exposure to the low dose rates or by reducing the high variability in background SCE frequencies.

Much effort has been directed toward understanding the mechanisms of CA and SCE formation and the relationships between the two types of alterations in genetic material (Gebhart, 1981; Carrano and Moore, 1982). It has been shown that chemicals attach to DNA and produce a variety of lesions that can vary from chemical to chemical (Wolff, 1982). With several chemicals, both SCEs and mutations increase linearly with the dose; thus, for a given chemical, the ratio of SCEs to mutations is constant over a large dose range (Carrano et al., 1978, 1980). Because this ratio changes for each chemical, it may indicate that of the multitude of lesions produced for a given chemical some could lead to SCEs and others to mutations, or that the lesions that lead to the induction of mutations are a subset of those that produce SCEs (Carrano and Thompson, 1982). Wolff (1982) states that, in any case, the induction of SCEs shows that DNA is being affected and that SCEs are an indicator of damage.

According to Wolff (1982),

> Most geneticists agree that induced mutations are detrimental and, therefore, that any general increase in the mutation rate will also be detrimental. The reasons for this are both empirical and theoretical. For instance, radiation-induced mutations in plants, fruit flies, or any other system that is favorable for genetic analysis usually lead to reduced fitness, i.e., are lethal or semilethal. This makes theoretical sense because all living organisms are the result of eons of evolution and have been selected to fit their particular ecological niche; mutations, which are random changes in the genetic constitution of the organisms, can upset the balance brought about by natural selection. There is, however, a problem in determining exactly how detrimental the effects of mutations will be and how much damage really will be done, especially after low doses.

Because of the uncertainties in our ability to predict the consequences of changes in chromosomes induced by low levels of irradiation, the presence of these changes should be used currently only to signal potential problems in a population. As more data on radiation effects at low doses become available, we may be able to relate changes in chromosomes to changes in populations and, in turn, to those in communities.

14.5. CONCLUSIONS

Nereid worms, because of their karyotype and life style, are a good cytogenetic model for studying radiation effects on benthic organisms. We used *N. arenaceodentata* larvae to characterize the rates of induction of CDs and SCEs from exposure to ionizing radiation.

The induction of CDs by X-rays delivered at 0.7 Gy min^{-1} was dose-related, but doses ≥ 2.0 Gy were required to obtain a response that was significantly different from background. Before this bioassay can be applied to conditions that exist at low-level radioactive waste disposal sites, increased sensitivity is required. We suggest that increased sensitivity of the response in the worm be achieved by identifying a long-lived cell system, similar to that of the lymphocytes in mammals, that can be used as an integrating dosimeter for CAs.

The induction of SCEs by ^{60}Co irradiation appeared to be dose-related at total doses ≤0.6 Gy, but appeared to be unrelated to higher doses. A response that was significantly different from controls was obtained at 0.6 Gy. The 0.6-Gy dose is lower than the 2.0-Gy dose required to obtain significant difference for CD frequencies, but higher than the dose that would be present at most radioactive waste disposal sites. Also, the dose over which SCE induction may be used as an indicator of environmental exposure is limited because of the decline in the dose response at higher doses. Before this bioassay is applied to field conditions, factors producing the decline of SCEs at higher doses need to be identified. Furthermore, we suggest that the required increased sensitivity of the SCE response be obtained either by decreasing the variability in the response at low doses or by identifying a cell system that can be used as an integrating dosimeter for SCEs.

ACKNOWLEDGMENTS

The authors thank Drs. Anthony Carrano and Irene Jones for their technical advice and Cathy Adams, Lynee Adams, Roger Martinelli, and Gay Story for their assistance in scoring slides. This work was funded by the U.S. Environmental Protection Agency Office of Radiation Programs and performed under the auspices of the U.S. Department of Energy by Lawrence Livermore National Laboratory under Contract W-7405-Eng-48.

REFERENCES

Abramovsky, I., G. Vorsanger, and K. Hirschhorn. 1978. Sister-chromatid exchange induced by X-ray of human lymphocytes and the effect of L-cysteine. *Mutation Research*, **50**, 93–100.

Blaylock, B. G., and T. R. Trabelka. 1978. Evaluating the effects of ionizing radiation on aquatic organisms. *Advances in Radiation Biology*, **7**, 103–152.

Carrano, A. V., and D. H. Moore. 1982. The rationale and methodology for quantifying sister chromatid exchange in humans. *In:* Mutagenicity: New Horizons in Genetic Toxicology. Academic Press, New York, pp. 267–304.

Carrano, A. V., and L. H. Thompson. 1982. Sister chromatid exchange and gene mutation. *Cytogenetics and Cell Genetics*, **33**, 57–61.

Carrano, A. V., L. H. Thompson, P. A. Lindl, and J. L. Minkler. 1978. Sister chromatid exchanges as an indicator of mutagenesis. *Nature*, **271**, 551–553.

Carrano, A. V., J. L. Minkler, D. G. Stetka, and D. H. Moore. 1980. Variation in the baseline sister chromatid exchange frequency in human lymphocytes. *Environmental Mutagenesis*, **2**, 325–337.

Dyer, R. S. 1976. Environmental surveys of two deep-sea radioactive waste disposal sites using submersibles. *In:* International Symposium on the Management of Radioactive Wastes from the Nuclear Fuel Cycle. IAEA-Su-207/65, International Atomic Energy Agency, Vienna, pp. 22–26.

Evans, H. J. 1977. Molecular mechanisms in the induction of chromosome aberrations. *In:* Progress in Genetic Toxicology, D. Scott, B. A. Bridges, and F. H. Sobels (Eds.). Elsevier, North-Holland Biomedical Press, New York, pp. 57–74.

Filliben, J. J. 1975. The probability plot correlation coefficient test for normality. *Technometrics*, **17**, 111–117.

Galloway, S. M. 1977. Ataxia Telangiectasia: the effects of chemical mutagens and X-rays on sister chromatid exchanges in blood lymphocytes. *Mutation Research*, **45**, 343–349.

Gatti, M., S. Pimpinelli, and G. Olivieri. 1974. The frequency and distribution of iso-labelling in Chinese hamster chromosomes after exposure to X-rays. *Mutation Research*, **23**, 229–238.

Geard, C. R., M. Rutledge-Freeman, R. C. Miller, and C. Borek. 1981. Antipain and radiation effects on oncogenic transformation and sister chromatid exchanges in Syrian hamster embryo and mouse C3H/10T-1/2 cells. *Carcinogensis*, **12**, 1229–1233.

Gebhart, E. 1981. Sister chromatid exchange (SCE) and structural chromosome aberration in mutagenicity testing. *Human Genetics*, **58**, 235–254.

Gibson, D. A., and D. M. Prescott. 1972. Induction of sister chromatid exchanges in chromosomes of rat kangaroo cells by tritium incorporated into DNA. *Experimental Cell Research*, **74**, 397–402.

Hagen, A. A. 1983. History of low-level radioactive waste disposal in the sea. *In.* Wastes in the Ocean, Vol. 3: Radioactive Wastes and the Ocean, P. K. Park, D. R. Kester, I. W. Duedall, and B. H. Ketchum (Eds.). Wiley-Interscience, New York, pp. 47–64.

Harrison, F. L., and D. W. Rice, Jr. 1981. Effects of Low ^{60}Co Dose Rates on Sister Chromatid Exchange Incidence in the Benthic Worm *Neanthes arenaceodentata*. UCRL-53205, Lawrence Livermore National Laboratory, Livermore, California, 23 pp.

Harrison, F. L., D. W. Rice, Jr., and D. H. Moore. 1984. Induction of Chromosomal Aberrations and Sister Chromatid Exchanges in the Benthic Worm, *Neanthes arenaceodentata* Exposed to Ionizing Radiation. UCRL-53524, Lawrence Livermore National Laboratory, Livermore, California, 57 pp.

Joseph, A. B., P. F. Gustafson, I. R. Russell, E. A. Schuert, H. L. Volchok, and A. Tamplin. 1971. Sources of radioactivity and their characteristics. *In:* Radioactivity in the Marine Environment. National Academy of Sciences, Washington, D.C., pp. 6–41.

Kato, H. 1974. Sponaneous sister chromatid exchanges detected by a BUdR-labelling method. *Nature,* **251,** 70–72.

Kato, H. 1979. Spontaneous and induced sister chromatid exchanges as revealed by the BUdR-labelling method. *International Review of Cytology,* **49,** 55–97.

Latt, S. A., J. Allen, S. Bloom, A. Carrano, E. Falke, D. Kram, E. Schneider, R. Schreck, R. Tice, B. Whitfield, and S. Wolff. 1981. Sister chromatid exchanges: a report of the GENE-TOX program. *Mutation Research,* **87,** 17–62.

Littlefield, L. G., S. P. Colyer, E. E. Joiner, and R. J. DuFrain. 1979. Sister chromatid exchanges in human lymphocytes exposed to ionizing radiation during G Zero. *Radiation Research,* **78,** 514–521.

Livingston, G. K., and L. A. Dethlefsen. 1979. Effects of hyperthermia and X-irradiation on sister chromatid exchange (SCE) frequency in Chinese hamster ovary (CHO) cells. *Radiation Research,* **77,** 512–520.

Marin, G., and P. M. Prescott. 1964. The freqency of sister chromatid exchanges following exposure to varying doses of H^3-thymidine or X-rays. *Journal of Cell Biology,* **21,** 159–167.

Miller, R. G. 1966. Simultaneous Statistical Inference. McGraw-Hill, New York, p. 62.

Minkler, J., D. Stetka, Jr., and A. V. Carrano. 1978. An ultraviolet light source for consistent differential staining of sister chromatids. *Stain Technology,* **53,** 359–360.

Morgan, W. F., and P. E. Crossen. 1980. X-irradiation and sister chromatid exchange in cultured human lymphocytes. *Environmental Mutagenesis,* **2,** 140–155.

Morgan, W. F., J. L. Schwartz, J. P. Murnane, and S. Wolff. 1983. Effect of 3-aminobenzamide on sister chromatid exchange frequency in X-irradiated cells. *Radiation Research,* **93,** 567–571.

Nagasawa, H., and J. B. Little. 1981. Induction of chromosome aberrations and sister chromatid exchanges by X-rays in density-inhibited cultures of mouse 10T-1/2 cells. *Radiation Research,* **87,** 538–551.

Nakanishi, Y., and E. L. Schneider. 1979. In vivo sister-chromatid exchange: a sensitive measure of DNA damage. *Mutation Research,* **60,** 329–337.

Noshkin, V. E., K. M. Wong, T. A. Jokela, R. J. Eagle, and J. L. Brunk. 1978. Radionuclides in the Marine Environment near the Farallon Islands. UCRL-52381, Lawrence Livermore Laboratory, Livermore, California, 17 pp.

Perry, P., and H. J. Evans. 1975. Cytological detection of mutagen–carcinogen exposure by sister chromatid exchange. *Nature,* **258,** 121–125.

Perry, P., and S. Wolff. 1974. New Giemsa method for the differential staining of sister chromatids. *Nature,* **251,** 156–158.

Pesch, G. G., and C. E. Pesch. 1980a. *Neanthes arenaceodentata* (Polychaeta: Annelida), a proposed cytogenetic model for marine genetic toxicology. *Canadian Journal of Fisheries and Aquatic Sciences,* **37,** 1225–1228.

Pesch, G. G., and C. E. Pesch. 1980b. Chromosome complement of the marine worm, *Neanthes arenaceodentata* (Polychaeta: Annelida). *Canadian Journal of Fisheries and Aquatic Sciences,* **37,** 286–288.

Pesch, G. G., J. S. Young, and M. Varela. 1981. Effects of Ionizing Radiation on the Chromosomes of the Marine Worm, *Neanthes arenaceodentata.* Contribution No. 198, Environmental Research Laboratory, U. S. Environmental Protection Agency, Narragansett, Rhode Island, 13 pp.

Popescu, N. C., S. C. Amsbaugh, and J. A. DiPaolo. 1981. Relationship of carcinogen-induced sister chromatid exchange and neoplastic cell transformation. *International Journal of Cancer,* **28,** 71–77.

Reish, D. J. 1957. The life history of the polychaetous annelid *Neanthes caudata* (delle chiaje), including a summary of development in the family Nereidae. *Pacific Science,* **11,** 216–228.

Reish, D. J. 1974. The establishment of laboratory colonies of polychaetous annelids. *Thalassia Jugoslavica,* **10,** 181–195.

Renault, G., A. Gentil, and I. Chouroulinkov. 1982. Kinetics of induction of sister-chromatid exchanges by X-rays through two cell cycles. *Mutation Research,* **94,** 359–368.

Snedecor, G. W., and W. G. Cochran. 1967. Statistical Methods, 6th Edition. The Iowa State University Press, Ames, Iowa, 593 pp.

Solomon, E., and M. Bobrow. 1975. Sister chromatid exchanges—a sensitive assay of agents damaging human chromosomes. *Mutation Research,* **30,** 273–278.

Taylor, J. H. 1958. Sister chromatid exchanges in tritium labeled chromosomes. *Genetics,* **43,** 515–529.

Templeton, W. L., M. Barnhard, B. G. Blaylock, C. Fisher, M. J. Holden, A. G. Klimov, P. Metalli, R. Mukherjee, O. Ravera, L. Sztanyik, and F. Van Hoeck. 1976. Effects of ionizing radiation on populations and ecosystems. *In:* Effects of Ionizing Radiation on Aquatic Organisms and Ecosytems. Technical Document 1972, International Atomic Energy Agency, Vienna, pp. 89–102.

U.S. Congress. 1972. The Marine Protection, Research and Sanctuaries Act of 1972. Public Law 92-532, 86 Stat. 1052, Washington, D. C.

U.S. National Academy of Sciences. 1980. The Effects on Populations of Exposure to Low Levels of Ionizing Radiation. U.S. National Academy of Sciences, Washington, D.C., 639 pp.

Walsh, J. E. 1962. Nonparametric confidence intervals and tolerance regions. *In:* Contribution to Order Statistics, A. E. Sarban and B. G. Greenberg (Eds.). Whey, New York, pp. 136–143.

Wolff, S. 1982. Difficulties in assessing the human health effects of mutagenic carcinogens by cytogenetic analyses. *Cytogenetics and Cell Genetics,* **33,** 7–13.

Woodhead, D. S. 1980. Marine disposal of radioactive wastes. *Helgoländer Meeresuntersuchungen,* **33,** 122–137.

Yu, L. 1976. Probing the Mechanism of Sister Chromatid Exchange Formation with the Fluorescent plus Giemsa Technique. Ph.D. Dissertation, University of California, San Francisco, 90 pp.

Chapter 15

Effects of Radiation Exposure on Fishes

Avril D. Woodhead and Virginia Pond

Biology Department
Brookhaven National Laboratory
Upton, New York

ABSTRACT

The use of oceans for disposal of radioactive wastes involves direct risks to the marine biota and indirect risks to humans through the food chain. This chapter addresses the somatic and genetic effects of ionizing radiation on fishes. A brief review of the information on radiation effects from a range of field and laboratory studies is presented. Also included are radionuclide-accumulation data derived from fish populations living in the vicinity of a nuclear reprocessing station located on the coast of the Irish Sea. There is a great deal of information on somatic damage caused by internal whole-body exposure of fishes to radionuclides or, more frequently, by exposure to external radiation sources. The histopathological work is detailed, and includes whole-body autoradiograms and evaluations of dose-rate effects. Most radiation injuries that lead to the death of the fishes are related to damage both to cell-renewal systems and to the central nervous system. There is evidence that fish can repair and recover from radiation damage and that ambient temperature affects the rate of repair. Although low-level irradiation may cause damage to DNA, resulting in mutation, there is little information on the genetic effects of exposure. The combined results of field observations and laboratory experiments have been interpreted as evidence that fishes are more resistant to radiation exposure than are mammals. It has been suggested that the disposal of radioactive wastes onto the seabed is unlikely to have any detrimental effects on fishes. We contend that laboratory studies have generally been inadequate and do not yet provide a wholly satisfactory background for this optimism.

15.1. INTRODUCTION

The development of nuclear energy has aroused concern for the protection of the environment through the safe disposal of radioactive wastes. The need for nuclear energy may result in an increase in exposure of humans and other organisms to ionizing radiation. In this chapter, we focus on the somatic and genetic effects that disposal of radioactive wastes into the ocean may have on marine fishes.

Over the years, information has accumulated describing the effects of exposure to radiation in a variety of invertebrates and fishes. Generally, the lethal dose range for fishes has been established at 10–50 Gy. These data are traditionally interpreted as evidence that fishes are more resistant to radiation exposure than are mammals (Rice and Angelovic, 1969; Chipman, 1972; Patel and Patel, 1979). Most of the few studies in which fishes have been exposed to chronic low-level radiation from external sources or from incorporated radionuclides have failed to show any marked detrimental effects either on individuals or on entire populations. It has been considered that this information is adequate to allow the prediction that radiation damage to fishes is unlikely to be a cause for public concern if the radiation dose is set at limits that protect humans (Purdom and Woodhead, 1973; Templeton, 1980). Others advocate the view that the ocean can be used safely as a repository for radioactive and other wastes (Osterberg, 1981, 1982).

Radiobiological research on mammals has been vigorously pursued over the years primarily in order to meet the practical needs of radiation protection for humans and for the development of radiation therapy and also to investigate the effects of radiation at the cellular and molecular levels. The major shift in emphasis has been away from descriptive work and toward quantitative evaluation of the dose-response relationships and away from studies of acute lethality toward the quantification of delayed effects of low doses over a lifetime. This shift has not been paralleled by studies of the cold-blooded vertebrates.

Therefore, it is somewhat difficult to evaluate the findings, particularly because the results are fragmentary and were obtained under different laboratory conditions.

15.2. ACUTE EXPOSURE

15.2.1. The $LD_{50/30}$

The $LD_{50/30}$ is widely used as a quantitative measure of the radiation sensitivity of laboratory animals. This value, defined as the dose of radiation required to kill 50% of an experimental group of organisms in 30 d, is readily obtained and easily quantified. Moreover, it is usually an appropriate measurement in laboratory mammals because it allows sufficient time for surviving cell-renewal systems, notably the hematopoietic system, to replace damaged cells. Acute lethal doses have been determined in mammals after exposure to X-rays, γ-rays, β-rays, and fast and slow neutrons. In closely related species, the $LD_{50/30}$ rarely differs by more than a factor of two or three.

This measurement has been widely adopted in radiation studies with aquatic organisms. The $LD_{50/30}$ values obtained, 8.8–44 Gy, suggest that fishes are less sensitive to radiation than are mammals. These values lend support to the view that if limits for exposure are set to protect humans, then fish populations will automatically be protected, and no serious ecological effects will be seen. There may be reason, however, to question whether this measurement is entirely satisfactory: the 30-d period is probably too short a time period in a life cycle for radiation effects to develop in cold-blooded vertebrates [International Atomic Energy Agency (IAEA), 1976, 1979]. Sparrow et al. (1970) reinforced the chromosome volume theory from the plant to the animal kingdom by showing that the amphibian *Necturus maculosus*, which has extremely large chromosomes, has an LD_{50} of only 98×10^{-2} Gy after 170 d. Extrapolation of the data suggests that this value would be further reduced to 67×10^{-2} Gy after 200 d. Willis and his colleagues recorded an $LD_{50/30}$ for the rough-skinned newt (*Taricha granulosa*) of 15 Gy, but they observed that deaths from acute radiation continued for several weeks thereafter. In fact, the LD_{50} after 160 d was between 1.9 and 2.9 Gy, a dose within the mammalian lethal range (Willis and Lappenbusch, 1976; Willis, 1980). Rees (1962) recorded an $LD_{50/30}$ of 14 Gy for the euryhaline grass shrimp (*Palaemontes pugio*) at a salinity of 30‰, but Engel et al. (1974) found that by 40 d the LD_{50} had decreased to 2.2 Gy. Similarly, mummichogs (*Fundulus heteroclitus*) kept at a temperature of 27°C and a salinity of 5‰ had an $LD_{50/30}$ of 12 Gy, but by 60 d it was as low as 3–3.5 Gy (Angelovic et al., 1969). Engel (1979) proposed that observations of mortality or any other physiological endpoint should be extended, in view of the slower rates of cell cycling in cold-blooded vertebrates, and that a value should be taken only when the mortality rate of the experimental animals is asymptotic with that of the controls.

15.2.1a. Age effects

The age of the animal at the time of exposure may modify the effects of acute and chronic radiation. Generally, embryos, juveniles, and pre-adults are more susceptible to radiation, probably due to the higher metabolic and proliferative state of their tissues. It is well established that the LD_{50} is lower for eggs, larvae, and juvenile fish than for adults (Donaldson and Foster, 1957; Egami and Ijiri, 1979). The most sensitive stage of the life cycle for the chinook salmon (*Oncorhynchus tshawytscha*) is the 32-cell-stage zygote (Wadley and Welander, 1971). Maximum sensitivity in the plaice (*Pleuronectes platessa*) occurs 24 h after fertilization (Ward et al., 1971). Zygotes at the 32-cell stage are most vulnerable in rainbow trout, chinook salmon, and medaka; in the latter species, this stage occurs ~5 h after fertilization (Welander, 1954; Wadley and Welander, 1971; Egami and Hama, 1975). The medaka may also have a period of increased sensitivity at gastrulation (Egami and Ijiri, 1979).

Hyodo-Taguchi et al. (1973) demonstrated the inhibitory effects of exposure to X-rays and neutrons on hatching in the medaka. Neutrons caused the greater inhibition, and again the earliest developmental stages were the most vulnerable. Egami and Ijiri (1979) proposed that radiation may cause lesions in the hatching gland, located in the endoderm of pregastrulation-stage embryos, thus adversely affecting the synthesis of the hatching enzyme.

15.2.1b. Temperature and salinity effects

Environmental factors considerably modify the radiation response of cold-blooded vertebrates. These modifying effects have been well demonstrated in studies where LD_{50} was the endpoint. Temperature and salinity, the *ecological master factors* described by Kinne (1964), are of prime importance both singly and in combination, but other factors such as pH and heavy-metal exposure also have a pronounced effect. Early demonstrations of the effects of temperature and salinity on the LD_{50} were provided by Angelovic et al. (1969), who studied the saltwater mummichog, and by Engel, who studied the euryhaline grass shrimp (Engel et al., 1974; Engel, 1979). Mummichogs kept at 5‰ salinity and 17°C were most resistant to radiation exposure, with a median lethal dose of 25 Gy for 40 d, the greatest sensitivity was seen at a salinity of 25‰ at 27°C, at which the LD_{50} was 3.3 Gy. Radiation sensitivity in the shrimp was primarily salinity-dependent, and mortality was lowest when the animals were held at a salinity of 15‰, which is close to the isosmotic point of the hemolymph. Similar modifications of the LD_{50} have been reported for rainbow trout (*Salmo gairdneri irideus*), medaka (*Oryzias latipes*), and the European eel (*Anguilla anguilla*) (Neuhold and Sharma, 1969; Egami, 1970; Hansen, 1975, 1980). When medaka were irradiated with a lethal dose at 23°C and maintained at that temperature, death occurred within 10 d, but similarly irradiated groups kept at 4°C survived for 200 d. After transfer to the higher temperature, however, the fish died rapidly.

The modifying effects of temperature on the radiation response have also been observed at the tissue level. Lockner (1972) showed that much larger doses of radiation were needed to cause a 50% depression of circulating lymphocytes, granulocytes, and erythrocytes in the tench (*Tinca vulgaris*) held at 10°C compared with those held at 18°C. Etoh (1968) also showed that the rate of development of radiation injury to the hematopoietic tissue of goldfish (*Carassius auratus*) was slower at low temperatures. Goldfish exposed to 70 Gy of X-rays survived for more than 100 d at 4°C, but most died within 10 d at 22°C. The low temperature inhibited the development of radiation damage to the epithelium of the gastrointestinal tract, but upon transfer to higher temperatures the goldfish rapidly developed lesions and died (Hyodo, 1964, 1965a, 1965b). Hyodo-Taguchi and Egami (1969) estimated that at 25°C the epithelial cell cycle times for G_1, S, G_2, and M phases were 5, 9, 2, and 2 h, respectively; lowering the temperature to 15°C prolonged these phases to 15, 20, 4, and 4 h. Replicon growth rates in cultured fish cells are two times as fast at 34°C than at 14°C (Grabar and Flickinger, 1977).

15.2.2. Pathological Effects

The physiology and pathology of acute-radiation injury in fishes is qualitatively similar to that seen in mammals despite the large differences in organ systems. The causes of acute-radiation death are damage to (1) the central nervous system at 88 Gy or more, (2) the gastrointestinal tract at 35–280 Gy, and (3) the hematopoietic system at 8.8–18 Gy. Latent effects such as shortened lifespan, cancer induction, and reduced growth may occur at 8.8 Gy or less. These values are not absolute and may vary with the environmental conditions under which the fishes are held. There are detailed accounts of the patterns of damage to the gastrointestinal tract and hematopoietic tissues; lesions to the central nervous system and sense organs have been described far less often. Amazon mollies (*Poecilia formosa*) exposed to acute doses of 60 Gy generally die within 10 d, and early in that period the fish show many signs of gross damage to the nervous system (Woodhead et al., 1979). These changes include extreme irritability to minor stimuli, visual and auditory impairment, and lack of neuromuscular coordination, often seen as corkscrew, or spinning, locomotion. Six days after irradiation, the erratic behavior generally ceases, and the fish lie immobile on the bottom of the tank until death (Woodhead et al., 1979).

We have seen no consistent pattern of histological changes in the brain of exposed mollies. In some individuals, there was moderate to conspicuous congestion in the blood vessels of the medulla oblongata; in others, the vacuolation of the optic tectum markedly increased. The brains of still other mollies dying from acute radiation injury appeared normal.

Gastrointestinal injury was the major cause of death in Amazon mollies at radiation doses of 40–60 Gy. Within 24 h, we saw a marked degeneration of the epithelial lining of the tract, and epithelial and goblet cells were shed in abundance, particularly those at the tips of the longitudinal folds (Fig. 15.1). A few days later, the longitudinal folds became flattened, the submucosal layer detached from the muscularis, and there were numerous hemorrhages within the submucosa. Mitotic activity in the intestinal stem cells at the bases of the villi was completely inhibited, and the fish ceased to feed; at death, the fish were bloated and edematous. Similar effects have been seen in goldfish (Hyodo, 1964, 1965a, 1965b; Hyodo-Taguchi, 1970) and silver salmon (Johnson et al., 1970). Hyodo-Taguchi (1970) also demonstrated a close correlation between probability of death and the extent of degeneration in the intestine of goldfish.

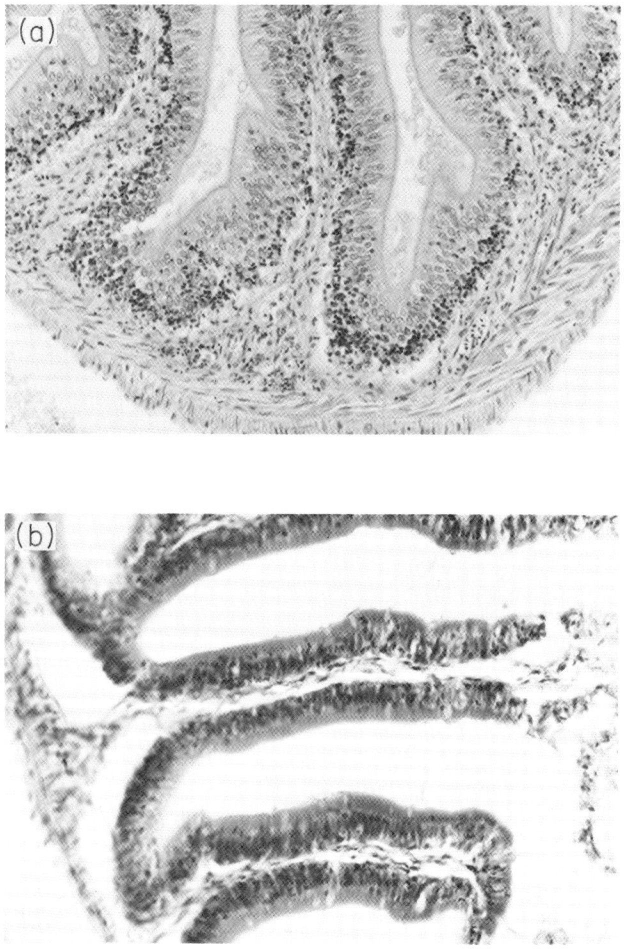

Figure 15.1. Section through the intestine of (a) an unirradiated Amazon molly and (b) an Amazon molly sampled 60 d after exposure to 40 Gy (145X). Note degeneration and atrophy of the intestine is apparent in the irradiated fish. (From Woodhead et al., 1979.)

Early and rapid depletion of the hematopoietic tissue of the kidney, spleen, and thymus follows radiation doses between 10 and 20 Gy. The depletion is grossly evident by a decrease in the weight of these various blood-forming organs and by examination of histological sections (Ellinger, 1940; Welander et al., 1948, 1949; Preston, 1959; Aoki, 1963; Egami, 1968; Cosgrove et al., 1975). Three hours after the exposure of the Amazon molly to 20 Gy, there was a large discharge of mature blood cells from the hematopoietic tissues of the kidney, spleen, and thymus, probably in response to the development of numerous small hemorrhages in the head, dorsal musculature, and swim bladder. By 12 h, the blood-forming cells showed nuclear pycnosis and cellular necrosis; considerable cell debris had accumulated in the interstitial spaces of the kidney and spleen (Fig. 15.2). There was no restoration of hematopoiesis, and over several weeks the fish became anemic, lethargic, and usually died from secondary bacterial or fungal infection (Woodhead et al., 1979; Woodhead and Setlow, 1980). Many investigators have reported a sharp decline in the number of erythrocytes and leucocytes in the blood, and lower packed-cell volumes and hemoglobin concentrations after acute and sublethal irradiations. Ghoneum and Egami (1980) found a marked drop in thymic volume in the medaka 1–3 d after radiation exposure; there was a subsequent increase in mitotic activity in the thymocytes of fish exposed to 9–18 Gy and a gradual regeneration of the gland, but no recovery occurred after exposure to 35–70 Gy. Much of this work has been reviewed by Patel and Patel (1979), who outline recent sophisticated techniques that can be used to detect radiation injury.

The gill epithelium, a monolayer of cells covering the gill lamellae, is also destroyed by high, acute doses of radiation, and the fish may die from osmotic imbalance (Etoh et al., 1968). Conte (1965) showed that juvenile coho salmon exposed to 8.8 Gy had malformed gill filaments; these fish were able to survive in freshwater but not in seawater, where they showed a marked elevation of plasma electrolytes. Bluegills (*Lepomis macrochirus*) exhibited a severe decrease in β globulins, α globulins, and albumins in the blood within 2–24 h after exposure to 8.8–26 Gy at a dose rate of 1.5 Gy min^{-1}. Part of these effects was due to changes in protein metabolism, but there was also a large loss of protein throughout the damaged gastrointestinal and gill epithelium, and also an osmotic imbalance (Ulrickson, 1971).

At doses ≤8.8 Gy, fishes may gradually recover from radiation damage to the hematopoietic tissue and the gastrointestinal tract. Autoradiographic studies with the Amazon molly have shown that many of the stem cells in both of these tissues are destroyed by radiation; mitosis is arrested in those few that escape death, but gradually mitotic activity is restored, and the tissue repopulated. We found little evidence of division in the hematopoietic stem cells of the molly for the first 16 d after exposure to 8.8 Gy; however, by 2 months hematopoiesis was well established and the blood was restored to a normal state. Recovery and restoration of the intestine were much faster and appeared to be complete between 1 and 2 weeks after irradiation (Woodhead and Setlow, 1980).

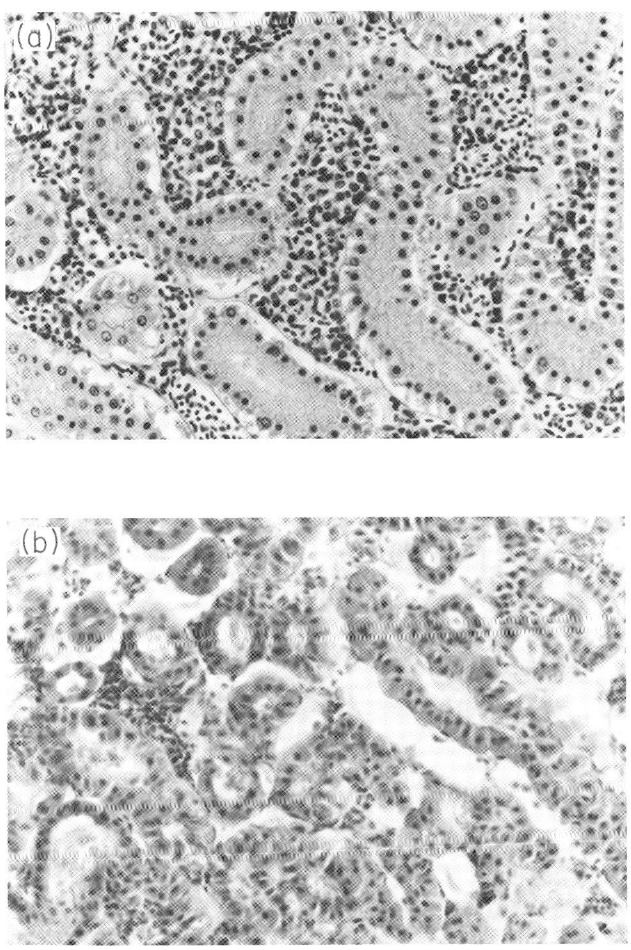

Figure 15.2. Section through the kidney tissue of (a) an unirradiated Amazon molly and (b) an Amazon molly sampled 60 d after exposure to 40 Gy (145X). Note the marked depletion of hematopoietic tissue in the irradiated fish. (From Woodhead et al., 1979.)

The idea that radiation has a nonspecific effect on the rate of aging has now been abandoned, and cancers are found to be the only specific diseases induced by low doses of radiation (Walburg, 1975). In mice, acute radiation doses of 2 Gy or more increased the incidence or advanced the age of onset of a large variety of rare and common neoplasms (Upton et al., 1960). We were unable to find any differences in lifespan or incidences of tumors in sublethally irradiated Amazon mollies. The experimental fish developed localized pigment spots in old age and were generally thinner than were the untreated fish. The growth check that they experienced after radiation, when damage to the intestine precluded feeding, did not seem to be accommodated later (Woodhead and Setlow, 1980).

15.2.3. Fertility

The effects of radiation exposure on fish gonads have been particularly well documented. A good criterion of whether a low dose of radiation is affecting a population adversely is whether the population is capable of reproducing at former levels. Much of this information comes from the efforts of Egami and his colleagues in Japan. Since 1955, they have published over 100 papers on this aspect of radiobiology (reviewed in Egami and Ijiri, 1979; Egami and Hama-Furukawa, 1981).

As with mammals, the most sensitive period in spermatogenesis occurs in the embryo just before hatching or birth and at an early stage of gonadal differentiation just before the proliferation of mature spermatogonia (Hamaguchi, 1980). The female germ-cell stage most susceptible to radiation damage is the immature arrested oocyte. Two critical periods have been identified: in the last premeiotic division and in the period between pachytene and diplotene (Egami and Ijiri, 1979). Similarly, mammalian oocytes are most sensitive to radiation exposure at the diplotene stage of prophase. Hamaguchi (1980) demonstrated that exposure of rainbow trout embryos to 6 Gy produces sterility in many individuals of both sexes; Konno (1980) found that all embryos are sterile after exposure to 6 Gy. In the adult medaka, multiplication of spermatogonia is slightly retarded by acute exposure to 1 Gy, but even after a dose as high as 20 Gy not all spermatogonia are damaged: some escape injury and eventually repopulate the testes (Egami and Ijiri, 1979). Autoradiographs show that such exposures inhibit further development of the actively dividing spermatogonia and spermatocytes at an early stage of meiotic prophase, whereas dormant spermatogonia, late-stage spermatocytes, spermatids, and spermatozoa are unaffected (Egami et al., 1964). In mice, although exposure to acute whole-body doses of 0.15 Gy causes the death of some spermatogonia, even a dose of 10 Gy fails to kill all of the cells or induce sterility (Committee on the Biological Effects of Ionizing Radiations, 1980). Temporary sterility was induced in adult female medaka after acute exposure to 20 Gy, but a massive dose of 80 Gy was required to destroy the ovary completely.

There are species differences in the gonadal responses of fishes to acute radiation exposure. Exposure of 3-month-old Amazon mollies to 10 Gy of radiation prevented subsequent ovarian maturation, even though the mollies were long past their most sensitive stage at the time of exposure (Woodhead and Setlow, 1980). Hyodo-Taguchi et al. (1970) and Egami and Hyodo-Taguchi (1971) showed that the marine goby (*Chasmichthys golusus*) was much more sensitive than the medaka; some medium-sized eggs were destroyed after treatment with 5 Gy, and almost all were killed after 10 Gy (Fig. 15.3). Egami and Hyodo (1965) believe, however, that reproduction in female fish may be adversely affected not so much from damage to the eggs as from a reduction in the secretion of gonadotrophins by the pituitary gland, which may be altered by exposure to ionizing radiation. Aoki et

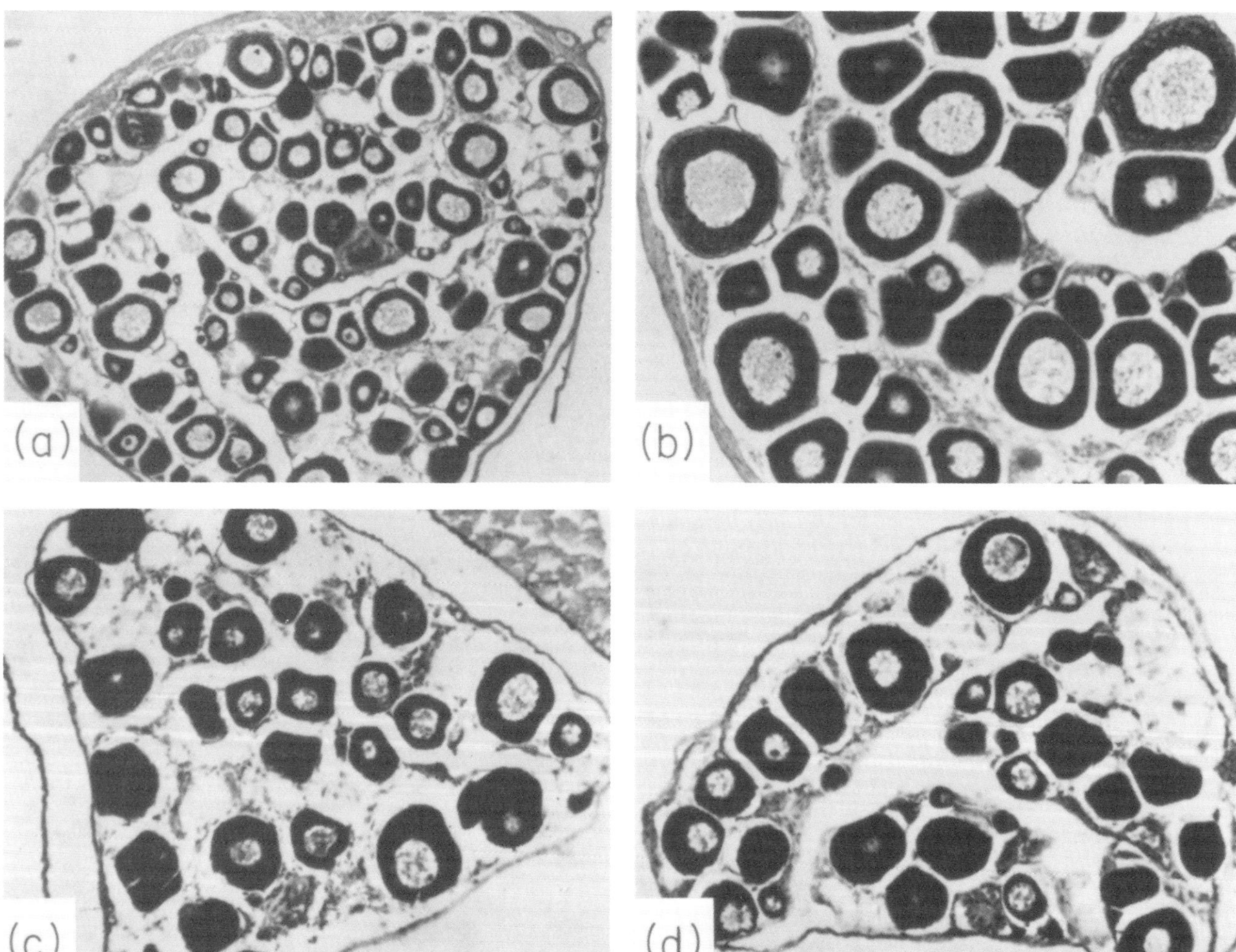

Figure 15.3. Ovaries of untreated marine gobies (*Chasmichthys gulosus*) sampled in winter (a and b) and the ovaries of marine gobies that were treated for 21 d with 3.7×10^{-2} Bq and 37×10^{-2} Bq of ^{90}Sr, respectively (c and d). Note the medium-sized oocyte damage in both cases. (From Hyodo-Taguchi et al., 1970.)

al. (1966) suggest that the marked hypertrophy of the interrenal gland of the goldfish after whole-body exposure to lethal and sublethal doses of X-rays is mediated by the hypophysis.

15.2.4. Fractionated Doses and Dose-Rate Effects

Etoh and Egami (1967) and Egami (1968) showed that fractionation of a high-radiation dose lessened mortality and had a sparing effect on radiation damage in tissues if the total radiation time was >12 h. Fractionation or reduction in dose rate allows time for the repair of the first lesions before the second are inflicted. Generally, survival is inversely related to the rate of delivery of the dose in the range of Gy s^{-1} to Gy d^{-1}. It is important to note the dose-rate or fractionating procedures, temperature, and salinity when describing the results of comparative experiments. Striking effects were shown by Chee et al. (1979), who exposed groups of fathead minnows (*Pimelphales promelas*) to the same dose of radiation, 30 Gy, delivered at different rates. The time to 50% mortality was >60 d at a dose of 2.6×10^{-2} Gy min^{-1}, 28 d at 38×10^{-2} Gy min^{-1}, and 17 d at 17 Gy min^{-1}. Egami and Hama (1975) also demonstrated the dose-rate effect by using the hatchability of irradiated embryos of *Oryzias latipes* as the endpoint. Two days after fertilization groups of eggs were exposed to 20 Gy at 2.5, 0.33, and 0.017 Gy min^{-1} (at the lowest rate, the entire dose was given in < 24 h). Hatching was reduced by ~50% at the two higher dose rates, but little effect was seen at the lowest. Michibata and Egami (1977) saw a similar sparing effect in the testes, a rapidly proliferating cell system, as well as in the tail fin, normally a slowly dividing tissue that retains the capability to divide quickly when necessary. Cell division in the tail fin of adult *Oryzias latipes* was inhibited when, immediately after amputation, the fish were exposed to 18 Gy at a rate of 2.2 Gy min^{-1}. At postoperative exposure rates of 0.24, 0.47, and 1.1 Gy d^{-1}, no effects were observed even when the experiments were continued longer and the final dose was well over 18 Gy (Hama-Furukawa and Egami, 1980).

Although radiation experiments show that the responses of fishes are qualitatively similar to those of mammals, fishes appear to be less sensitive to radiation by at least one order of magnitude. This finding would be reassuring when considering the disposal of radioactive wastes into the oceans were it not for the growing realization that many studies may have given overestimates of the resistance of fish to radiation doses.

15.3. CHRONIC EXPOSURE

15.3.1. External Sources

In many laboratory experiments fishes have been irradiated at low dose rates for prolonged periods in an attempt to duplicate the conditions of exposure to a contaminated environment. Low dose-rate exposures also reduce some of the problems of extrapolation of data from a high dose to a low dose (Setlow, 1983). Generally, when fishes are exposed continuously to low radiation levels, the total dose necessary to cause mortality (or to reach any endpoint) is increased in comparison with a single high exposure. Egami and his colleagues extended many of their earlier studies of acute-radiation effects on gametogenesis to studies of chronic radiation effects (Egami and Hama, 1977, 1979; Michibata and Egami, 1977; Egami and Hama-Furukawa, 1980, 1981). They exposed embryos or fry of *Oryzias latipes* to a gamma field for periods of up to 86 d and then reared them in an area without radiation for 1 y or more. In all experiments there was a good inverse correlation between dose rate and fecundity or fertility in females and males. At dose rates of ~0.53 Gy d^{-1}, no differences from control fish were seen, whereas at rates ≥1.2 Gy d^{-1} the medaka were sterile. At values in between, a balance was established in males between destruction and restoration of spermatogenesis, a balance in which spermatogonial stem cells existed at about 20–40% of the nonirradiated value. Similarly, in mice given a total dose >2.6 Gy at a rate of 8.8×10^{-6} Gy min^{-1}, there was an equilibrium established in the spermatogonial population of the testes at about 80% of the control value. Hyodo-Taguchi (1980) made a detailed study of the fertility of male *Oryzias latipes* given doses of 1.3×10^{-2} to 84×10^{-2} Gy d^{-1} for periods up to 200 d (Table 15.1). The response of the testes was principally dependent on dose rate rather than on accumulated dose, as is also the case in mammals (Oakberg, 1968; Erickson and Martin, 1976). Fish germ cells, however, appear to be less sensitive to radiation exposure than those of mammals. At these levels of radiation, ovotestes frequently occurred in the fish. Sex determination is quite labile in the medaka, and the formation of ovotestes can be initiated by any factor causing temporary testicular degeneration (Egami and Ijiri, 1979).

In an attempt to combine the advantages of controlled laboratory experiments with field studies, Donaldson and his colleagues made a series of large-scale experiments in which they irradiated the eggs of the anadromous chinook salmon with doses ranging from 4.4×10^{-3} to 180×10^{-3} Gy d^{-1}. The embryos were exposed shortly after fertilization until feeding began. From each experiment, 96 × 10^3 to 260 × 10^3 fingerlings were reared and then allowed to migrate to sea. Precocious males returned during the second year, and ripening males and females returned during the third and fourth years. There was no apparent damage to the reproductive capacity of the spawning stock over several generations after exposures of 4.4 × 10^{-3} to 44 × 10^{-3} Gy d^{-1} (a total dose of 3.1 Gy). There was a threefold increase in abnormalities in young fish at exposures of 17 × 10^{-2} Gy d^{-1}, but the number of returning adult salmon was not significantly different from the control. Radiation damage affecting the subsequent growth of the fingerlings was seen at doses of 8.8 × 10^{-2} Gy d^{-1} after a total dose of 7.1 Gy (Donaldson and Bonham, 1964, 1970; Bonham and Donaldson, 1966; Donaldson, 1970). In view of the problems and the prohibitive costs of large-scale field experiments and the problems in extrapolating from laboratory experiments, combined laboratory–field tests such as these should be encouraged [International Atomic Energy Agency, (IAEA), 1979]. Experiments with irradiated fish placed in cages in their natural environment might also be effective.

15.3.2. Radionuclides in Water

Contradictions in the literature on the effects of acute, high doses of radiation are even more apparent in studies of fish exposed to low doses of radionuclides in water. This is not surprising in view of the complicated dosimetry involved. The impact of a radionuclide uptake and assimilation depends in part on its physical, chemical, and biochemical properties and on the physiology of the animal. Cell-turnover rate, organ size, and immune surveillance are among the physiological factors affecting the numbers and lifespan of the target cells and the distribution of radiation-induced effects. Polikarpov (1966, 1979) noted that environmental and biological factors together greatly affect the uptake of radionuclides so that accumulation may differ by six orders of magnitude for different radionuclides and for different organisms.

Soviet workers in the 1960s made detailed studies of the effects of the long-lived ^{90}Sr and its short-lived daughter ^{90}Y on the developing eggs and larvae of several species of marine fish from the Black Sea. They exposed the eggs to low doses of the radioisotopes, ranging from 3.7 × 10^{-4} to 3.7 × 10^6 Bq $liter^{-1}$. Their findings suggested that young stages of fishes are very sensitive to radiation (IAEA, 1976). For example, reduced hatching and early mortality were seen at concentrations ranging from 3.7 × 10^3 to 3.7 × 10^6 Bq $liter^{-1}$, whereas malformed larvae were seen with consistency at concentrations of 3.7 to 3.7 × 10^3 Bq $liter^{-1}$. In one experiment, they found that 35% of the larvae were atypical at 3.7 × 10^{-9} Bq $liter^{-1}$, 50% at 3.7 × 10^2 Bq $liter^{-1}$, and 70% at 3.7 × 10^4 Bq $liter^{-1}$. Twenty percent of the control was atypical, and the highest concentration that did not cause any significant difference from the unexposed fish was 3.7 × 10^6 Bq $liter^{-1}$. Water containing ^{137}Cs, ^{144}Ce, and mixtures of nuclear fission products also produced larval deaths and deformities at similarly low concentrations (Polikarpov and Ivanov, 1962; Kosheleva, 1973a, 1973b). It became apparent early in their work that there were differences in the results of similar experiments. The increased mortality seen in loach (*Misgurnus fossilis*) eggs kept in a solution of ^{90}Sr–^{90}Y at 11 Bq $liter^{-1}$ by Shekhanova and Pechkurenkov (1968) was not confirmed in later studies (Pechkurenkov et al., 1972).

Brown and Templeton (1964) did not observe any reduction in the hatching of plaice (*Pleuronectes platessa*) and brown trout (*Salmo trutta*) eggs exposed to ^{90}Sr in their water at

Table 15.1. Fecundity of Irradiated Male *Oryzias latipes*[a]

Dose Rate (Gy d^{-1}) (10^{-2})	Accumulated Dose (10^{-2})	Number of Fish Examined	Irradiation Period (d)	Oviposition Frequency (%)[b]	Unfertilized Eggs (%)[b]	Infertile Fish (%)[b]
0	0	13	200	88 ± 5	14 ± 4	0
1.3	78	13	190	90 ± 6	13 ± 7	0
2.9	170	11	160	69 ± 11	27 ± 12	18
6.8	410	7	100	53 ± 16[c]	42 ± 17[c]	29
16	940	3	51	85 ± 8	5 ± 3	0
24	1400	5	72	36 ± 19[c]	48 ± 17[c]	60
84	5100	8	110	28 ± 16[c]	78 ± 15[c]	75

[a] Adapted from Hyodo-Taguchi (1980).
[b] Percentage ± standard deviation.
[c] Results signficantly different from the control value (p<0.05).

levels of 3.7 to 3.7 × 10^6 Bq liter^{-1} or to an external radiation source at levels up to 5 Gy delivered at a rate of 1 × 10^{-2} Gy h^{-1}. Till et al. (1976) showed that the minimum concentrations of ^{238}Pu and ^{232}U that produce abnormalities in the carp (*Cyprinus carpio*) and fathead minnow (*Pimephales promelas*), respectively, are 59 and 44 × 10^6 Bq liter^{-1} for ^{238}Pu and 2.8 and 1.9 × 10^6 Bq liter^{-1} for ^{232}U. Accumulated doses of more than 20 Gy during embryogenesis were needed to produce any significant differences in hatching and survival between exposed and unexposed groups of fishes. Abnormalities in the larvae were seen, however, when the total dose exceeded 7.5 Gy.

In spite of the many studies that have been made with mammals, there are serious gaps in knowledge of the dosimetry of internalized radionuclides and of their metabolism in different tissues (Woodard, 1980). Early studies of incorporated radionuclides in fish quickly revealed the problems of dosimetry. Thus Shekhanova and Pechkurenkov (1968) compared the dynamics of accumulation of ^{90}Sr and its daughter ^{90}Y. They reported that ^{90}Y was taken up by loach eggs much more readily than ^{90}Sr (an accumulation factor >100, compared to <1). However, about one-fourth to one-third of the ^{90}Sr passes through the membrane of the egg to the embryo and continues to accumulate up to the blastula stage, at which point saturation is reached. On the other hand, ^{90}Y accumulates principally on the surface of the egg and does not appear in embryonic tissues until near the end of gastrulation, and then it continues to accumulate until near the end of embryogenesis. Egami and Ijiri (1979) discussed the critical periods of embryogenesis and differences in radiation sensitivity in fishes. The considerable differences in the dynamics of accumulation and the half-lives of ^{90}Y and ^{90}Sr might explain any differences in the effect induced by these two elements. Till et al. (1976) also reported differences in the uptake and concentration of ^{238}Pu and ^{233}U in carp (*Cyprinus carpio*) eggs. The concentration of ^{238}Pu in the embryo at hatching was about three times higher than that of the surrounding water, but the concentration of ^{233}U in the embryo was only a fraction higher than that of the water. However, of the total ^{238}Pu in the egg, 67% was accumulated on the chorion, whereas 60% of the total ^{233}U was contained in the embryo. Polikarpov (1979), who initiated many of the early radiation studies, recently discussed sources of variability and suggested ways to improve dosimetry and standardize experimental procedures. In the same report, Woodhead (1979) described computational methods to provide a best estimate of radiation dose in aquatic organisms living in polluted waters and also outlined methods to evaluate absorbed radiation dose from an external source.

Measurements of accumulated radioactivity in the tissues of fishes from polluted environments, or after injection of a radionuclide, have been better documented than have been exposure experiments; however, the sample sizes in some experiments were very small. Pentreath and Fowler (1979) have given some values for the accumulation in fish of the α-ray emitters ^{232}U, ^{210}Po (Table 15.2), ^{237}Pu, and ^{238}Pu, and also for several β- and soft X-ray emitters, including ^{3}H, ^{90}Sr, ^{90}Y, and the isotopes of ^{55}Fe, ^{59}Fe, and ^{131}I. Their studies and those by Geldiay and Uysal (1976) showed that the distribution is not uniform, and, in many cases, more of the radionuclides accumulate in the liver (or hepatopancreas) than in tissues such as muscle. A detailed investigation of the concentration of several elements in the organs of young plaice (*Pleuronectes platessa*) revealed that Mn, Cr, and Co were highest in bony tissues, Cu was highest in the liver, Fe in the blood, and Hg in the kidneys. The distribution of Ag was fairly even throughout the body. Plutonium radionuclides accumulated in the bones of fishes, as they do in mammals.

The differential accumulation of radionuclides in developing ova is particularly interesting in relation to fertility and genetic effects. Tarrant (1971) showed that 70% of a dose of injected ^{131}I was taken up by the developing oocytes of the channel catfish (*Ictalurus punctatus*). In freshwater pike (*Esox lucius*) and tench (*Tinca tinca*) ~10% of a dose of injected Cs was in the ova; however, Cs had a lower concentration factor in the marine environment (Pentreath and Fowler, 1979).

Pentreath and Fowler (1979) discussed the large differences in the accumulation of particular elements in different species. For example, Ag is accumulated and retained in the liver of the thornback ray (*Raja clavata*) but is poorly retained and rapidly excreted by the plaice (*Pleuronectes platessa*). They also emphasized the importance of the entry route of the radionuclide, because its subsequent distribution depends on whether it is in the food or the surrounding water. Food is usually the main route of entry of radionuclides even in young fishes, and only in the case of inorganic Hg and ^{137}Cs does the contribution of water approach that of the food (Hewett and Jefferies, 1976; Pentreath, 1976). Absorption from the gut does not appear to be an efficient process; the gut content of ^{238}Pu in plaice is usually far greater than the tissue content (Pentreath, 1980). In goldfish the retention of ingested ^{90}Sr depends also on food composition and the specific activity of the radionuclide. For food containing Ca at about 0.6 × 10^{-3} mg g^{-1}, ~27% of the ingested ^{90}Sr is retained, but the addition of either Ca, Mn, or Sr lowers retention by ~95% (Ophel and Judd, 1967). Ophel (1979) found two examples from the freshwater environment that show how the content of K in the water affects the body burden of ^{137}Cs and how Ca affects uptake of ^{90}Sr. Sakaizumi and Egami (1980) have shown that the uptake of Hg by medaka eggs is reduced in water with a high salinity.

15.4. GENETIC EFFECTS

Early in the 1920s it was realized that exposure of humans to radiation could cause genetic defects in future generations. At that time the interests of geneticists centered on the mechanisms of heredity, and radiation-induced variants were widely used in basic research. By the 1950s, concern over nuclear fallout from above-ground testing of atomic bombs and increasing use of radiation in medical practice resulted in a need for experiments designed to assess the genetic hazards to humans from radiation exposure. Many studies using rodents as test organisms have been made of mutations and chromosomal aberrations after acute and lifetime exposures to radiation. By contrast, there have been few studies with cold-blooded vertebrates.

15.4.1. Chromosome Aberrations

Woodhead (1976) noted that the spectrum of chromosomal damage seen in fishes after radiation exposure was similar to that in mammals. He established cultures from the embryonic tissue of a small tropical fish (*Ameca splendens*), which has a diploid chromosome number of 26. Cells exposed to ^{60}Co radiation showed centric rings and dicentric, tricentric, and acentric fragments. Since the start of the 1960s, a number of Soviet investigators have made extensive studies of the effects of exposure to high doses of X-rays on fish chromosomes (reviewed by Kligerman, 1979). Irradiated tissues were fixed in Carnoy's fixative and squash preparations were made in aceto-carmine stain. Criteria such as the occurrence of chromosome fragments, anaphase bridges, and multipolar mitoses were used to assess the damage. In loach (*Misgurnus fossilis*) and Atlantic salmon (*Salmo salar*) there were dose-dependent increases in the number of aberrations per cell that persisted for many generations and were correlated with morphological malformations and early death of the embryo (Kligerman, 1979; Tsytsugina, 1979). Tsytsugina (1972, 1973, 1979) and Migalovskaya (1973) reported chromosome damage in several species of fish living with low concentrations of ^{90}Sr–^{90}Y in their aquaria. Suyama et al. (1980) also studied the effects of low levels of radionuclides (^{3}H and ^{90}Sr–^{90}Y) on plaice (*Limanda yokohamae*) eggs, and found that tritiated water causes an increase in the frequency of chromosome bridges directly related to the concentration of the radionuclide, but no effects were seen with ^{90}Sr–^{90}Y

Table 15.2. Percentage of ^{210}Po and ^{232}U in Organs of Fishes[a]

	^{210}Po			^{232}U		
Organ	**Yellow Fin Tuna (*Thunnus albacares*)**	**Broomtail Grouper (*Mycteroperca xenarcha*)**	**Lantern Fish (*Myctophum nitidulum*)**	**Mullet (*Mullus barbatus ponticus*)**	**Garfish (*Belone vulgaris*)**	**Scorpion Fish (*Scorpaena porcus*)**
Caeca	52	42				
Caeca and liver			34			
Intestine	6.5		40	2.0		2.8
Intestinal content	6.5	58				
Contents of stomach	1.7					
Stomach and stomach content			3.3			
Liver	8.5	0.3		6.7	6.8	8.9
Spleen	3.0	0.3		2.6		
Kidney	0.6		14			
Gall bladder	18					
Heart	0.9	0.1	1.9	0.2		
Gill	1.4	0.1	1.6			
Muscle	0.2	0.01		0.9	4.0	12
Gonad	0.9	0.01	5.5			
Gonad and eggs				1.4		
Skeletal bones				35	26	26
Head				51	47	50
Skin					16	

[a]Adapted from Pentreath and Fowler (1979).

up to a concentration of 3.7×10^7 Bq liter^{-1}. Medaka eggs incubated in freshwater containing the same radionuclides at 3.7×10^5 Bq liter^{-1}, however, showed much higher absorbed doses than those kept in seawater. Suyama et al. (1980) concluded that the decreased absorption in the chromosome-aberration study with plaice eggs could have been due to reduced uptake in seawater.

Studies of radiation-induced chromosome aberrations may be valuable in surveying fishes caught in contaminated environments. These studies can be used to identify damages rapidly, even in fishes with large numbers of chromosomes, such as is the case with the two species used most often in these studies: *Misgurnus fossilis,* in which $2n = 100$ chromosomes, and *Salmo salar,* in which $2n = 53$–60 chromosomes (Roberts, 1970; Raicu and Taisescu, 1972). Usually these surveys cannot provide precise, quantitative determinations of the genetic damage caused by exposure to low-level radiation because of the large numbers of chromosomes and because the individual chromosomes are so small. Kligerman (1979, 1980) has attempted to put fish cytogenetics on a quantitative basis and has listed the criteria important for any *in vivo* fish model. The principal requirement is that the fish karyotype consist of small numbers of large chromosomes with easily identified centromeric regions. Most fishes have large numbers of small, ill-defined chromosomes (Fig. 15.4). So far, Kligerman has found only 10 suitable, easily maintained species with chromosome complements ranging from $2n = 16$ to $2n = 26$. Endo and Ingalls (1968) proposed that the two-day-old zebra fish (*Brachydanio rerio*), which has been cloned (Streisinger et al., 1981), might be useful, but it has a diploid chromosome number of 50. Other species, such as the black ghost knifefish (*Apteronotus albifrons*), which has 24 chromosomes, have suitable karyotypes but are difficult to keep in the laboratory (Howell, 1972). Kligerman has suggested the use of the central mudminnow (*Umbra limi*) because it is a hardy species with 22 large, meta- and submetacentric chromosomes (Kligerman and Bloom, 1976; Kligerman, 1979, 1980) (Fig. 15.5). Kligerman's studies showed chromosomal breaks and gaps in cells from the gills, intestine, and kidneys after the mudminnows were exposed to 2.8 Gy.

Metalli (1979) has questioned the value of chromosome studies, considering both the time and effort required and the detailed background knowledge of the karyotype and spontaneous mutation rates needed as a basis for comparisons. He believes, however, that studies of chromosome aberrations may be useful as indicators in the more sensitive species of any particular environment. Chromosome aberrations may, in fact, be the first indication of the early stages of perturbation in an ecosystem.

15.4.2. Mutations

15.4.2a. Morphological

Fishes have been used as models in basic genetic research for over 50 y, and a great deal is known about the regulation and expression of genes responsible for color polymorphism, melanoma formation, and sex determination (Schröder, 1973). Purdom and Woodhead (1973) studied the effects of acute doses of radiation (8.8–18 Gy) on the F_1 offspring of irra-

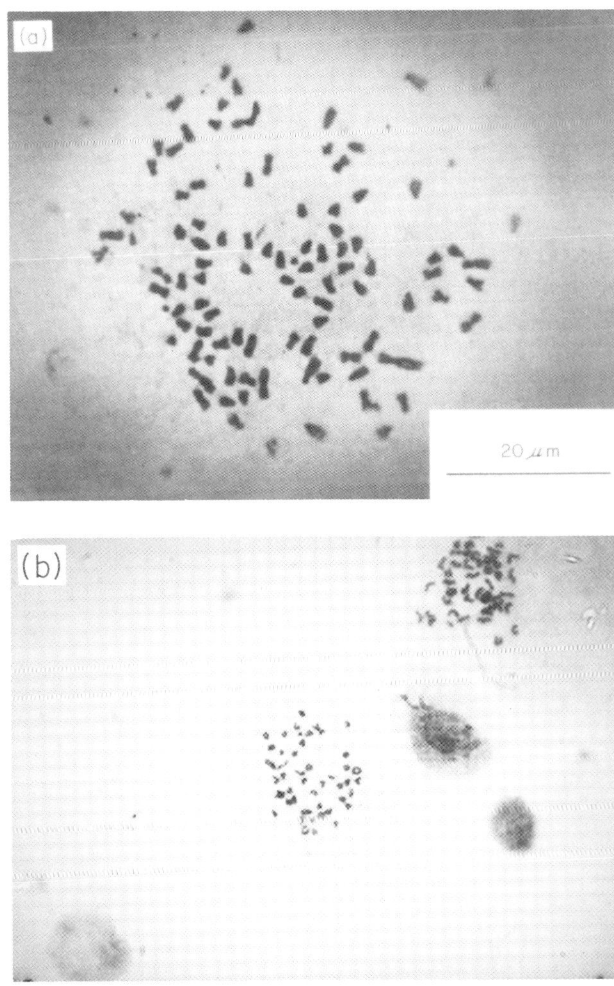

Figure 15.4. Typical metaphase chromosome spread from cultured cells of (a) the goldfish (CAF-MM1) with 98 chromosomes and (b) the Amazon molly with 46–48 chromosomes. The photograph of the goldfish chromsomes is reproduced from Mitani et al. (1982) with permission of the author.

diated male guppies (*Lebistes reticulatus*). They found three mutations out of the 852 fish examined. They concluded that although the number of mutations was lower than predicted when compared to similar mouse specific-locus assays, radiation did not necessarily have a lower mutagenic effect in fishes. Anders et al. (1973a, 1973b) found that X-ray irradiation disturbed the regulatory elements controlling the expression of macromelanophore genes in xiphophorid fishes, leading to the development of premelanomas and melanomas. Although hereditary changes in color pattern are well documented in these small aquarium fishes, which are ideal for laboratory studies, few of the commercially important species of the northern hemisphere have the marked and bright color patterns of tropical species. Other morphological characteristics may be evaluated qualitatively and quantitatively in the field as well as in the laboratory. Skeletal deformities (dominant mutations) have been widely employed as a criterion of chemically and radiation-induced mutations in mice; they are also an excellent criterion in fish. Spinal and vertebral deformities caused by degenerative changes in vertebral connective tissues lead to conditions similar to lordosis, scoliosis, and kyphosis found in mammals; these conditions in fishes have been described in detail by Rosenthal and Rosenthal (1950), Gordon (1954), Takeushi (1966), and Weis and Weis (1976).

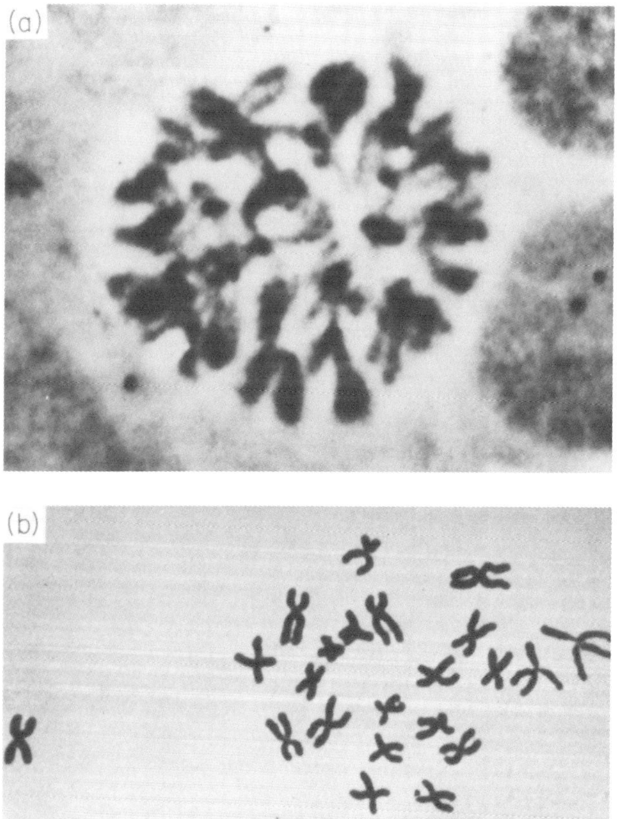

Figure 15.5. The 22 metaphase chromosomes of the mudminnow (*Umbra limi*): (a) in a squash preparation from gill tissue treated with aceto-carmine (phase contrast, 1668X) and (b) in a solid-tissue preparation from intestinal tissue exposed to colcemid *in vivo* and treated with hypotonic solution before fixation (Giemsa-stained preparation, 870X). Note the well-defined morphology of the chromosomes and the lack of clumping in the latter preparation. (From Kligerman, 1979; with permission of the author.)

Schröder (1979), in his review of radiation-induced mutations in fishes put skeletal deformities at the top of his list of traits that can be evaluated qualitatively and quantitatively. Spinal curvatures are prominent among the offspring in the F_2 generation of guppies and *Xiphophorus helleri* exposed to 4–10 Gy of X-rays (Penners, 1959, 1962; Schröder, 1969). By appropriately outcrossing lordotic guppies, Schröder showed that recessive mutation of polygenes is responsible for some of these deformities. Some severely affected fishes demonstrated reduced reproductive functioning; male fetuses with vertebral abnormalities have a higher probability of dying before emergence than do their female counterparts. Irradiation of the eggs and sperm of rainbow trout (*Salmo gairderi*) with 1.8–18 Gy causes spinal deformation and anomalies of lateral symmetry in the F_1 generation (Newcombe and McGregor, 1967; McGregor and Newcombe, 1968, 1972a, 1972b). Radiographs of the affected fish reveal extensive fusion and duplication of ribs and changes in the numbers of vertebrae. Mutations causing congenital swim-bladder defects in guppies (Schröder, 1979) and eye and head malformations due to chromosome aberrations in rainbow trout (McGregor and Newcombe, 1968, 1972a, 1972b) have also been studied extensively; in rainbow trout there is a linear relationship at doses from 2.5 to 4 Gy.

Viability is often used as an indicator of radiation-induced, dominant, lethal mutations in fish and is measured in terms of the number of stillbirths, brood size, frequency of postnatal deaths, and altered sex ratios (Schröder, 1979). The rate of dominant lethal mutations increases with the dose of radiation in the salmon, guppy, and medaka (McGregor and Newcombe, 1972a; Schröder and Holzberg, 1972; Egami and Hyodo-Taguchi, 1973). Purdom and Woodhead (1973) exposed guppies to chronic, long-term radiation at dose rates of 17×10^{-3}, 5×10^{-3}, and 2.5×10^{-3} Gy h^{-1} for 2 y. At the highest dose rate, all of the guppies were sterile within 4 months. Fecundity was reduced by 50% at the lower doses, possibly as a result of induction of dominant lethal mutations.

Woodhead (1977) also reported a marked reduction in fecundity in guppies continuously exposed at rates of 4.1 $\times 10^{-2}$, 9.6 $\times 10^{-2}$, and 30 $\times 10^{-2}$ Gy d^{-1}. The reduction was exhibited as a decrease in mean brood size and an increase in temporary or permanent sterility. Using the direct relationship between the haploid DNA content and induced mutation rate [demonstrated for a range of organisms by Abrahamson et al. (1973)], together with an actual value obtained in experiments with guppies (Purdom, 1966), it was estimated that the natural mutation rate for fishes (with a modal value of 1 pg DNA) would be in the order of 7 $\times 10^{-10}$ Gy per locus per zygote (IAEA, 1976). There are, however, many uncertainties in extrapolating information on mutation rate from one species to all fishes; furthermore, these data yield only a first-generation effect, not an equilibrium condition (Committee on the Biological Effects of Ionizing Radiations, 1980).

15.4.2b. Behavioral

The use of behavioral responses as an assessment of radiation damage in fishes is of recent origin; however, the studies do not lag too far behind those in which behavioral changes have been used to monitor effects of chemical contaminants (Eisler, 1979). To measure the mutagenic effects of radiation, Schröder (1980) recorded changes in the innate behavior patterns of the postirradiation F_1 generation. Activity, aggressiveness, social cohesiveness, and courtship have been defined under standardized conditions in nonirradiated cichlids and poeciliids, and quantitative tests are sufficiently developed to focus on radiation-induced changes in behavior. Holzberg and Schröder (1972) and Holzberg (1973) compared the activity and aggressive behavior of F_1 groups of nonmated and mated male convict cichlids (*Cichlasoma nigrofasciatum*) from irradiated parents (10 Gy to both parents) and the appropriate control males from sham-exposed parents. Aggressiveness was measured as the number of bites to subadult male conspecifics during a 15-min observation period; activity was recorded as the number of 90° turns made in the same time period. Locomotory activity did not differ between the two groups, but the level of aggression in the nonmated and the mated irradiated F_1 fish was significantly lower than that of the nonirradiated fish (by one-third and one-half, respectively). The aggressiveness of the F_2 males was also lower than that of the control F_2 males, and the mean value for any particular male tended to correspond with that of his father. Schröder (1979) suggested that mutations can affect genes controlling innate behavior and that these mutations are then transmitted to later generations. Thus, behavioral manifestations can be assessed just as satisfactorily as inherited morphological traits.

Social cohesiveness was determined in convict cichlids by counting the distribution of 10 male or female fish in a tank subdivided into 12 equal parts, each of which could contain the entire group. Schröder (1980) found an increase in cohesiveness of postirradiation F_1 males. This cohesiveness was most pronounced after exposure of the parents to 4.4 Gy, and he interpreted this change to be the result of an inherited decline in aggression. The results with females were the opposite and were difficult to explain. Schröder also subdivided the strain, and kept and bred one group in freshwater and the other in seawater. A male courtship activity test was devised that involved four separate traits. It was used to compare the effects of parental exposure to 8.8 Gy of X-rays on spermatogonia and oogonia of neonatal fish. There was an increase in all four parameters in the postirradiation F_1 male offspring kept in seawater, but there was no change in those held in freshwater; in the postirradiation F_2 seawater group the differences disappeared (Schröder, 1980). The predator–prey interaction technique developed by Goodyear (1972) for assessing the lethal effects of toxic compounds and the effects of exposure to ionizing radiation also holds great promise.

So far, behavioral studies have been of limited value in the estimation of risk to fish populations because the results are equivocal and difficult to explain. However, once the details of the innate behavior patterns of particular species are documented and the tests rigorously standardized, behavioral studies may demonstrate sublethal effects of low-level, long-term radiation exposure before changes in morphology or fertility can be identified.

15.5. MOLECULAR STUDIES

Thirty years ago there was a rapid growth of knowledge in the field of animal-tissue culture. Since then, many cell strains have been established and maintained by the American Type Culture Collection. Tissue culture has enabled researchers to study the effects of radiation on cellular macromolecules. Lower vertebrates have received far less attention; the Collection houses only 6 fish cell lines, although 61 cell lines exist (Wolf and Mann, 1980). The maintenance of lower vertebrate cell lines is not very different from the handling of homeothermic tissues. Fish cells have the advantage that they can be grown at room temperature (20–25°C), and the growth medium can go unchanged between successive passages. A line in culture has been established from the embryonic tissue of the Amazon molly by R. W. Hart, and we have used this cell line in studies of radiation resistance and repair of lesions.

15.5.1. Radiation Resistance of Cultured Fish Cells

We were interested in seeing whether the resistance of whole fishes to radiation is reflected at the cellular level. Survival parameters of cultured cells are usually estimated by means of the colony-formation assay, in which a known density of cells is inoculated into flasks of growth media, the cells are irradiated, the medium is replaced, and the cells are left to divide for several days. Colonies, usually consisting of 32 or more cells, are considered survivors. Amazon molly cells in culture do not form discrete colonies, and their plating efficiency is low. Even the certified fish lines in the American Type Culture Collection have a plating efficiency of $<1\%$. Consequently, no one has studied the sensitivity of fish cells to radiation until recently. Etoh and Suyama (1980) succeeded in establishing a cell line that has a plating efficiency of $\sim 10\%$ derived from the caudal fin of goldfish. At passage 70, these cells were irradiated with 2.2, 4.4, 8.8, 18, or 26 Gy of X-rays at a dose rate of 0.7 Gy min^{-1}. A dose-response curve was obtained in which the mean lethal dose (D_0) was approximately 5.7 Gy. Mitani et al. (1982) confirmed these results (Fig. 15.6) and concluded that the duration of mitotic arrest in fish cells is shorter than that in mammalian cells, and that fishes are more resistant to radiation exposure at the cellular and whole-body levels. Shima et al. (1980) established a second cell line of fibroblasts from the original goldfish fin line and found that the new cell line was also less sensitive to ionizing and ultraviolet (UV) radiation than were mammalian cells. Shima's experiments were also conducted with cells at passage 70 and later.

We studied the resistance of fish cells to radiation at an early passage when the cells still retained their diploid chromosome number. Cells kept in culture through many passages often show an increase in chromosome number. We used a procedure similar to the one used to study mammalian cells (Erickson et al., 1980). Cells in the exponential growth phase were exposed to doses of X-rays ranging from 1.5 to 10 Gy, and their growth was observed for the following 10 weeks. This time period allowed the cells to pass through several divisions and allowed for expression of radiation damage. Doses of 1.5 and 3 Gy had no effect on the rate of replication. Over 75% of the cell population survived exposure to 5 Gy and subsequently divided normally; some cells survived and eventually divided even after 6 Gy (Fig. 15.7). Much lower doses of radiation adversely affected the survival and growth of cultured mammalian cells (Elkind and Whitmore, 1967). The doubling time of human fibroblasts increased from a control value of 24 h to 27 h (an increase of 12%) after a dose of 0.4 Gy, and a dose as low as 0.2 Gy caused a small, but reproducible, decline in growth rate (Cho, 1977).

15.5.2. Repair

Fishes, like other organisms, are better able to survive a particular dose of radiation when it is fractionated or given at a low dose rate. Splitting the dose allows time for damaged cells to be repaired, to resume division, and to replace damaged tissue. At very low dose rates, repair may keep pace with damage, and no lesions are seen. At slightly higher dose rates, a new steady state may be established between the rates of damage and repair so that the tissue functions at a level below its normal capacity, as Egami and Ijiri (1979) showed in their studies of spermatogenesis in irradiated medaka.

The integrity of nuclear DNA is essential to normal replication of cells; damage to DNA molecules is the source of cell reproductive death and mutation. Ionizing radiation induces single-strand and double-strand breaks in DNA as well as base damages. The particular lesion responsible for radiation effects has not yet been identified (Setlow, 1982). Healthy mammalian cells repair single-strand breaks very efficiently with little disruption of the genetic code (Omerod, 1976). Double-strand breaks are repaired less efficiently, and there may be disturbance of the genetic code leading to cell death and somatic mutation (Leenhouts and Chadwick, 1978). Excision repair of base damage is excellent in mammalian cells (Mattern et al., 1975).

There has been no study of double-strand repair in irradiated fish cells. There are, however, several other well-documented mechanisms of DNA repair, such as photoreactivation, nucleotide excision, and DNA adduct removal, which we looked for in fish tissues. Cells from the Amazon molly exposed to UV light develop cyclobutylpyrimidine dimers in their DNA, which, if left unrepaired, may lead to mutation and cancer. When the damaged cells are exposed to the longer wavelengths of visible light, however, the dimers are reversed *in situ* by an enzymatic process called photoreactivation. We found high levels of photoreactivating activity in cells of the Amazon molly, the blue shark (*Prionace glauca*), and the blind cave fish (*Anophtichthys jordani*) (Woodhead et al., 1978; Achey et al., 1979; Woodhead and Achey, 1979). Efficient photoreactivating ability has also been found in cultured cells from the grunt (*Haemulon sciu-*

rus) and the goldfish (*Carassius auratus*) (Cook and McGrath, 1967; Regan and Cook, 1967; Shima et al., 1981). Mano et al. (1982) found that cultured cells from goldfish fin are more sensitive to UV light than are mouse *L* cells, and that about one-half of the dimers are photoreactivated in visible light shortly after exposure. On the other hand, cells from the Amazon molly and the rainbow trout show little nucleotide excision repair of induced dimers (Woodhead et al., 1980), and defects in excision repair were also reported in grunt and goldfish cells (Regan and Cook, 1967; Shima et al., 1981; Mano et al., 1982). Ishikawa et al. (1978) demonstrated, with autoradiography, that the highly differentiated ganglion cells of the adult medaka efficiently carry out unscheduled DNA synthesis (a measure of DNA repair) *in vivo* in response to chemically induced DNA damage. Extracts of Amazon molly cells from brain, kidney, and liver are also able to remove O^6-methylguanine adducts in exogenous DNA. It seems apparent that fish cells recognize lesions and repair them in ways similar to those used by mammalian cells. All of the findings with cultured cells, including mammalian cells, however, are open to the valid objection that the mechanisms and rates of repair may differ from those *in vivo*.

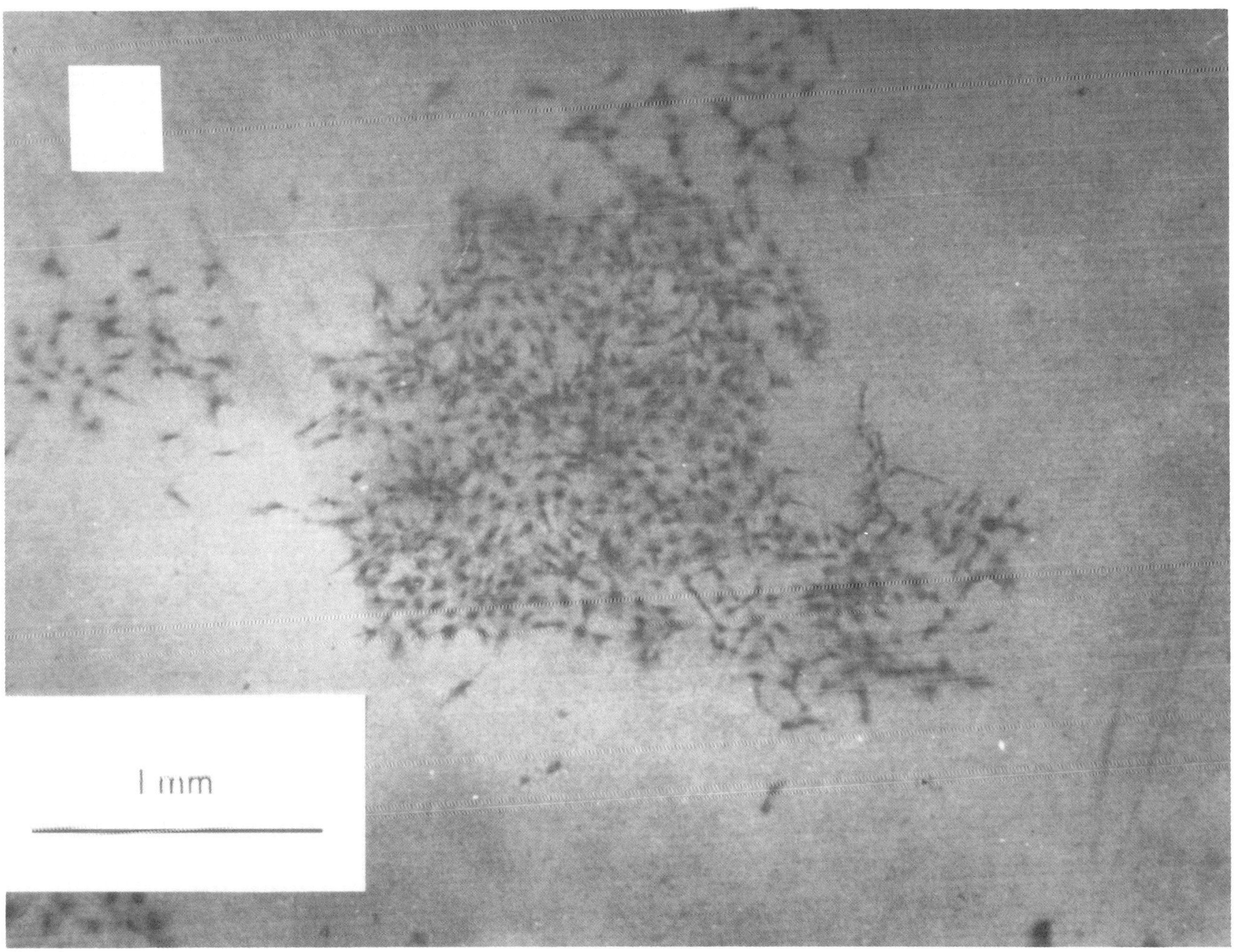

Figure 15.6. The morphology of a colony of cultured goldfish cells (CAF-MM1) at the 90th passage. (From Mitani et al., 1982; with permission of the author.)

15.6. RADIATION EFFECTS ON FISH POPULATIONS

Much of what is known about the effects of radiation on fishes comes from laboratory studies. The contrasts between laboratory findings and those from field experiments or sampling in nature both for the effects of radiation and for chemical pollution have been brought out in many experiments and discussed in many reports (IAEA, 1976, 1979; Cole, 1979). The problems of extrapolation of radiation effects on fishes have been reviewed by Patel (1975). Young (1977) dramatically demonstrated laboratory and field differences by showing that the rate of excretion of ^{65}Zn in the dogwhelk (*Nucella lapillus*) in a polluted environment was only about one-tenth of the value obtained in a laboratory study. With only a few notable exceptions (Bonham and Donaldson, 1966; Donaldson, 1970), there have been few attempts to study radiation effects in large-scale marine experimental ecosystems, although the need for such semi-natural experiments has been stressed repeatedly (IAEA, 1976, 1979). Studies with simplified ecosystems would undoubtedly highlight and clarify certain relationships, such as aspects of bioaccumulation and biomagnification. Large-scale field studies, however, are expensive and technically difficult; the observation periods must be long, and such experiments are difficult to replicate (Steele, 1979).

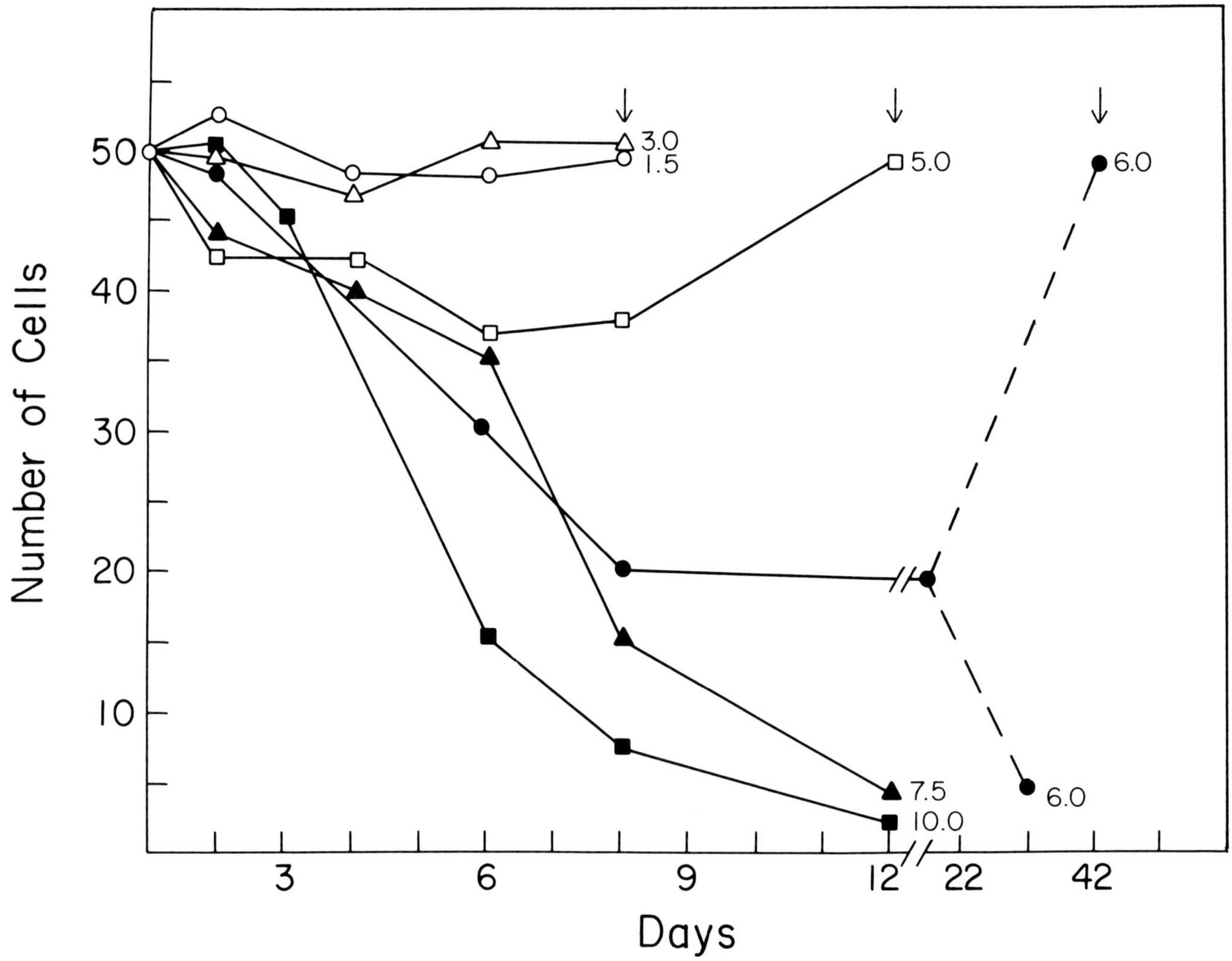

Figure 15.7. The growth and survival of cultured cells of the Amazon molly after exposure to various doses of X-rays, ranging from 1.5 to 10 Gy. Arrows denote cells that have reached confluency.

In assessing the effect of radiation on fishes, we have the advantage of a large volume of data from other sources. One is the extensive information on stock and recruitment relationships in plaice (*Pleuronectes platessa*) populations living in radiation-polluted areas of the northeastern Irish Sea. Another is a limited number of tissue samples analyzed from these same populations. The Sellafield nuclear reprocessing station releases more than 832.5×10^{13} Bq y^{-1} of low-level waste into the Irish Sea. In the vicinity of the outlet out to a distance of 10 km, the concentrations of Pu in the water are higher than is the background level. According to Templeton (1980), plaice eggs are exposed to about 7×10^{-9} Gy h^{-1} from 3.7×10^{-2} Bq of Pu per liter of seawater. A release rate of 3.7×10^{10} Bq d^{-1} from ^{137}Cs produces a mean annual concentration of 15.5×10^{-2} Bq liter^{-1} of seawater, which is diluted 30-fold at the northern exit of the Irish Sea and 300-fold at the southern exit (Jefferies et al., 1982). So far, there seem to have been no detrimental effects on the plaice stocks in this area, as measured by the number of plaice caught per 100 h of fishing (IAEA, 1976). Fisheries experts generally agree that under operating conditions as they stand now, the radioactivity in the discharge from a nuclear power plant is unlikely to have any adverse effects on coastal marine ecosystems (IAEA, 1976, 1979). These conclusions are based on many factors, including the physical and chemical properties of the radionuclides, their bioaccumulation and biomagnification through the food chain, and the physiology of the fishes.

A large part of the resistance of populations to radiation exposure resides in their stock and recruitment relationships. Most exploited stocks of fish are highly fecund, and there is high larval mortality. In addition, humans may cause a fish mortality as high as 50% per year for all age classes. Very few larvae, however, need to survive in order to maintain stock size (perhaps only one out of 10,000). Furthermore, in highly fecund species, there is no direct relationship between spawning stock size and recruitment. It seems unlikely, therefore, that any increase in mortality due to low levels of radiation would have a perceptible effect. Woodhead (1979) estimated that, if all mutations result in the death of the embryo, then less than 1 out of every 1000 fish would die as a result of exposure of each parent to 5×10^{-3} Gy. In considering the effects of radiation on fish populations, the sensitivity of their prey must also be taken into account, as fish population size is dependent on the amount of available food. Although the dynamics of populations of other aquatic animals have not received the attention given to fishes, evidence indicates that many organisms have density-dependent regulatory mechanisms that enable population size to reflect environmental changes. Furthermore, fish eggs are among the most sensitive components of an aquatic ecosystem: invertebrates and microorganisms are much less sensitive to ionizing radiation (IAEA, 1976). Any damage to fish populations as a result of radiation exposure is likely to result from direct effects on the fishes rather than from alterations in the food web (Templeton, 1980).

Exploited species with low fecundities, such as sharks and rays, for which there is a direct relationship between adult stock size and recruitment, may be affected by the additional stress of radiation-induced mortality or sterility, particularly if the stocks are already at the limit of exploitation. Nonetheless, Templeton (1980) believes that the levels of radiation to which species of fish with low fecundity are presently exposed are unlikely to have any adverse effects on their stocks. The possible effects on populations with low fecundity have been extrapolated, rather than demonstrated. Such effects, however, have been demonstrated in lizards by Turner et al. (1971), who monitored populations of lizards under continuous radiation in the southern Nevada desert. Populations of leopard and horned lizards rapidly declined as the females became sterile; soon they were all extinct. In contrast, coexisting populations of whiptail lizard and *Uta* spp. were unaffected by the same radiation exposure. The population responses appeared to be due to the dynamics of the populations rather than to any differences in the sensitivity of the ovaries. *Uta* spp. and whiptails mature early in life, are highly fecund, and die off rapidly, so there is a fast turnover of individuals in the population. Leopard and horned lizards, on the other hand, take much longer to mature, have a long lifespan, and lay fewer eggs; the long lifespan maintains the level of recruitment, but it also allows time for the accumulation of much larger total doses of radiation.

The beneficial effects of radiation exposure are often discussed (IAEA, 1976, 1979; Luckey, 1980). Donaldson and Bonham (1964, 1970), in their decade-long study of radiation effects in the migratory chinook salmon, reported that groups exposed to low level radiation early in life return from the sea to their home stream in greater numbers than do the controls and produce a greater number of viable eggs. A lower rate of mortality was observed in the developing eggs of loach reared with exposure to moderate concentrations of Sr and Y. Adult *Tilapia mossambica* exposed under similar conditions spawn earlier than do control fish, and both the frequency of spawning and the number of eggs shed increases (Shekhanova and Pechkurenkov, 1968; Shekhanova and Voronina, 1971).

A number of experiments have been cited as evidence of an induced, triggered repair system for radiation damage. Kulikov (1970) proposed that exposure to a low dose of radiation increases resistance to larger radiation exposures. The work of McGregor and Newcombe (1972a, 1972b) and Newcombe and McGregor (1972) shows that exposure of rainbow trout sperm to low levels of radiation increases the production and the viability of embryos. They favored the viewpoint that low doses of radiation stimulate repair mechanisms in the cell, and then cause overcompensation and repair of other damage. Apparent stimulation of growth, survival, and fertility by low doses of radiation have been reported for a wide variety of aquatic organisms, including the salmon and trout (IAEA, 1976). Experiments with protozoa and invertebrates have produced additional indications of the beneficial effects of radiation. For example, postradiation exposure of the protozoan *Tetrahymena* to UV light after exposure to α-particles enhanced survival, indicating two possible repair mechanisms: constitutive and induced (Calkins, 1973; Pittock and Calkins, 1973). Laszlo and Horsley (1973) suggest that the enhanced capacity for growth in the irradiated green alga *Oedogonium cardiacum* depends on the induction of protein synthesis. It has also been postulated that the growth stimulation seen in some invertebrates may result from the action of intracellular enzymes scavenging for radiation-induced free radicals that then indirectly stimulate the oxidative pentose phosphate pathway, RNA synthesis, and growth (Croute et al., 1982).

The repair of DNA damage caused by ionizing radiation is not inducible in mammalian systems, and there is no evidence that repair in response to lesions caused by ionizing radiation goes on to repair other damages. The suggestion that a modest increase in radiation may be beneficial to populations by increasing the mutation rate, thereby providing more material upon which natural selection can work, is also debatable. A committee convened to evaluate the biological effects of ionizing radiation strongly disagreed with this suggestion (Committee on the Biological Effects of Ionizing Radiation, 1980). They considered that there are excellent reasons for believing that the rate of evolutionary adaptation is not limited by the rate of mutation; moreover, they found that detectable mutations are almost always deleterious in their effects. We take the view that the concept of radiation hormesis, a word used by Luckey (1980) to mean stimulation by low doses of any potentially harmful agent, requires careful reconsideration. Undoubtedly, some of the earlier experiments cited by Luckey (1980) that demonstrated beneficial effects were rather naively designed or recorded. There appears to be a divergence of opinion on this question between radiation biologists working with mammals and those working with lower vertebrates and invertebrates.

15.7. CONCLUSIONS

At present few of the adverse effects of radiation exposure observed in the laboratory or predicted from laboratory findings appear to have occurred in fish populations exposed to radiation in their natural surroundings. Populations of plaice in the shallow waters of the Irish Sea were exposed to output from an operating nuclear reprocessing plant and appear unaffected by radiation exposure; this conclusion is drawn from detailed statistics of fisheries' landings for over two decades (IAEA, 1976). Therefore, is there anything to be gained from further pursuit of radiation studies with fish? A National Research Council panel believes that the present information concerning environmental effects of radiation may be adequate to formulate standards for radioprotection under normal conditions (Committee on Federal Research on the Biological and Health Effects of Ionizing Radiation, 1981). The data may not, however, be sufficient to predict with confidence the long-term effects of accidental, large-scale discharges into coastal waters, or to predict what might happen to deepwater organisms in the event of leakage or disruption of a container of high-level radioactive waste disposed of onto or into the seabed. Predictions will have to be made rather than actual sampling or monitoring of the seabed, because there are still great technical difficulties involved in sampling the sparse and patchily distributed animals.

A number of reports have emphasized the importance of chronic low-level exposure experiments for the purposes of environmental assessment of radiation effects (IAEA, 1975, 1976, 1979). It has been suggested that, unless laboratory experiments are made at doses in the range of 10^{-6}–10^{-3} Gy h^{-1}, they will be of little value because it is virtually impossible to extrapolate from the response of a single species exposed to acute high doses of radiation in the controlled conditions of the laboratory to the functional behavioral responses of natural populations that are exposed continuously (or intermittently) to low doses of radiation. Furthermore, the sublethal doses that elicit effects in the laboratory are often greater than the doses to which any population in the sea is currently exposed. It might seem, therefore, that studies of acute exposures in the laboratory should be relegated to the past, but we suggest that such studies are still a powerful investigatory tool. They are invaluable in providing definitive data on the nature of radiation-induced lesions, differential sensitivity of species and developmental stages, as well as dose-rate effects and the mechanisms of repair of radiation-induced damage in fishes. Cole (1970), in discussing the effects of chemical pollution in the seas and the ways to assess the problem, suggested that "a different approach may be preferable by which the effects of a potentially toxic pollutant on the processes of growth, physiology, behaviour and breeding are established in controlled laboratory conditions and then looked for in the field." Perkins (1979) believes that the field situation may be confusing and discusses the problems attendant on the collection and interpretation of field data. It has been questioned how far existing fisheries' statistics can demonstrate the sublethal effects of pollution, and it is not universally accepted that the effects of overfishing are strictly comparable to sublethal stress on a population (Morris, 1979). Species may respond to a decline in population by lowering the age of first maturity, but sublethal stress may result in a general loss of physiological efficiency and a lowering of reproductive potential (Morris, 1979).

It would be useful to continue radiation studies under standardized conditions so that findings from different laboratories can be compared. It would also be useful to attempt to identify particularly vulnerable species in a specific ecosystem. A series of relevant, precise, and carefully controlled laboratory studies together with shallow-water analogies might put to rest any concerns regarding the disposal of radioactive wastes into the ocean.

ACKNOWLEDGMENT

Research was carried out at Brookhaven National Laboratory under the auspices of the U.S. Department of Energy.

REFERENCES

Abrahamson, S., M. A. Bender, A. D. Conger, and S. Wolff. 1973. Uniformity of radiation-induced mutation rates among different species. *Nature*, **245**, 460–462.

Achey, P. M., A. D. Woodhead, and R. B. Setlow. 1979. Photoreactivity of pyrimidine dimers in DNA from thyroid cells of the teleost, *Poecilia formosa*. *Photochemistry and Photobiology*, **29**, 305–310.

Anders, A., F. Anders, and K. Klinke. 1973a. Regulation of gene expression in the Gordon-Kosswig melanoma system. I. The distribution of the controlling genes in the genome of the xiphophorin fish, *Platypoecilus maculatus* and *Platypoecilus variatus*. *In:* Genetics and Mutagensis of Fish, J. H. Schröder (Ed.). Springer-Verlag, New York, pp. 33–52.

Anders, A., F. Anders, and K. Klinke. 1973b. Regulation of gene expression in the Gordon–Kosswig melanoma system. II. The arrangement of chromatophore determining loci and regulating elements in the sex chromosomes of xiphophorin fish, *Platypoecilus maculatus* and *Platypoecilus variatus*. *In:* Genetics and Mutagensis of Fish, J. H. Schröder (Ed.). Springer-Verlag, New York, pp. 53–63.

Angelovic, J. W., J. R. White, Jr., and E. Davis. 1969. Interactions of chronic gamma radiation, salinity and temperature in the estuarine fish *Fundulus heteroclitus*. *In:* Symposium on Radioecology, D. J. Nelson and F. C. Evans (Eds.). Conf-670503, U.S. Atomic Energy Commission, Oak Ridge, Tennessee, pp. 131–141.

Aoki, K. 1963. The effects of whole body x-irradiation on the hematopoietic tissue in the goldfish, *Carassius auratus*. *Zoological Magazine*, **72**, 283–288.

Aoki, K., N. Egami, and R. Arai. 1966. Histological changes in the interrenal tissue of the goldfish, *Carassius auratus*, following x-irradiation. *Journal of Radiation Research*, **7**, 83–90.

Bonham, K., and L. R. Donaldson. 1966. Low-level chronic irradiation of salmon eggs and alevins. *In:* Disposal of Radioactive Wastes into Seas, Oceans and Surface Waters. International Atomic Energy Agency, Vienna, pp. 869–884.

Brown, V. M., and W. L. Templeton. 1964. Resistance of fish embryos to chronic irradiation. *Nature*, **203**, 1257–1259.

Calkins, J. 1973. The fine structure of radiation response of *Tetrahymena* as demonstrated by the use of repair inhibitors. *Radiation Research*, **55**, 594–596.

Chee, P. C., R. L. Vettner, and G. J. Atchison. 1979. Effect of dose rate on survival time in the fathead minnow, *Pimephelas promelas*. *Radiation Research*, **77**, 98–102.

Chipman, W. A. 1972. Ionizing radiation. *In:* Marine Ecology, Vol. 1., Part 3: A Comprehensive Integrated Treatise on Life in Oceans and Coastal Waters, O. Kinne (Ed.). John Wiley & Sons, London, pp. 1579–1657.

Cho, S. O. 1977. The Effect of Low Doses of X-Rays on Human Diploid Fibroblasts. M.S. Thesis, Wright State University, Dayton, Ohio, 118 pp.

Cole, H. A. 1970. North Sea Pollution. *In:* FAO Technical Conference on Marine Pollution and Its Effects on Living Resources and Fishing (Abstract). FAO/FIR-MP70/R20, FAO Fisheries Report 99, Food and Agricultural Organization of the United Nations, Rome, p. 112.

Cole, H. A. 1979. Summing-up: deficiencies and future needs. *Philosophical Transactions of the Royal Society of London*, **286**(B), 625–633.

Committee on the Biological Effects of Ionizing Radiations, Division of Medical Sciences, Assembly of Life Sciences, National Research Council. 1980. The Effects on Populations of Exposure to Low Levels of Ionizing Radiation. National Academy Press, Washington, D.C., 524 pp.

Committee on Federal Research on the Biological and Health Effects of Ionizing Radiation, Division of Medical Sciences, Assembly of Life Sciences, National Research Council. 1981. Federal Research on the Biological and Health Effects of Ionizing Radiation. National Academy Press, Washington, D. C., 800 pp.

Conte, F. P. 1965. Effects of ionizing radiation on osmoregulation in fish *Oncorhynchus kisutch*. *Comparative Biochemistry and Physiology*, **15**, 293–302.

Cook, J. S., and J. R. McGrath. 1967. Photoreactivating-enzyme activity in Metazoa. *Proceedings of the National Academy of Sciences of the United States of America*, **58**, 1359–1365.

Cosgrove, G. E., B. F. Blaylock, G. U. Ulrickson, and P. H. Cohan. 1975. Radiation-induced hematopoietic lesions in fish. *In:* The Pathology of Fishes, W. E. Ribelin and G. Migaki (Eds.). University of Wisconsin Press, Madison, pp. 463–466.

Croute, F., J. P. Soleilhavoup, S. Vidal, D. Dupouy, and H. Panel. 1982. *Paramecium tetraurelia*. Growth stimulation under low-level chronic irradiation: Investigations on a possible mechanism. *Radiation Research*, **92**, 560–567.

Donaldson, L. R. 1970. Low-Level Chronic Irradiation of Salmon. Annual Progress Report of United States Atomic Energy Commission. No. RLO-2225-72-1, U.S. Atomic Energy Commission, Washington, D.C., 22 pp.

Donaldson, L. R., and K. Bonham. 1964. Effects of low-level chronic irradiation of chinook and coho salmon eggs and alevins. *Transactions of the American Fisheries Society,* **93,** 333–341.

Donaldson, L. R., and K. Bonham. 1970. Effects of chronic exposure of chinook salmon eggs and alevins to gamma radiation. *Transactions of the American Fisheries Society,* **99,** 112–119.

Donaldson, L. R., and R. F. Foster. 1957. Effects of radiation on aquatic organisms. *In:* The Effects of Atomic Radiation on Oceanography and Fisheries. Publication 551, National Research Council, National Academy Press, Washington, D.C., 96–102.

Egami, N. 1968. Kinetics of recovery from injury after whole-body x-irradiation of the fish, *Oryzias latipes,* at different temperatures. *Radiation Research,* **37,** 192–201.

Egami, N. 1970. Effects of cysteamine given before and after x-irradiation under different temperature conditions on mortality of the fish, *Oryzias latipes. International Journal of Radiation Biology,* **18,** 391–394.

Egami, N., and A. Hama. 1975. Dose-rate effects on the hatchability of irradiated embryos of the fish, *Oryzias latipes. International Journal of Radiation Biology,* **28,** 273–278.

Egami, N., and A. Hama. 1977. Effects of chronic γ-irradiation on the fish, *Oryzias latipes.* I. Mortality of fish and hatchability of eggs laid by the irradiated fish. *Journal of Radiation Research,* **18,** 54.

Egami, N., and A. Hama. 1979. Effects of chronic γ-irradiation on the fish, *Oryzias latipes.* V. Further study of the sterilized fish exposed to γ-rays during development. *Journal of Radiation Research,* **20,** 15.

Egami, N., and A. Hama-Furukawa. 1980. Late effects of continuous γ-irradiation of the developmental stage on the gonads in *Oryzias latipes. In:* Radiation Effects on Aquatic Organisms, N. Egami (Ed.). Japan Scientific Societies Press, Tokyo, pp. 105–117.

Egami, N., and A. Hama-Furukawa. 1981. Response to continuous γ-irradiation of germ cells on embryos and fry of the fish, *Oryzias latipes. International Journal of Radiation Biology,* **40,** 563–568.

Egami, N., and Y. Hyodo. 1965. Effects of x-irradiation on the oviposition of the teleost, *Oryzias latipes. Annotationes Zoologicae Japonenses,* **38,** 171–181.

Egami, N., and Y. Hyodo-Taguchi. 1971. Species difference in radiosensitivity of germ cells among teleostean fishes. *Journal of Radiation Research,* **13,** 23.

Egami, N., and Y. Hyodo-Taguchi. 1973. Dominant lethal mutation rates in the fish, *Oryzias latipes,* irradiated at various stages of gametogenesis. *In:* Genetics and Mutagenesis of Fish, J. H. Schröder (Ed.). Springer-Verlag, New York, pp. 75–81.

Egami, N., and K-I. Ijiri. 1979. Effects of irradiation on germ cells and embryonic development in teleosts. *International Review of Cytology,* **59,** 195–248.

Egami, N., Y. Hyodo-Taguchi, and K. Konno. 1964. Autoradiographical studies on spermatogenesis in the fish, *Oryzias latipes,* with special reference to radiation effect on the fertility of fish. *Gunman Symposia on Endocrinology,* **4,** 147–159.

Eisler, R. 1979. Behavioural responses of marine poikilotherms to pollutants. *Philosophical Transactions of the Royal Society of London,* **286**(B), 507–521.

Elkind, M., and G. F. Whitmore. 1967. The Radiobiology of Cultured Mammalian Cells. Gordon and Breach, New York, 615 pp.

Ellinger, F. 1940. The goldfish as a new biologic test object in experimental radiation therapy. *Radiology,* **35,** 563–574.

Endo, A., and T. H. Ingalls. 1968. Chromosomes of the zebra fish. *Journal of Heredity,* **59,** 382.

Engel, D. W. 1979. Experimental approaches to demonstrate the interactions of radiation with other environmental stresses. *In:* Methodology for Assessing Impacts of Radioactivity on Aquatic Ecosytems. Technical Report Series No. 190, International Atomic Energy Agency, Vienna, pp. 301–316.

Engel, D. W., M. G. Shelton, and J. C. White, Jr. 1974. The effect of ionizing radiation and salinity on the grass shrimp, *Palaemontes pugio. Radiation Research,* **59,** 72–80.

Erickson, B. H., and P. G. Martin. 1976. Effects of continuous prenatal radiation on the pig and rat. *In:* Biological and Environmental Effects of Low-Level Radiation Pertinent to Protection of Man and His Environment, Vol. 1. International Atomic Energy Agency, Vienna, pp. 111–118.

Erickson, L.C., M. D. Bradley, J. M. Ducore, R. A. G. Ewig, and K. W. Kohn. 1980. DNA crosslinking and cytotoxicity in normal and transformed human cells treated with antitumor nitrosoureas. *Proceedings of the National Academy of Sciences of the United States of America,* **77,** 467–471.

Etoh, H. 1968. Changes in the incorporation of ^{3}H-thymidine into hematopoietic cells of goldfish following X-irradiation. *Zoological Magazine,* **77,** 213.

Etoh, H., and N. Egami. 1967. Damage accumulation and recovery in the fish *Oryzias latipes* exposed to fractionated or protracted radiation at different temperatures. *Radiation Research,* **32,** 884–891.

Etoh, H., and I. Suyama. 1980. Effects of radiation on cultured fish cells. *In:* Radiation Effects on Aquatic Organisms, N. Egami (Ed.). Japan Scientific Societies Press, Tokyo, pp. 195–203.

Etoh, H., Y. Hyodo-Taguchi, and T. Maruyama. 1968. Effects of x-irradiation of a part of the body on mortality rate and histological changes in the goldfish, *Carassius auratus. Journal of Radiation Research,* **9,** 141–151.

Geldiay, R., and H. Uysal. 1976. Accumulation and distribution of trace metals and radionuclides in marine organisms (particularly *Tapes decussatus* L.) in the Izmir Bay Area, Turkey. *In:* Trace Contaminants of Agriculture, Fisheries and Food in Developing Countries. International Atomic Energy Agency, Vienna, pp. 81–85.

Ghoneum, M. M. H., and N. Egami. 1980. Effect of γ-irradiation of adult and embryo of *Oryzias latipes* on thymus size. *In:* Radiation Effects on Aquatic Organisms, N. Egami (Ed.). Japan Scientific Societies Press, Tokyo, pp. 135–137.

Goodyear, C. P. 1972. A simple technique for detecting effects of toxicants or other stresses on a predator–prey interaction. *Transactions of the American Fisheries Society,* **101,** 367–370.

Gordon, M. 1954. The genetics of fish diseases. *Transactions of the American Fisheries Society,* **83,** 229–240.

Grabar, F., and R. A. Flickinger. 1977. Replicon number and growth rate in fathead minnow cells cultured at 14 and 34°C. *Cell and Tissue Kinetics,* **10,** 505–507.

Hama-Furukawa, A., and N. Egami. 1980. Effect of irradiation on the fin regeneration of the fish, *Oryzias latipes,* with special reference to the dose rate at different temperatures. *In:* Radiation Effects on Aquatic Organisms, N. Egami (Ed.). Japan Scientific Societies Press, Tokyo, pp. 139–146.

Hamaguchi, S. 1980. Differential radiosensitivity of germ cells according to the developmental stages in the teleost, *Oryzias latipes. In:* Radiation Effects on Aquatic Organisms, N. Egami (Ed.). Japan Scientific Societies Press, Tokyo, pp. 119–128.

Hansen, H. J. M. 1975. Changes in the lipid fraction of eel gills after ionizing irradiation *in vivo* and a shift from fresh to sea water. *Radiation Research,* **62,** 216–224.

Hansen, H. J. M. 1980. Changes in the lipid fraction of eel (*Anguilla anguilla*) gills after ionizing radiation and a shift from fresh to sea water: further influence of temperature and treatment with actinomycin D. *In:* Radiation Effects on Aquatic Organisms, N. Egami (Ed.). Japan Scientific Societies Press, Tokyo, pp. 147–155.

Hewett, C. J., and D. F. Jefferies. 1976. The accumulation of radioactive caesium from water by the brown trout (*Salmo trutta*) and its comparison with plaice and rays. *Journal of Fish Biology,* **9,** 479–485.

Holzberg, S. 1973. Change of aggressive readiness in post-irradiation generations of the convict cichlid fish, *Cichlasoma nigrofasciatum. In:* Genetics and Mutagenesis of Fish, J. H. Schröder (Ed.). Springer-Verlag, New York, pp. 173–176.

Holzberg, S., and J. H. Schröder. 1972. Behavioural mutagenesis in the convict cichlid fish, *Cichlasoma nigrofasciatum* Guenther. I. The reduction of male aggressiveness in the first post-irradiation generation. *Mutation Research,* **16,** 289–296.

Howell, W. M. 1972. Somatic chromosomes of the black ghost knifefish, *Apteronotus albifrons* (Pisces, Apteronotidae). *Copeia,* **1972** (1), 191–193.

Hyodo, Y. 1964. Effect of x-irradiation in the intestinal epithelium of the goldfish, *Carassius auratus.* I. Histological changes in the intestine of irradiated fish. *Annotationes Zoologicae Japonenses,* **37,** 104–111.

Hyodo, Y. 1965a. Development of intestinal damage after x-irradiation and ^{3}H-thymidine incorporation into intestinal epithelial cells of irradiated goldfish, *Carassius auratus,* at different temperatures. *Radiation Research,* **26,** 383–394.

Hyodo, Y. 1965b. Effect of x-irradiation on the intestinal epithelium of the goldfish, *Carassius auratus.* II. Influence of temperature on the development of histopathological changes in the intestine. *Radiation Research,* **24,** 133–141.

Hyodo-Taguchi, Y. 1970. Effect of x-irradiation on DNA synthesis and cell proliferation in the intestinal epithelial cells of goldfish at different temperatures with special reference to recovery process. *Radiation Research,* **41,** 568–578.

Hyodo-Taguchi, Y. 1980. Effects of chronic γ-irradiation on spermatogenesis in the fish, *Oryzias latipes,* with special reference to regeneration of testicular stem cells. *In:* Radiation Effects on Aquatic Organisms, N. Egami (Ed.). Japan Scientific Societies Press, Tokyo, pp. 91–104.

Hyodo-Taguchi, Y., and N. Egami. 1969. Development of intestinal radiation injury and recovery at different temperatures in fish. *In:* Comparative Cellular and Species Radiosensitivity, V. P. Bond and T. Sugahara (Eds.). Igaku-Shoin, Tokyo, pp. 244–254.

Hyodo-Taguchi, Y., N. Egami, and N. Yoshimura. 1970. Histological effects of X-ray irradiation on the ovary of the marine goby, *Chasmichthys gulosus. Journal of the Faculty of Science, University of Tokyo, Zoology,* **12,** 47.

Hyodo-Taguchi, Y., H. Etoh, and N. Egami. 1973. RBE of fast neutrons for inhibition of hatchability in fish embryos irradiated at different developmental stages. *Radiation Research,* **53,** 385–391.

International Atomic Energy Agency (IAEA). 1975. Design of Radiotracer Experiments in Marine Biological Systems. Technical Report Series No. 167, International Atomic Energy Agency, Vienna, 289 pp.

IAEA. 1976. Effects of Ionizing Radiation on Aquatic Organisms and Ecosystems. Technical Report Series No. 172, International Atomic Energy Agency, Vienna, 131 pp.

IAEA. 1979. Methodology for Assessing Impacts of Radioactivity on Aquatic Ecosystems. Technical Report Series No. 190, International Atomic Energy Agency, Vienna, 416 pp.

Ishikawa, T., S. Takayama, and T. Kitagawa. 1978. Autoradiographic demonstration of DNA repair synthesis in ganglion cells of aquarium fish at various ages *in vivo. Virchows Archiv B. Zellpathologie,* **28,** 235–242.

Jefferies, D. J., A. K. Steele, and A. Preston. 1982. Further studies on the distribution of ^{137}Cs in British coastal waters. I. Irish Sea. *Deep-Sea Research,* **29,** 713–738.

Johnson, T. S., R. E. Nakatani, and F. P. Conte. 1970. Influence of temperature and x-irradiation on the cellular dynamics of the intestinal epithelium in coho salmon, *Oncorhynchus kisutch. Radiation Research,* **42,** 129–140.

Kinne, O. 1964. The effects of temperature and salinity on marine and brackish water animals. II. Salinity and temperature–salinity combinations. *In:* Oceanography and Marine Biology, Vol. 2, H. Barnes (Ed.). Allen and Unwin, London, pp. 281–339.

Kligerman, A. D. 1979. Cytogenetic methods for the detection of radiation-induced chromosome damage in aquatic organisms. *In:* Methodology for Assessing Impacts of Radioactivity on Aquatic Ecosystems. Technical Report Series No. 190, International Atomic Energy Agency, Vienna, pp. 349–367.

Kligerman, A. D. 1980. The use of aquatic organisms to detect mutagens that cause cytogenetic damage. *In:* Radiation Effects on Aquatic Organisms, N. Egami (Ed.). Japan Scientific Societies Press, Tokyo, pp. 241–252.

Kligerman, A. D., and S. E. Bloom. 1976. Sister-chromatid differentiation and exchanges in adult mudminnows (*Umbra limi*) after *in vivo* exposure to 5-bromodeoxyuridine. *Chromosoma,* **56,** 101–109.

Konno, K. 1980. Effects of irradiation on the gonads of the rainbow trout, *Salmo gairdnerii irideus,* during embryonic stages. *In:* Radiation Effects on Aquatic Organisms, N. Egami (Ed.). Japan Scientific Societies Press, Tokyo, pp. 129–133.

Kosheleva, V. V. 1973a. Accumulation of radioactive isotopes by the developing eggs of the Atlantic salmon. *In:* Effect of Ionizing Radiation on the Organism, B. P. Sorokin (Ed.). AEC-tr-7418, U.S. Atomic Energy Commission, Washington, D.C., pp. 8–15.

Kosheleva, V. V. 1973b. Change in peripheral blood in the embryos and larvae of Atlantic salmon under the influence of radioactive contamination of the water and X-rays. *In:* Effect of Ionizing Radiation on the Organism, B. P. Sorokin (Ed.). AEC-tr-7418, U.S. Atomic Energy Commission, Washington, D.C., pp. 75–88.

Kulikov, N. V. 1970. Radiosensitivity of tench (*Tinca tinca* L.) embryos at early stages of development. *Radiobiologiya,* **10,** 127–132.

Laszlo, A., and R. J. Horsley. 1973. Additional recovery in x-irradiated *Oedogonium cardiacum* can be suppressed by cycloheximide. *International Journal of Radiation Biology,* **23,** 201–204.

Leenhouts, H. P., and K. H. Chadwick. 1978. The crucial role of DNA double-strand breaks in cellular radiobiological effects. *Advances in Radiation Biology,* **7,** 56–102.

Lockner, D. 1972. Radiosensitivity of haemopoiesis in fish. II. Studies at 18° and 10°C. *International Journal of Radiation Biology,* **21,** 559–572.

Luckey, T. D. 1980. Hormesis with Ionizing Radiation. CRC Press, Boca Raton, Florida, 222 pp.

Mano, Y., K. Kunihiko, and N. Egami. 1982. Photoreactivation and excision repair of thymine dimers in ultraviolet-irradiated cultured fish cells. *Radiation Research,* **90,** 501–508.

Mattern, M. R., P. V. Harihan, and P. A. Cerutti. 1975. Selective excision of gamma ray damaged thymine from the DNA of cultured mammalian cells. *Biochimica et Biophysica Acta,* **395,** 48–55.

McGregor, J. F., and H. B. Newcombe. 1968. Major malformations in trout embryos irradiated prior to active organogenesis. *Radiation Research,* **35,** 282–300.

McGregor, J. F., and H. B. Newcombe. 1972a. Decreased risk of embryo mortality following low doses of radiation to trout sperm. *Radiation Research,* **51,** 536–544.

McGregor, J. F. and H. B. Newcombe. 1972b. Dose-response relationships for yields of major eye malformations following low doses of radiation to trout sperm. *Radiation Research,* **49,** 155–169.

Metalli, P. 1979. An outline of methodological problems related to the assessment of genetic effects of radiation. *In:* Methodology for Assessing Impacts of Radioactivity on Aquatic Ecosystems. Technical Report Series No. 190, International Atomic Energy Agency, Vienna, pp. 403–412.

Michibata, H., and N. Egami. 1977. Effect of chronic gamma-irradiation on the fish, *Oryzias latipes.* III. Effects on the gonad with special reference to spermatogonial population. *Journal of Radiation Research,* **18,** 54–55.

Migalovskaya, V. N. 1973. Chronic effects of strontium-90 + yttrium-90 on the frequency of chromosomal aberrations in the embryonic cells of the Atlantic salmon. *In:* Effect of Ionizing Radiation on the Organism, B. P. Sorokin (Ed.). AEC-tr-7148, U.S. Atomic Energy Commission, Washington, D.C., pp. 89–99.

Mitani, H., H. Etoh, and N. Egami. 1982. Resistance of a cultured fish cell line (CAF-MM1) to γ-irradiation. *Radiation Research,* **89,** 334–347.

Morris, R. J. 1979. Discussion of D. H. Cushing's paper, "The monitoring of biological effects: the separation of natural changes from those induced by pollution." *Philosophical Transactions of the Royal Society of London,* **286**(B), 597–609.

Neuhold, J. M., and R. K. Sharma. 1969. Interacting effects of gamma irradiation and sodium halide concentrations on rainbow trout. *In*: Symposium on Radioecology, D. J. Nelson and F. C. Evans (Eds.). U.S. Atomic Energy Commission Conf-670503, U.S. Atomic Energy Commission, Oak Ridge, Tennessee, pp. 142–151.

Newcombe, H. B., and J. F. McGregor. 1967. Major congenital malformations from irradiations of sperm and eggs. *Mutation Research,* **4,** 663–673.

Newcombe, H. B., and J. F. McGregor. 1972. Increased embryo production following low doses of irradiation to trout sperm. *Radiation Research,* **52,** 536–544.

Oakberg, E. F. 1968. Mammalian gametogenesis and species comparisons in radiation response of the gonads. *In:* Effects of Radiation on Meiotic Systems. Report of a Study Group on the Effects of Radiation on Meiotic Systems (1967). International Atomic Energy Agency, Vienna, pp. 3–15.

Omerod, M. G. 1976. Radiation-induced strand breaks in the DNA of mammalian cells. *In:* Biology of Radiation Carcinogenesis, J. M. Yuhas, R. W. S. Tennant, and J. D. Regan (Eds.). Raven Press, New York, pp. 67–84.

Ophel, I. L. 1979. Radiation effects in contaminated aquatic ecosystems and their relation to laboratory studies. *In:* Methodology for Assessing Impacts of Radioactivity on Aquatic Ecosystems. Technical Report Series No. 190, International Atomic Energy Agency, Vienna, pp. 211–216.

Ophel, I. L., and J. M. Judd. 1967. Experimental studies of radiostrontium accumulation by fresh water fish from food and water. *In:* Radioecological Concentration Processes, B. Aberg and F. P. Hungate (Eds.). Pergamon Press, London, pp. 859–865.

Osterberg, C. L. 1981. The seas: to waste or not. *The New York Times,* 9 August 1981.

Osterberg, C. L. 1982. Why not in the ocean? *International Atomic Energy Agency Bulletin,* **24,** 30–34.

Patel, B. 1975. Field and laboratory comparability of radioecological studies. *In:* Design of Radiotracer Experiments in Marine Biological Systems. Technical Report Series No. 167, International Atomic Energy Agency, Vienna, pp. 211–240.

Patel, B., and S. Patel. 1979. Techniques for detecting radiation effects on biochemical and physiological systems in aquatic organisms. *In:* Methodology for Assessing Impacts of Radioactivity on Aquatic Ecosystems. Technical Report Series No. 190, International Atomic Energy Agency, Vienna, pp. 237–265.

Pechkurenkov, V. L., I. A. Shekhanova, and I. G. Telysheva. 1972. The effect of chronic exposure to small doses of irradiation on the embryonic development in fish and the validity of various assessment methods. *Vorposy Ikthiologii,* **12,** 84.

Penners, R. 1959. Durch Röntgenstrahlen verursachte biologische Schaden. Untersuchungen mit und an dem Schwertträger. *Zeitschrift für Naturforschung,* **5,** 403–407.

Penners, R. 1962. Experimentel erzeugte Missbildung bei Fischen. *Jahrbuch der Fürsorge für Körperbehinderte,* pp. 35–38.

Pentreath, R. J. 1976. The accumulation of mercury from food by the plaice, *Pleuronectes platessa* L. *Journal of Experimental Marine Biology and Ecology,* **25,** 51–65.

Pentreath, R. J. 1980. Nuclear Power, Man, and the Environment. Wykeham Publications, London, 225 pp.

Pentreath, R. J., and S. W. Fowler. 1979. Irradiation of aquatic animals by radionuclide incorporation. *In*: Methodology for Assessing Impacts of Radioactivity on Aquatic Ecosystems. Technical Report Series No. 190, International Atomic Energy Agency, Vienna, pp. 129–171.

Perkins, E. J. 1979. The need for sublethal studies. *Philosophical Transactions of the Royal Society of London,* **286**(B), 425–442.

Pittock, J. W., and J. Calkins. 1973. Reactivation of protozoa irradiated with americium-241 alpha particles. *Radiation Research,* **55,** 597.

Polikarpov, G. G. 1966. Radioecology of Aquatic Organisms, V. Schultz and A. W. Klement, Jr. (Eds.). Reinhold, New York, 314 pp.

Polikarpov, G. G. 1979. Experimental methods for radiobiological investigations with developing fish eggs. *In:* Methodology for Assessing Impacts of Radioactivity on Aquatic Ecosystems. Technical Report Series No. 190, International Atomic Energy Agency, Vienna, pp. 173–194.

Polikarpov, G. G., and V. N. Ivanov. 1962. The harmful effect of ^{90}Sr-^{90}Y in the early development of the red mullet, the green wrasse, the horse mackerel and the Black Sea anchovy. *Doklady Akademii Nauk SSSR, Seriya Biologiya,* **144,** 219.

Preston, A. 1959. Leucocytosis in response to bacteraemia as a feature of the radiation syndrome in the plaice. *Nature,* **183,** 832–833.

Purdom, C. E. 1966. Radiation and mutation in fish. *In:* Disposal of Radioactive Wastes into Seas, Oceans and Surface Waters. Proceedings of the Symposium, International Atomic Energy Agency, Vienna, pp. 861–867.

Purdom, C. E., and D. S. Woodhead. 1973. Radiation damage in fish. *In:* Genetics and Mutagenesis of Fish, J. H. Schröder (Ed.). Springer-Verlag, New York, pp. 67–73.

Raicu, P., and E. Taisescu. 1972. *Misgurnus fossilis,* a tetraploid fish species. *Journal of Heredity,* **63,** 92–97.

Rees, G. H. 1962. Effects of gamma radiation on two decapod crustaceans, *Palaemonetes pugio* and *Uca pugnax. Chesapeake Science,* **3,** 29–34.

Regan, J. D., and J. S. Cook. 1967. Photoreactivation in an established vertebrate cell line. *Proceedings of the National Academy of Sciences of the United States of America,* **58,** 2274–2279.

Rice, T. R., and J. W. Angelovic. 1969. Radioactivity in the sea: effects of fisheries. *In:* The Encyclopedia of Marine Resources, F. E. Firth (Ed.). Van Nostrand, New York, pp. 574–576.

Roberts, F. L. 1970. Atlantic salmon (*Salmo salar*) chromosomes and speciation. *Transactions of the American Fisheries Society,* **99,** 105–111.

Rosenthal, H. L., and R. S. Rosenthal. 1950. Lordosis, a mutation in the guppy. *Journal of Heredity,* **41,** 217–218.

Sakaizumi, M., and N. Egami. 1980. Enhancement of mercury compound toxicity to the embryos of *Oryzias latipes* by halogens. *In:* Radiation Effects on Aquatic Organisms, N. Egami (Ed.). Japan Scientific Societies Press, Tokyo, pp. 73–77.

Schröder, J. H. 1969. Inheritance of radiation-induced spinal curvatures in the guppy, *Lebistes reticulatus. Canadian Journal of Genetics and Cytology,* **11,** 937–947.

Schröder, J. H. (Ed.). 1973. Genetics and Mutagenesis of Fish. Springer-Verlag, New York, 356 pp.

Schröder, J. H. 1979. Methods for screening radiation-induced mutations in fish. *In:* Methodology for Assessing Impacts of Radioactivity on Aquatic Ecosystems. Technical Report Series No. 190, International Atomic Energy Agency, Vienna, pp. 381–402.

Schröder, J. H. 1980. Radiation-induced mutations in fish. *In:* Radiation Effects on Aquatic Organisms, N. Egami (Ed.). Japan Scientific Societies Press, Tokyo, pp. 217–222.

Schröder, J. H., and S. Holzberg. 1972. Population genetics of *Lebistes (Poecilia) reticulatus* Peters (Poeciliidae, Pisces). I. Effects of radiation-induced mutations on the segregation ratio in postirradiation F_2. *Genetics,* **70,** 621–630.

Setlow, R. B. 1982. Dose response relations: The effects of DNA repair. *In:* Genetic Toxicology: An Agricultural Perspective, R. A. Fleck and A. Hollaender (Eds.). Plenum Press, New York, pp. 391–401.

Setlow, R. B. 1983. Quantitative assessment of risks. *In:* Short-Term Tests for Environmentally Induced Chronic Health Effects, A. D. Woodhead and M. D. Waters (Eds.). U.S. Environmental Protection Agency, Washington, D.C., pp. 1–10.

Shekhanova, I. A., and V. L. Pechkurenkov. 1968. Accumulation of strontium-90 and yttrium-90 dissolved in the water and its effect on the embryonic development of loach. *Journal of Ichthyology,* **8,** 551–561.

Shekhanova, I. A., and E. A. Voronina. 1971. Effect of radioactive contamination on the reproductive capacity of fish. *Trudy Vsesoyuznyi Nauchno-Issledovatel'skogo Instituta Morskogo Rybnogo Khozyaistva Okeanografii,* **70,** 95.

Shima, A., O. Nikaido, S. Shinohara, and N. Egami. 1980. Effect of ultraviolet light on fish cells *in vivo* and *in vitro* as revealed by Feulgen–DNA cytofluorometry. *In:* Radiation Effects on Aquatic Organisms, N. Egami (Ed.). Japan Scientific Societies Press, Tokyo, pp. 183–193.

Shima, A. M., M. Ikenaga, O. Nikaido, H. Takebe, and N. Egami. 1981. Photoreactivation of ultraviolet light-induced damage in cultured fish cells as revealed by increased colony forming ability and decreased content of pyrimidine dimers. *Photochemistry and Photobiology,* **33,** 313–316.

Sparrow, A. H., C. H. Nauman, G. M. Donnelly, D. L. Willis, and D. G. Baker. 1970. Radiosensitivities of selected amphibians in relation to their nuclear and chromosome volumes. *Radiation Research,* **42,** 353–371.

Steele, J. H. 1979. The uses of experimental ecosystems. *Philosophical Transactions of the Royal Society of London,* **286**(B), 583–595.

Streisinger, G., C. Walker, N. Dower, D. Knauber, and F. Singer. 1981. Production of clones of homozygous diploid zebra fish (*Brachydario rerio*). *Nature,* **291,** 293–296.

Suyama, I., H. Etoh, and R. Ichikawa. 1980. Effects of ionizing radiation on the development of *Limanda* eggs. *In:* Radiation Effects on Aquatic Organisms, N. Egami (Ed.). Japan Scientific Societies Press, Tokyo, pp. 205–207.

Takeushi, K. 1966. 'Wavy-fused' mutants in the medaka, *Oryzias latipes. Nature,* **211,** 866–867.

Tarrant, R. M. 1971. Seasonal variation in the accumulation and loss of ^{131}I by tissues of adult female channel catfish, *Ictalurus punctatus* (Rafinesque). *Transactions of the American Fisheries Society,* **100,** 237–246.

Templeton, W. L. 1980. Ecological effects of transuranics in the marine environment. *In:* Radiation Effects on Aquatic Organisms, N. Egami (Ed.). Japan Scientific Societies Press, Tokyo, pp. 3–12.

Till, J. E., S. V. Kaye, and J. R. Trabalka. 1976. The Toxicity of Uranium and Plutonium to the Developing Embryos of Fish. Report ORNL-5160, Environmental Sciences Division No. 884, Oak Ridge, Tennessee, 193 pp.

Tsytsugina, V. G. 1972. Effect of incorporated radionuclides on the chromosome apparatus of ocean fish. *In:* Marine Radioecology, G. G. Polikarpov (Ed.). AEC-tr-7299, U.S. Atomic Energy Commission, Washington, D.C., pp. 157–165.

Tsytsugina, V. G. 1973. Effect of incorporated radionuclides on the chromosomal apparatus of marine fish. *In:* Effects of Ionizing Radiation on the Organism, B. P. Sorokin (Ed.). AEC-tr-7418, U.S. Atomic Energy Commission, Washington, D.C., pp. 157–165.

Tsytsugina, V. G. 1979. Methods of demonstrating cytogenetic radiation damage to aqueous organisms *in vivo* or in systems of tissue culture. *In:* Methodology for Assessing Impacts of Radioactivity on Aquatic Ecosystems. Technical Report Series No. 190, International Atomic Energy Agency, Vienna, pp. 369–380.

Turner, F. B., P. Licht, J. D. Thrasher, P. A. Medica, and J. R. Lannom, Jr. 1971. Radiation-induced sterility in natural populations of lizards (*Crotaphytus wislizenii* and *Cnemidorus tigris*). *In:* Radionuclides in Ecosystems, D. J. Nelson (Ed.). Proceedings Third National Symposium on Radioecology, Conf-710501, Oak Ridge National Laboratory, Oak Ridge, Tennessee, pp. 1131–1143.

Ulrickson, G. U. 1971. Radiation effects on serum proteins, hematocrits, electrophoretic patterns and protein components in the blue gill (*Lepomis macrochirus*). *In:* Radionuclides in Ecosystems, D. J. Nelson (Ed.). Proceedings Third National Symposium on Radioecology, Conf-710501, Oak Ridge National Laboratory, Oak Ridge, Tennessee, pp. 1100–1105.

Upton, A. C., A. W. Kimball, J. Furth, K. W. Christenberry, and W. E. Benedict. 1960. Some delayed effects of atom-bomb radiations in mice. *Cancer Research,* **20,** 1–62.

Wadley, G. W., and A. D. Welander. 1971. X-rays and temperature: combined effects on mortality and growth of salmon embryos. *Transactions of the American Fisheries Society,* **100,** 267–274.

Walburg, H. E., Jr. 1975. Radiation-induced life-shortening and premature aging. *Advances in Radiation Biology,* **5,** 147–179.

Ward, E., S. A. Beach, and E. D. Dyson. 1971. The effect of acute x-irradiation on the development of the plaice *Pleuronectes platessa* L. *Journal of Fish Biology,* **3,** 251–258.

Weis, P., and J. S. Weis. 1976. Abnormal locomotion associated with skeletal malformations in the sheephead minnow, *Cyprinodon variegatus,* exposed to malathion. *Environmental Research,* **12,** 196–200.

Welander, A. D. 1954. Some effects of x-irradiation of different embryonic stages of the trout (*Salmo gairdnerii*). *Growth,* **18,** 227–255.

Welander, A. D., L. R. Donaldson, R. F. Foster, K. Bonham, and A. H. Seymour. 1948. The effects of roentgen rays on the embryos and larvae of the chinook salmon. *Growth,* **12,** 203–242.

Welander, A. D., L. R. Donaldson, R. F. Foster, K. Bonham, A. H. Seymour, and F. G. Lowman. 1949. The Effects of Roentgen Rays on Adult Rainbow Trout. United States Atomic Energy Commission Report, Publication No. 17, University of Washington Applied Fisheries Laboratory, Seattle, Washington, 21 pp.

Willis, D. L. 1980. The effects of temperature on the radiation response of the rough-skinned newt, *Taricha granulosa*. *In:* Radiation Effects on Aquatic Organisms, N. Egami (Ed.). Japan Scientific Societies Press, Tokyo, pp. 157–167.

Willis, D. L., and W. L. Lappenbusch. 1976. The radiosensitivity of the rough-skinned newt (*Taricha granulosa*). *In:* Radioecology and Energy Resources, C. E. Cushing, Jr. (Ed.). Dowden, Hutchinson and Ross, Stroudsburg, Pennsylvania, pp. 363–375.

Wolf, K., and J. A. Mann. 1980. Poikilotherm vertebrate cell lines and viruses: a current listing for fishes. *In Vitro,* **16,** 168–179.

Woodard, H. Q. 1980. Radiation Carcinogenesis in Man: A Critical Review. EML-380, U.S. Department of Energy, Washington, D.C., 137 pp.

Woodhead, A. D., and P. M. Achey. 1979. Photoreactivating enzyme in the blind cave fish, *Anoptichthys jordani*. *Comparative Biochemistry and Physiology,* **63**(B), 73–76.

Woodhead, A. D., and R. B. Setlow. 1980. Response to and recovery from acute sublethal gamma-radiation in the Amazon molly, *Poecilia formosa*. *In:* Radiation Effects on Aquatic Organisms, N. Egami (Ed.). Japan Scientific Societies Press, Tokyo, pp. 171–182.

Woodhead, A. D., P. M. Achey, R. B. Setlow, and E. Grist. 1978. Photoenzymatic repair of ultraviolet-irradiated DNA in the cells of a shark, *Prionace glauca*. *Comparative Biochemistry and Physiology,* **60**(B), 205–208.

Woodhead, A. D., R. B. Setlow, and R. W. Hart. 1979. Genetically uniform strains of fish as laboratory models for experimental studies of the effects of ionizing radiation. *In:* Methodology for Assessing Impacts of Radioactivity on Aquatic Ecosystems. Technical Report No. 190, International Atomic Energy Agency, Vienna, pp. 317–333.

Woodhead, A. D., R. B. Setlow, and E. Grist. 1980. DNA repair and longevity in three species of cold-blooded vertebrates. *Experimental Gerontology,* **15,** 301–304.

Woodhead, D. S. 1976. Influence of acute irradiation on induction of chromosome aberrations in cultured cells of the fish *Ameca splendens*. *In:* Biological and Environmental Effects of Low-Level Radiation, Vol. 1. International Atomic Energy Agency, Vienna, pp. 67–76.

Woodhead, D. S. 1977. The effects of chronic irradiation on the breeding performance of the guppy, *Poecilia reticulata* (Osteichthyes: Teleostei). *International Journal of Radiation Biology,* **32,** 1–22.

Woodhead, D. S. 1979. Methods of dosimetry for aquatic organisms. *In:* Methodology for Assessing Impacts of Radioactivity on Aquatic Ecosystems. Technical Report No. 190, International Atomic Energy Agency, Vienna, pp. 43–96.

Young, M. L. 1977. The roles of food and direct uptake from the water in the accumulation of zinc and iron in the tissues of the dogwhelk, *Nucella lapillus* (L.). *Journal of Experimental Marine Biology and Ecology,* **30,** 315–325.

PART V: ORGANIC ENRICHMENT FROM SEWAGE-SLUDGE DISPOSAL IN THE MARINE ENVIRONMENT

Chapter 16

Potentially Pathogenic Protozoa in Sediments from Oceanic Sewage-Disposal Sites

Thomas K. Sawyer

National Marine Fisheries Service
Northeast Fisheries Center
Oxford, Maryland

Thomas A. Nerad and Pierre-Marc Daggett

American Type Culture Collection
Rockville, Maryland

Shawn M. Bodammer

Environmental Concern, Inc.
St. Michaels, Maryland

ABSTRACT

Free-living amoebae belonging to the genus *Acanthamoeba* and capable of causing granulomatous amoebic encephalitis in man and animals were isolated from sediments contaminated with sewage sludge. Strains of *Acanthamoeba* capable of infecting laboratory mice were isolated from the New York Bight dumpsite, Philadelphia–Camden dumpsite, Puerto Rico outfall, the Apalachicola River, Florida, and the Gulf of Mexico seaward from Mobile Bay, Alabama. Seaward from the New York disposal site, amoebae were recovered from sediments that tested negative for fecal bacteria but were contaminated with organic pollutants such as coprostanol and polychlorinated biphenyls. Biochemical analyses of the amoebae enzyme systems (acid phosphatase, leucine aminopeptidase, and propionyl esterase) provided enzyme patterns that showed that certain morphologically similar species differed enzymatically and, therefore, probably represent new species. Electron microscope and isoenzyme studies employed to compare two similar species, one pathogenic and the other nonpathogenic, showed that the species were probably the same and differed only in pathogenicity and temperature tolerance. Characteristics such as cyst morphology, pathogenicity, or temperature tolerance taken alone are not always sufficient for determining species identifications.

16.1. INTRODUCTION

Two of the most serious consequences of sewage-sludge disposal in the marine environment are (1) accumulation of contaminants in fish and shellfish to levels that render them unfit for human consumption, and (2) the spread of pollutants that pose public health hazards in recreational waters. Also, the long-term use of designated disposal sites may lead to the accumulation of sludge wastes at depths that smother the sea bottom and eliminate the settling of commercially valuable, sessile shellfish. Disposal sites in the New York Bight apex have been studied extensively and were found to be contaminated with fecal bacteria (Babinchak et al., 1977), polychlorinated biphenyls (West and Hatcher, 1980), and heavy metals (O'Connor and Rachlin, 1982; Steimle et al, 1982). The Philadelphia–Camden disposal site, although used for a period of only 7.5 y, was similarly contaminated by fecal bacteria (O'Malley et al., 1982), organic pollutants (Lear et al., 1981), and heavy metals (Lear et al., 1982). Free-living protozoans (Amoebida: Acanthamoebidae), some of them capable of causing fatal disease in man and animals, were found in sewage-contaminated oceanic sediments less than 10 y ago (Sawyer, 1975, 1980; Sawyer et al., 1977). More recently, Goyal et al. (1982) published a preliminary account of the isolation of human enteric viruses from water and sediment taken at the New York Bight and Philadelphia–Camden dumpsites.

The recovery of pathogenic protozoans and human enteroviruses, in addition to well-known fecal bacteria from contaminated sediments, emphasizes the importance of continued studies on the introduction of microbial pathogens into marine ecosystems by way of sewage-sludge disposal at sea. Prior to the discovery of the amoebae in marine sediments, nine species of *Acanthamoeba* were recognized, and only two of them were pathogenic to experimentally infected mice. Approximately 18 new species have been described during the past 10 y, bringing the total number of known species to 27, 6 of them proving to be pathogenic after intranasal inoculation into mice. Eight of the 27 recognized species have been isolated from marine sediments. Other strains tentatively identified as *A. culbertsoni* and *A. glebae* in earlier publications (Sawyer, 1975, 1980; Sawyer et al., 1977) were found to differ from the type strains when compared by using isoenzyme electrophoresis (Daggett et al., 1982).

Sawyer and Griffin (1975) established the family Acanthamoebidae to resolve taxonomic problems brought about by frequent reference to many species of *Acanthamoeba* as species of *Hartmannella*. The type species of the genus *A. castellanii* was originally named *Hartmannella castellanii* by Douglas (1930) but later was transferred to *Acanthamoeba* by Volkonsky (1931) when he established the genus *Acanthamoeba* and placed it within the family Hartmannellidae. Thus, prior to 1975, the two genera *Hartmannella* and *Acanthamoeba* were included in the family Hartmannellidae. The inclusion of both genera within the same family was acceptable until Culbertson et al. (1959) isolated a spiny filose amoeba from tissue culture cells and found that it killed experimentally infected laboratory animals. Singh and Das (1970) considered the strain to be a new species and called it *Hartmannella culbertsoni*. After the family Acanthamoeba was introduced, *H. culbertsoni* became known as *A. culbertsoni* and this change has been accepted by most investigators.

Data to be presented in this chapter illustrate the value of starch-gel electrophoresis, mouse inoculation experiments, and ultrastructural morphology in resolving some of the taxonomic problems associated with morphologically similar species. Reliable species identifications are especially important in cases of human death when clinical procedures fail to provide a specific identification of causative amoebic agents.

16.2. EXPERIMENTAL

Sediment samples tested for the presence of *Acanthamoeba* were collected at the New York, Philadelphia–Camden, and Puerto Rico sewage-disposal sites, and from the Apalachicola River, Florida, and the Gulf of Mexico. Bacterial most-probable-numbers (MPNs) in the Puerto Rico sediments were provided by investigators from the University of Maryland, College Park, Maryland. Most-probable-numbers from sediments at the New York, Philadelphia–Camden sites, and the Apalachicola River were provided by investigators from the U.S. Food and Drug Administration, Davisville, Rhode Island. Bacteriological studies were not conducted at the Gulf of Mexico stations.

Sediment samples were collected with a Smith-McIntyre, Ekman, or Ponar grab device (Kahl Scientific Instrument, El Cajon, California), and subsampled with sterile wooden applicators, and stored under refrigeration in sterile plastic dishes until culture preparation. Cultivation on bacterized agar plates and identification procedures were carried out as described by Sawyer et al. (1982) and Sawyer and Bodammer (1983). After several weeks or months of continuous cultivation, most amoeba strains were tested for heat tolerance at 37 to 40°C. Temperature-tolerant strains were tested for pathogenicity in laboratory mice as described by Daggett et al. (1982). Test mice that did not die were killed 10–35 d after inoculation, and small pieces of brain tissue were placed on bacterized agar plates to determine whether amoebae could be recultured from exposed animals. Extracts of selected amoeba strains were prepared for starch-gel isoenzyme electrophoresis as described by Nerad and Daggett (1979).

Two strains of amoebae characterized as *A. hatchetti* and *A. triangularis*, both producing triangular-shaped cysts and having nearly identical isoenzyme patterns, were processed for ultrastructural studies. A third strain, also pathogenic to mice and previously identified as *A. culbertsoni*, was also processed for electron microscopy. Amoebae were grown in 50-ml screw-capped glass tubes for 3–5 d, fixed in cold 2% glutaraldehyde prepared in 0.1 *M* phosphate buffer (pH 7.4) and then transferred to conical centrifuge tubes for three rinsings in the buffer solution. Pelletized amoebae were post-fixed in 1% osmium tetraoxide:phosphate buffer, dehydrated through ethanol and propylene oxide, and embedded in Epon-Araldite. Thin sections were stained with uranyl acetate and lead citrate and examined with a Zeiss Model EM 9S2 electron microscope (Carl Zeiss, Inc., Thornwood, New York).

16.3. RESULTS

Nine strains of amoebae grew at 40°C or higher and were designated as heat-tolerant species. Four were identified as *A. hatchetti, A lenticulata, A. royreba,* and *A. comandoni,* and the other five were provisionally identified as *A.* spp. 1–5 (Table 16.1). Each of these nine species yielded distinctly different zymograms (isoenzyme patterns) in all three enzyme systems tested: acid phosphatase, leucine aminopeptidase, and propionyl esterase. Four of the nine species (*A. hatchetti, A. lenticulata, A.* sp.1, and *A.* sp. 5) either killed mice after nasal inoculation or yielded cultures of amoebae from brain tissue. The remaining five strains require further testing under conditions in which mice are maintained for longer than 35 d. Including the 5 undesignated species, 14 taxonomically different species were recovered from the sediment samples, and 9 were identified as well-known species (Table 16.2). Not all strains of the pathogenic species killed mice or were recovered from brain tissue (Table 16.1).

16.3.1. New York Bight Dumpsite

Forty-eight stations, closed to commercial shellfishing, in the New York Bight area were sampled during July 1978–May 1981 (Fig. 16.1). Thirty-seven (77%) of the stations were positive for *Acanthamoeba*; there were 35 stations (73%) with coliform bacterial MPNs up to >160,000, and 22 stations (46%) with fecal coliform MPNs up to 3300. Seven strains of the amoebae grew at temperatures of 40°C or higher, and two of them (*A. lenticulata* and *A.* sp.1) either killed inoculated mice or were recultured from brain tissue (Table 16.1). The other heat-tolerant, apparently nonpathogenic strains were identified as *A. comandoni, A. royreba, A.* sp. 2, and *A.* sp. 3. Each of the seven heat-tolerant strains produced distinctly different enzyme patterns when tested by starch-gel electrophoresis. Species that did not grow at elevated temperatures included *A. griffini, A. polyphaga, A. castellanii,* and *A. rhysodes* (Table 16.1).

16.3.2. Philadelphia–Camden Dumpsite

One hundred and twenty-five stations in the area affected by sludge disposal at the Philadelphia–Camden site (Fig. 16.2) were sampled for *Acanthamoeba* during the period April 1978–May 1981 (Table 16.2). Within a radius of 8.5 km from the center of the dumpsite, 55% of the stations were positive for *Acanthamoeba*, while 15% of the stations outside of this zone, extending ~33 km to the northeast and southwest, yielded amoebae. Coliform bacteria were present at 64% and fecal coliforms at 47% of the stations at the dumpsite center, and were also present at 10% and 6%, respectively, of the stations outside of the center. Coliform MPNs were sometimes >2400, while fecal coliform MPNs were usually <2400; most stations without coliform bacteria were outside the center of the site. Four strains of amoebae (Table 16.1) grew at temperatures of 40°C, and *A. hatchetti* were recultured from brain tissue. A second heat-tolerant strain tentatively identified as *A. lenticulata* did not kill mice, although a similar strain from the New York site was pathogenic. The remaining temperature-tolerant strains, designated *A.* spp. 3 and 4, were not pathogenic. All four heat-tolerant strains yielded distinctly different enzyme patterns. Three other species that did not grow at 40°C, *A. polyphaga, A. castellanii,* and *A. rhysodes*, also were present at the Philadelphia–Camden site (Tables 16.1 and 16.2).

16.3.3. Apalachicola River, Florida

Six stations in the Apalachicola River and Apalachicola Bay, Florida, were sampled for amoebae during February 1981

Table 16.1. Distribution, Temperature Tolerance, and Pathogenicity of *Acanthamoeba* Species by Location

Species[a]	Source	Maximum Growth Temperature (°C)	Time Required to Kill Mice (d)	*Acanthamoeba* Recultured from Mouse Brain
A. hatchetti	New York Bight Philadelphia–Camden[b] Apalachicola River	40		+
A. lenticulata	*New York Bight*[b] *Philadelphia–Camden* *Puerto Rico* *Apalachicola River*[b]	*40–43*[c]	*17–35*[c]	+
A. royreba	New York Bight	40		
A. griffini	New York Bight	35		
A. polyphaga	New York Bight Philadelphia–Camden Puerto Rico Apalachicola River Gulf of Mexico	35		
A. castellanii	New York Bight Philadelphia–Camden Apalachicola River	35		
A. rhysodes	New York Bight Philadelphia–Camden Puerto Rico Apalachicola River	37		
A. astronyxis	Puerto Rico	35		
A. comandoni	New York Bight	40		
A. sp.1	New York Bight[b]	40–45[c]	32–35[c]	+
A. sp. 2	New York Bight	40		
A. sp. 3	New York Bight Philadelphia–Camden	40		
A. sp. 4	Philadelphia–Camden Puerto Rico Apalachicola River	40		
A. sp. 5	Gulf of Mexico[b]	40	9–10	+

[a] Certain strains of *A. hatchetti* and *A. lenticulata* were re-established from cultured mouse brain and others were not. Strain *A.* sp. 5 was highly pathogenic.

[b] Location yielded pathogenic strains; however, not all strains were tested for pathogenicity, and of the strains tested not all were pathogenic.

[c] From Daggett et al. (1982).

(Fig. 16.3). The stations ranged from fresh to saline (26.7 ‰) water (Table 16.3). The six stations yielded coliform MPNs up to 23,000 and fecal coliform MPNs up to 700. All six stations had one or more species of *Acanthamoeba*. Three strains grew at 40°C, and two of them yielded amoebae from cultured brain tissue (Table 16.1). The three temperature-tolerant strains were designated *A. hatchetti*, *A. lenticulata*, and *A.* sp. 4, and all had distinct enzyme patterns. Species present that did not grow at 40°C were *A. polyphaga*, *A. castellanii*, and *A. rhysodes* (Table 16.1).

Table 16.2. *Acanthamoeba* Species Present in Sediments[a]

Species	New York Bight	Philadelphia–Camden	Puerto Rico	Apalachicola River	Gulf of Mexico
A. hatchetti	+	+	+	+	
A. lenticulata	+	+	+	+	
A. royreba	+				
A. griffini	+				
A. polyphaga	+	+	+	+	+
A. castellanii	+	+		+	
A rhysodes	+	+	+		
A. astronyxis			+		
A. comandoni	+				
A. sp. 1	+				
A. sp. 2	+				
A. sp. 3	+	+			
A. sp. 4		+	+	+	
A. sp. 5					+
Total	11	7	6	6	2

[a] Amoebae identified as *A.* sp. 1–5 produced unique zymograms as did the nine identified species.

16.3.4. Gulf of Mexico

Seven stations were sampled in the coastal waters of Florida, Alabama, and Mississippi in the Gulf of Mexico during February 1980. The stations were not located near ocean

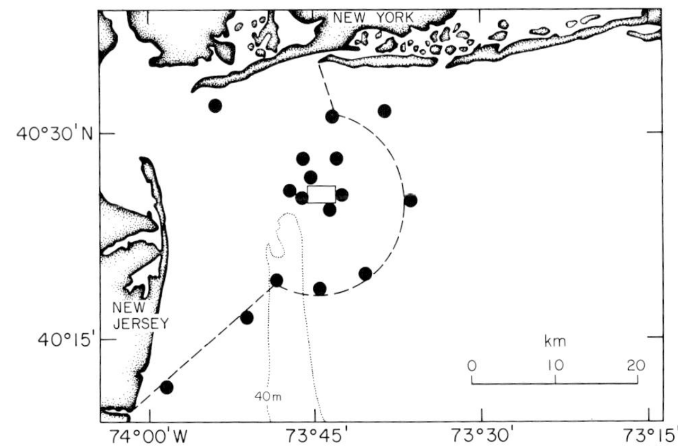

Figure 16.1. Representative *Acanthamoeba* sediment sampling sites (solid circles) in the New York Bight apex from which 48 samples were collected. The dashed line represents the area closed to shell fishing, and the square represents the sewage-sludge dumpsite.

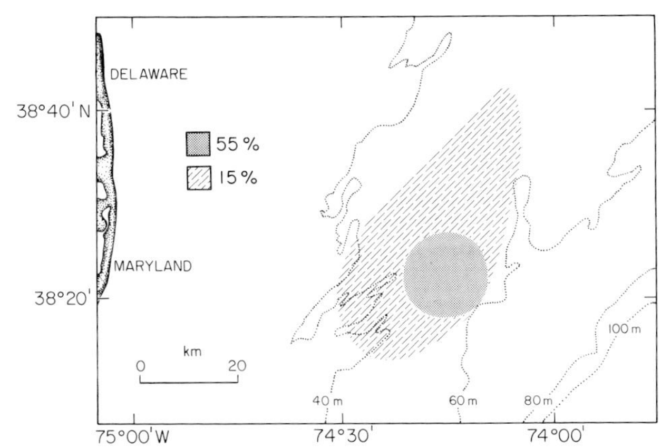

Figure 16.2. Percentage of stations at the Philadelphia–Camden sewage dumpsite that were positive for *Acanthamoeba*. The dotted area is within an 8.5 km radius of the dumpsite, and the striped area extends ~33 km to the northeast and southwest.

Table 16.3. Environmental Data for the Apalachicola River and Bay, Florida, Stations Sampled for *Acanthamoeba*.

Station[a]	Salinity (‰)	Total Coliform MPN[b]	Fecal Coliform MPN[b]	Number of *Acanthamoeba* Species
Near Jackson River	0	490	33	5
Apalachicola River	0	23,000	700	2
Near Scipio Creek	13.3	23,000	330	2
Apalachicola Bay	26.2	230	45	2
St. Vincent Island	26.7	45	20	2
St. George Island	23.0	78	20	1

[a] See Fig. 16.3 for station locations.

[b] Unpublished data, U.S. Food and Drug Administration Laboratory, Davisville, Rhode Island (courtesy Cpt. John Musselman). MPN is most-probable-number.

disposal sites, and bacteriological studies were not conducted on the sediments. One station (Fig. 16.4) yielded a temperature-tolerant strain of *Acanthamoeba* that killed two of six mice within 10 d after infection. The strain had a unique enzyme pattern and was designated *A.* sp. 5 (Tables 16.1 and 16.2). The station that yielded the pathogenic strain was located 33 km offshore from Mobile Bay, Alabama, and was the only station in which sediments were mixed with dark brown sand and silt.

16.3.5. Puerto Rico Outfall

Two stations to the east and northeast of a land-based sewage outfall between Arecibo and San Juan, Puerto Rico, were sampled in February 1982 (Fig. 16.5). One sediment sample was taken at a depth of 40 m and another at 100 m. One strain identified as *A. lenticulata* grew at 40°C; however, it did not kill mice and was not recovered from brain tissue. Five strains that were not temperature-tolerant were identified as *A polyphaga, A. rhysodes, A. astronyxis, A. hatchetti,* and *A.* sp. 4. Bacterial MPNs were not determined for the sites, but water samples taken from nearby stations contained both coliforms and fecal coliforms. Coliform MPNs from these water samples ranged up to 115, fecal coliforms ranged up to 174, and fecal streptococci ranged up to 35.

16.3.6. Amoeba Identifications

Nine species of *Acanthamoeba* were readily identified on the basis of cyst morphology, temperature tolerance, and pathogenicity to mice (Table 16.1). Five strains of amoebae, all forming spherical or subspherical cysts and growing at temperatures of 40°C or higher, produced unique zymograms that did not correspond to those produced by known strains of *Acanthamoeba*. Each of these five strains require further study to determine whether or not they are new species. The

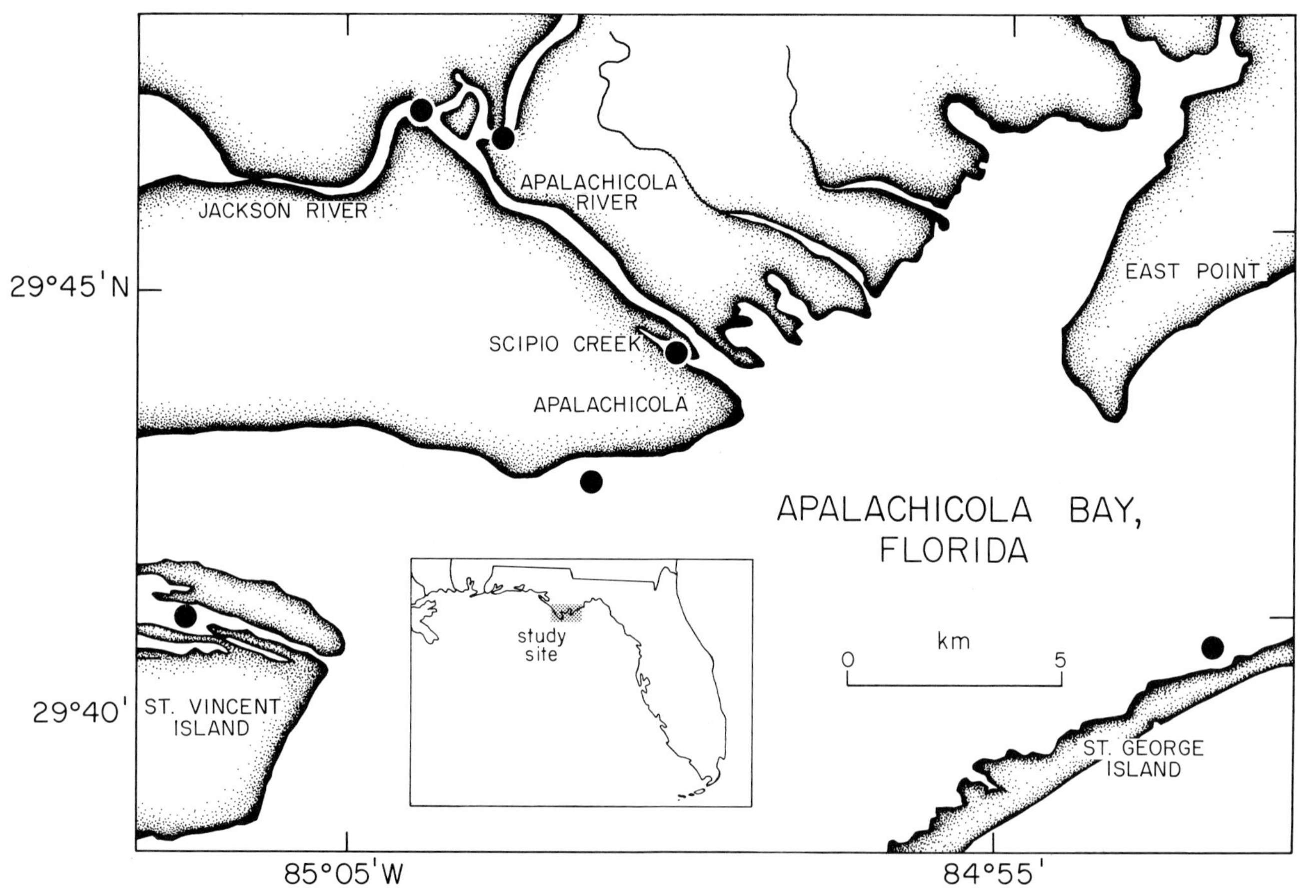

Figure 16.3. Stations sampled for *Acanthamoeba* in and around Apalachicola Bay, Florida.

strains identified as *A. lenticulata* varied in pathogenicity: some killed mice, some yielded amoebae only from cultured brain tissue, and some were nonpathogenic. Initially, strains that grew at 40°C were identified as *A. hatchetti*, and those that did not as *A. triangularis*. Tests on strains from different collections in this study showed that *A. hatchetti* did not always grow at elevated temperatures, and that limited growth sometimes occurred with *A. triangularis*. Enzyme analyses on seven strains, including the type strains of both species, showed that all produced nearly identical zymograms. *Acanthamoeba triangularis* is therefore considered a junior synonym of *A. hatchetti*, pending the outcome of further studies. Acid phosphatase zymograms for strains of *A. triangularis, A. hatchetti,* and pathogenic *A. hatchetti,* and differences between the three strains and a 45°C strain (*A.* sp. 1) are shown in Fig. 16.6. Propionyl esterase zymograms for all three strains of *A. hatchetti* were similar (Fig. 16.7). Distinct differences between acid phosphatase and propionyl esterase zymograms of *A. hatchetti* and *A.* sp. 1 were apparent (Fig. 16.8). Comparative ultrastructural features between the rounded cysts and trophozoites of *A.* sp. 1 and triangular cysts of *A. hatchetti* and *A. triangularis* are shown in Figs. 16.9 and 16.10. Amoebae of *A.* sp. 5 that killed mice in 10 d are shown in a section of brain tissue (Fig. 16.11).

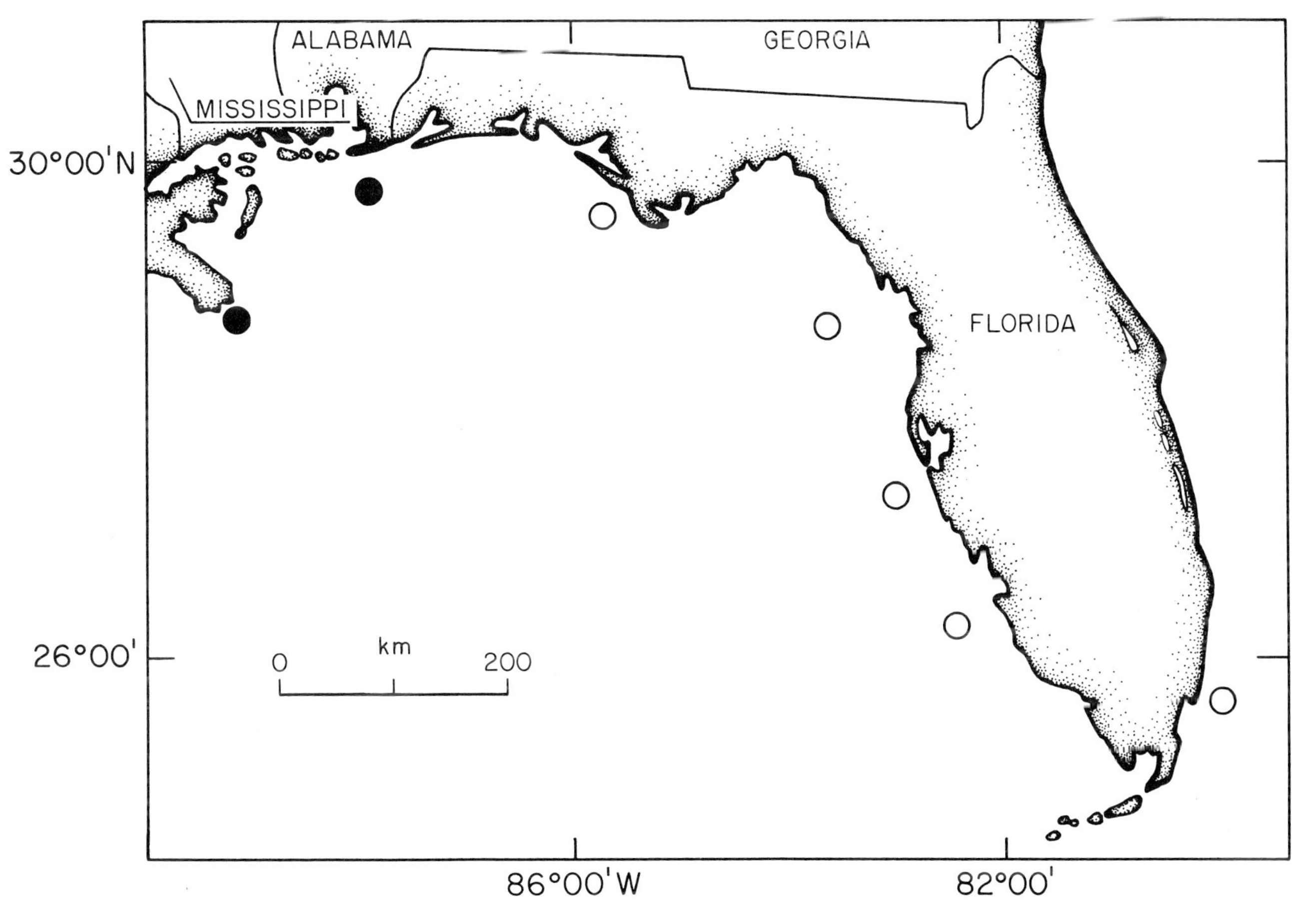

Figure 16.4. Gulf of Mexico sampling sites: Stations negative for *Acanthamoeba* are represented by open circles, and stations positive for *Acanthamoeba* are represented by solid circles. The station below Alabama (AL) yielded an unidentified pathogenic species (*A.* sp. 5, Table 16.1) and the station below Mississippi (MS) yielded *A. polyphaga* in previous studies (Sawyer, 1980).

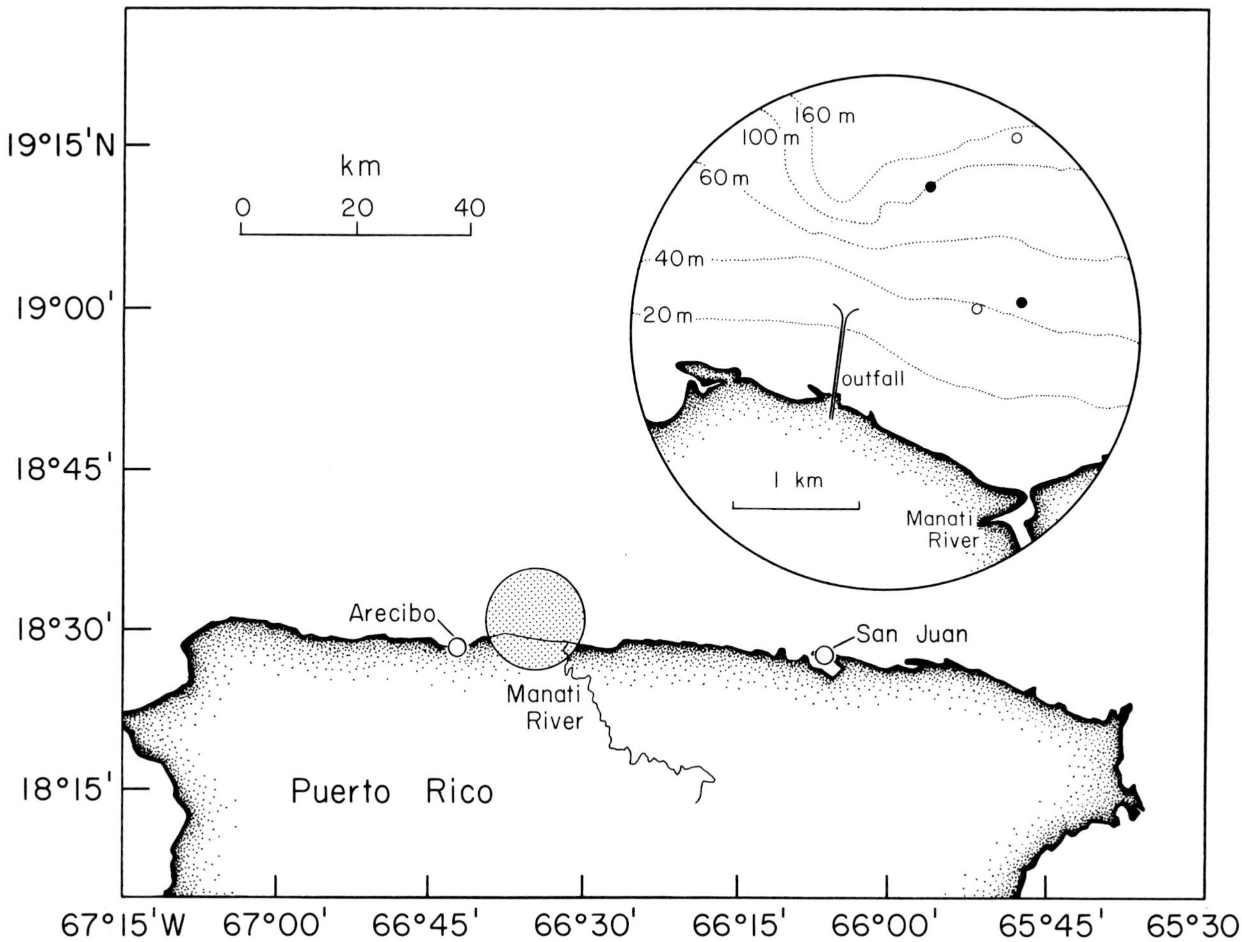

Figure 16.5. Puerto Rico sampling sites: Stations represented by solid circles were positive for *Acanthamoeba* in sediments, and stations represented by open circles were positive for coliform and fecal coliform bacteria in water.

16.3.7. Bacteriological Data

Coliform and fecal coliform MPN determinations by other investigators show that the highest numbers of sewage-associated bacteria in sediment samples were found at the New York Bight dumpsite. The second highest numbers were found in Apalachicola River sediments. These two geographically separated sampling areas showed that high bacterial MPNs in bottom sediments were associated with recovery of *Acanthamoeba* from 75 to 100% of the stations sampled. In contrast, the Philadelphia–Camden dumpsite, 67 km offshore and geographically isolated from sources of land runoff and sewage outfalls, had relatively low bacterial MPNs and fewer stations that yielded species of *Acanthamoeba*. At the Puerto Rico outfall, the recovery of amoebae from sediments and coliform and fecal coliform bacteria from the water column indicated a relationship between these two types of indicator organisms.

16.4. DISCUSSION AND CONCLUSIONS

Precise identifications for species of *Acanthamoeba* have become important because several species of the amoeba have been recovered from tissues and body fluids of humans and animals that died from granulomatous amoebic encephalitis (GAE) (Griffin, 1978). Species identifications usually have been made on the basis of cyst morphology, temperature tolerance, pathogenicity to mice, and antigenic properties. Costas and Griffiths (1980), however, compared 32 strains of *Acanthamoeba* by using starch-gel electrophoresis and

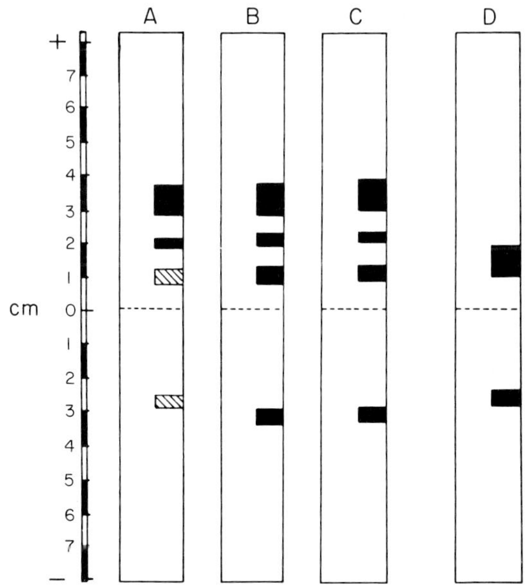

Figure 16.6. Zymograms (isoenzyme patterns) for acid phosphatase in four strains of *Acanthamoeba*: (A) *A. hatchetti* from Baltimore Harbor, Maryland; (B) *A. hatchetti* from the Philadelphia–Camden site; (C) *A. triangularis* from a human fecal specimen; and (D) *A.* sp. 1, *A. culbertsoni* from the New York Bight sewage-sludge dumpsite.

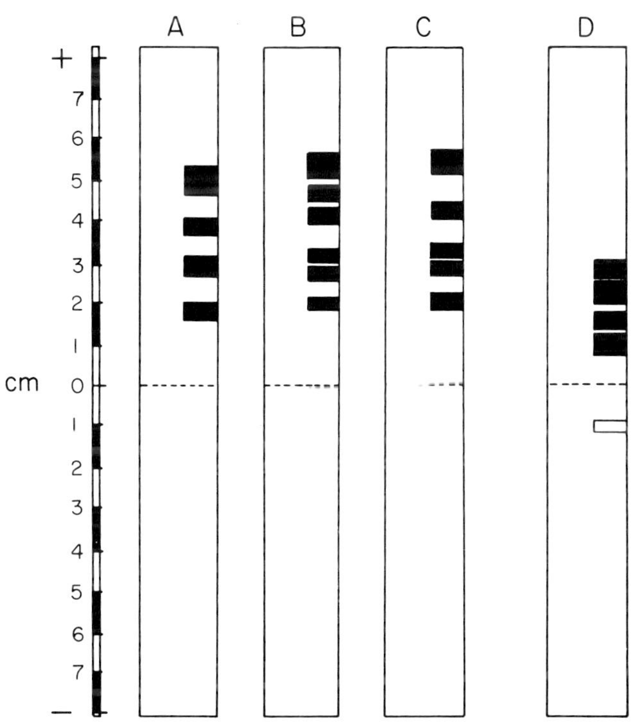

Figure 16.7. Zymograms for propionyl esterase. See Fig. 16.6 for definitions of A, B, C, and D.

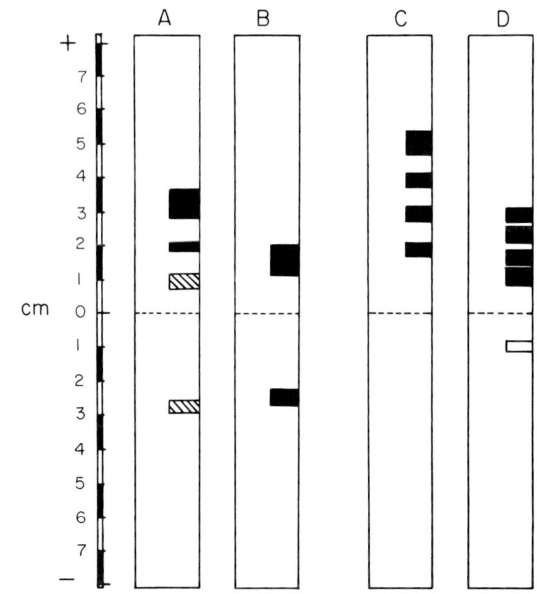

Figure 16.8. Zymograms for acid phosphatase (A and B) and propionyl esterase enzymes (C and D) in two pathogenic species of *Acanthamoeba:* (A, C) *A. hatchetti,* and (B,D) *A.* sp. 1.

found that zymograms of strains of the same species did not always correspond. Daggett et al. (1982) found that strains isolated from marine environments and sharing the same identifying morphologic characteristics did not necessarily produce the same zymograms.

Strains that grew at temperatures above 40°C killed laboratory mice and were tentatively identified as *A. culbertsoni* in earlier publications (Sawyer et al., 1977; Daggett et al., 1982). Now these strains are known to have zymograms that differ from those of the type species *A. culbertsoni*. Further studies are in progress to determine whether more than one pathogenic species is involved or whether there are geographically distinct races that form a species complex. Our studies also show that *A. hatchetti* and *A. triangularis* are probably members of a single species that have both pathogenic and nonpathogenic variants. It is of interest that pathogenic *A. hatchetti* has been isolated from highly contaminated bottom sediments in Baltimore Harbor (Sawyer et al., 1977), and that nonpathogenic *A. triangularis* has been isolated from a human fecal specimen (Pussard and Pons, 1977). Similarly, *A. lenticulata* isolated from a medicinal pool in France (Molet and Ermolieff-Braun, 1976), has both pathogenic and nonpathogenic strains (De Jonckheere, 1980). Although we isolated *A. lenticulata* from all locations except the Gulf of Mexico, only those from the New York dumpsite killed laboratory mice.

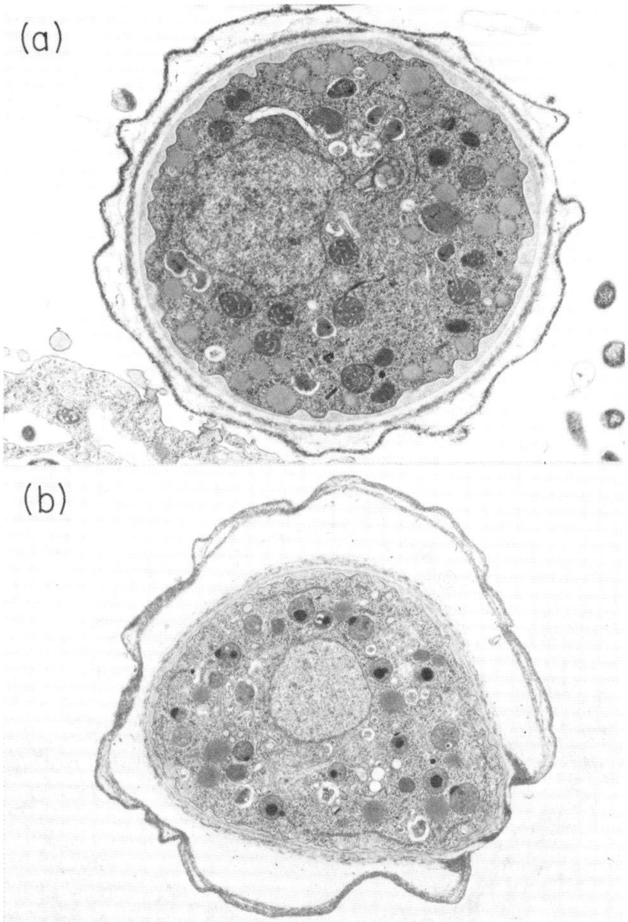

Figure 16.9. Electron photomicrographs of (a) the spherical cyst of *A.* sp. 1 (12,500X) and (b) the triangular cyst of *A. hatchetti* (9600X). Note the wrinkled ectocyst of both species.

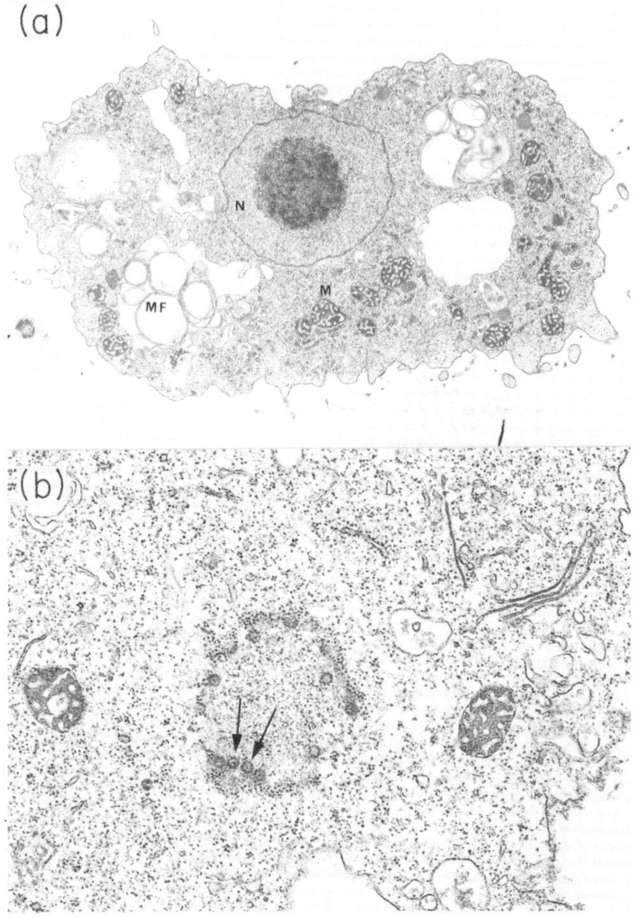

Figure 16.10. Electron photomicrographs of (a) *A.* sp. 1 showing nucleus (N) with two membranes, mitochondria (M), and myelin figures (MF) (9025X); and (b) the nucleus of *A. hatchetti* showing distinct pores through the nuclear membrane (arrows) (21,800X).

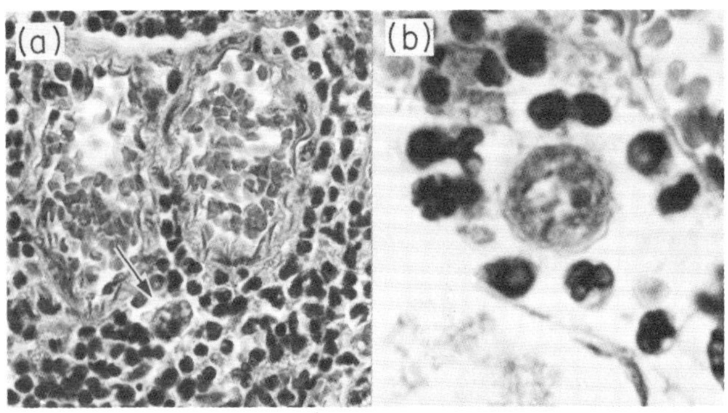

Figure 16.11. Unidentified strain (*A* sp. 5) in the brain of an experimentally infected mouse: (a) shows a trophozoite (arrow) within a large granuloma (stained with Harris' hematoxylin-eosin (640X), and (b) a cyst in the same brain tissue section (note wrinkled ectocyst wall) (1600X).

The discovery that typically free-living amoebae sometimes were capable of causing disease or death in humans and animals necessitated a new term to clearly distinguish them from obligately parasitic amoebae such as *Entamoeba histolytica*. Page (1974) resolved the problem of terminology by proposing the term "amphizoic" to characterize free-living amoebae known to cause opportunistic infections in humans and animals. Martinez (1980) considered *Acanthamoeba* an opportunistic pathogen because in most cases infection occurred in chronically ill, immunosuppressed, or immunologically deficient individuals. Infections in humans were probably overlooked prior to the discoveries reported by Culbertson et al. (1959) since many recently reported cases consider GAE disease as secondary to chronic diseases such as leukemia, Hodgkin's disease, and alcoholism (Martinez, 1982). *Acanthamoeba* infections are probably more widespread than is generally recognized since all known species form resistant cysts, and many of them are well-known contaminants in soil, water, dust, sewage sludge, and cooling towers (Daggett et al., 1982). Environmental factors that influence temperature tolerance and pathogenicity are not clearly understood, but may explain why some strains of a given species infect mice and others do not. De Jonckheere (1981) and Janitschke et al. (1982) isolated pathogenic strains of *Acanthamoeba* from thermally polluted waters. In contrast, Brown et al. (1982) isolated *Acanthamoeba* from Antarctic soil and water; some strains grew at 37°C, but none were pathogenic to mice.

The results of our studies clearly indicate that pathogenic and nonpathogenic strains of *Acanthamoeba* are useful indicators of sewage pollution in freshwater and marine ecosystems. Bacteriological studies (O'Malley et al., 1982) have shown that there is good agreement between the distribution of fecal bacteria and the amoebae in contaminated oceanic sediments. Present observations, however, show that sediments that are negative for such bacteria may yield cultures of *Acanthamoeba* when the sea bottom is known to be contaminated with other sewage pollutants such as human enteroviruses, coprostanol, PCBs, and heavy metals. Our unpublished studies of the Philadelphia–Camden dumpsite showed that the amoebae persisted in the sediments for as long as 2.5 y after the cessation of dumping. Studies at the Puerto Rico outfall and in the Apalachicola River, although limited to a single collection, yielded results consistent with our more extensive findings from the New York and Philadelphia–Camden dumpsites. *Acanthamoeba* spp. were isolated from all of the study areas contaminated by sewage bacteria. Although bacteriological studies were not made on sediments from the Gulf of Mexico, one pathogenic species was recovered from sediments near Mobile Bay. Previous studies reported the isolation of *A. polyphaga* at one station located near the outflow of the Mississippi River (Sawyer, 1980). New studies should be made in the Gulf of Mexico to test our hypothesis that river systems may be a second source of protozoan contamination in coastal waters.

Our studies show that pathogenic and nonpathogenic amoebae along with sewage bacteria, human enteroviruses, and coprostanol, are useful monitoring indicators, and that the resistant cysts of the amoebae persist in contaminated sediments when other organisms are no longer viable. In addition, the cysts may serve as markers for the dispersion of sewage pollutants away from the actual dumpsite. The pathogenic properties of certain amoeba strains suggest that their potential public health threat may be greater than previously appreciated.

ACKNOWLEDGMENTS

The authors gratefully acknowledge Cpt. John Musselman, U.S. Public Health Service, Food and Drug Administration, Davisville, Rhode Island, for providing the sediment samples and environmental data from the Apalachicola River, Florida, system and Dr. Jay Grimes, University of Maryland, College Park, Maryland, for the sediments and bacteriological data from the Puerto Rico outfall. Mr. Mark Galasso, Environmental Concern, Inc., collected the Gulf of Mexico samples and prepared the graphs and maps used in the text. We thank Dr. Joel Bodammer and Ms. Jane Wade for assistance with electron microscope preparations. Drs. Michael Devine and P. Kilho Park, of the U.S. National Oceanic and Atmospheric Administration, Rockville, Maryland, provided continuous interest and support throughout the course of our studies.

REFERENCES

Babinchak, J. A., J. T. Graikoski, S. Dudley, and M. F. Nitkowski. 1977. Distribution of fecal coliforms in bottom sediments from the New York Bight. *Marine Pollution Bulletin*, **8**, 150–153.

Brown, T. J., R. T. M. Cursons, and E. A. Keys. 1982. Amoebae from Antarctic soil and water. *Applied and Environmental Microbiology*, **44**, 491–493.

Costas, M., and A. J. Griffiths. 1980. The suitability of starch-gel electrophoresis of esterases and acid-phosphatases for the study of *Acanthamoeba* taxonomy. *Archiv für Protistenkunde*, **123**, 272–279.

Culbertson, C. G., J. W. Smith, H. K. Cohen, and J. R. Minner. 1959. Experimental infection of mice and monkeys by *Acanthamoeba*. *American Journal of Pathology*, **35**, 185–197.

Daggett, P.-M., T. K. Sawyer, and T. A. Nerad. 1982. Distribution and possible interrelationships of pathogenic and nonpathogenic *Acanthamoeba* from aquatic environments. *Microbial Ecology*, **8**, 371–386.

De Jonckheere, J. F. 1980. Growth characteristics, cytopathic effect in cell culture, and virulence in mice of 36 type strains belonging to 19 different *Acanthamoeba* spp. *Applied and Environmental Microbiology*, **39**, 681–685.

De Jonckheere, J. F. 1981. Pathogenic and nonpathogenic *Acanthamoeba* spp. in thermally polluted discharges and surface waters. *Journal of Protozoology*, **28**, 56–59.

Douglas, M. 1930. Notes on the classification of the amoeba found by Castellani in cultures of yeast-like fungus. *Journal of Tropical Medicine and Hygiene*, **33**, 258–259.

Goyal, S. M., K. Goyal, and J. K. Melnick. 1982. Impact of ocean disposal of sewage sludge on the distribution and cycling of human enteric viruses. Abstract No. Q3, March 1982. American Society for Microbiology, Atlanta, Georgia.

Griffin, J. L. 1978. Pathogenic free-living amoebae. *In:* Parasitic Protozoa, Vol. II: Intestinal Flagellates, Histomonads, Trichomonads, Amoebas, Opalinids, and Ciliates, J. P. Kreier (Ed.). Academic Press, New York, pp. 507–549.

Janitschke, K., S. Lichy, and C. Westphal. 1982. Examination of thermally polluted water for free living amoebae and testing for their possible pathogenic properties. *Zentrablatt fuer Bakteriologie und Hygiene I Abt Original B*, **176**, 160–166.

Lear, D. W., M. L. O'Malley, S. K. Smith. 1981. Effects of ocean dumping on a temperate midshelf environment. *In:* Ocean Dumping of Industrial Wastes, B. H. Ketchum, D. R. Kester, and P. K. Park (Eds.). Plenum Press, New York, pp. 458–502.

Lear, D. W., M. L. O'Malley, W. C. Muir, and G. Pence. 1982. Environmental effects of sewage sludge at the Philadelphia dumping site. *In*: Ecological Stress and the New York Bight: Science and Management, G. F. Mayer (Ed.). Estuarine Research Federation, Columbia, South Carolina, pp. 481–493.

Martinez, A. J. 1980. Is *Acanthamoeba* encephalitis an opportunistic infection? *Neurology*, **30**, 567–574.

Martinez, A. J. 1982. Acanthamoebiasis and immunosuppression. *Journal of Neuropathology and Experimental Neurology*, **41**, 548–557.

Molet, B., and G. Ermolieff-Braun. 1976. Description d'une amibe d'eau douce: *Acanthamoeba lenticulata*, sp. nov. (Amoebida). *Protistologica*, **12**, 571–576.

Nerad, T. A., and P.-M. Daggett. 1979. Starch gel electrophoresis: an effective method for separation of pathogenic and nonpathogenic *Naegleria* strains. *Journal of Protozoology*, **26**, 613–615.

O'Connor, J. M., and J. W. Rachlin. 1982. Perspectives on metals in New York Bight organisms: factors controlling accumulation and body burdens. *In:* Ecological Stress and the New York Bight: Science and Management, G. F. Mayer (Ed.). Estuarine Research Federation, Columbia, South Carolina, pp. 655–673.

O'Malley, M. L., D. W. Lear, W. N. Adams, J. Gaines, T. K. Sawyer, and E. J. Lewis. 1982. Microbial contamination of continental shelf sediments by wastewater. *Journal Water Pollution Control Federation*, **54**, 1311–1317.

Page, F. C. 1974. *Rosculus ithacus* Hawes, 1963 (Amoebida, Flabelluidae) and the amphizoic tendency in amoebae. *Acta Protozoologica*, **13**, 134–154.

Pussard, M., and R. Pons. 1977. Morphologie de la paroi kystique et taxonomie du genre *Acanthamoeba* (Protozoa, Amoebida). *Protistologica*, **13**, 557-598.

Sawyer, T. K. 1975. Description of a heat-tolerant species of *Acanthamoeba* (Amoebida) from ocean sediment. (Abstract). *Transactions of the American Microscopical Society*, **94**, 437.

Sawyer, T. K. 1980. Marine amoebae from clean and stressed bottom sediments of the Atlantic Ocean and Gulf of Mexico. *Journal of Protozoology*, **27**, 13–32.

Sawyer, T. K., and S. M. Bodammer. 1983. Marine amoebae (Protozoa: Sarcodina) as indicators of healthy or impacted sediments in the New York Bight apex. *In:* Wastes in the Ocean, Vol. 1: Industrial and Sewage Wastes in the Ocean, I. W. Duedall, B. H. Ketchum, P. K. Park, and D. R. Kester (Eds.). Wiley-Interscience, New York, pp. 338–352.

Sawyer, T. K., and J. L. Griffin. 1975. A proposed new family, Acanthamoebidae n. fam. (order Amoebida), for certain cyst-forming filose amoebae. *Transactions of the American Microscopical Society*, **94**, 93–98.

Sawyer, T. K., G. S. Visvesvara, and B. A. Harke. 1977. Pathogenic amoebas from brackish and ocean sediments, with a description of *Acanthamoeba hatchetti*, n. sp. *Science*, **196**, 1324–1325.

Sawyer, T. K., E. J. Lewis, M. Galasso, D. W. Lear, M. L. O'Malley, W. N. Adams, and J. Gaines. 1982. Pathogenic amoebae in ocean sediments near wastewater sludge disposal sites. *Journal Water Pollution Control Federation*, **54**, 1318–1323.

Singh, B. N., and S. R. Das. 1970. Studies on pathogenic and nonpathogenic small free-living amoebae and the bearing of nuclear division on the classification of the order Amoebida. *Philosophical Transactions of the Royal Society of London*, **259** (B), 435–476.

Steimle, F., J. Caracciolo, and J. B. Pearce. 1982. Impacts of dumping on New York Bight apex benthos. *In:* Ecological Stress and the New York Bight: Science and Management, G. F. Mayer (Ed.). Estuarine Research Federation, Columbia, South Carolina, pp. 213–223.

Volkonsky, M. 1931. *Hartmannella castellanii* Douglas et classification des Hartmannelles. *Archives de Zoologie Experimental et General*, **72**, 317–339.

West, R. H., and P. G. Hatcher. 1980. Polychlorinated biphenyls in sewage sludge and sediments of the New York Bight. *Marine Pollution Bulletin*, **11**, 126–129.

Chapter 17

Benthic Ecology in an Accumulating Sludge-Disposal Site

Thomas H. Pearson

Dunstaffnage Marine Research Laboratory
Oban, Argyll, Scotland

ABSTRACT

The distribution of benthic fauna in the area of the sludge-disposal site in the Firth of Clyde, Scotland, is related to the gradient of increasing organic enrichment created by dumping. Faunal succession in relation to enrichment is related to changing productivity along the gradient. Community changes are ascribed to progressive maximization of carbon utilization as input levels increase; this involves responses at many levels of organization. It is concluded that the benthic ecosystem has fully adapted to the continuously high organic carbon input.

17.1. INTRODUCTION

Deposition of sewage sludge at Garroch Head in the Firth of Clyde, Scotland, has been taking place for many years. Currently the operating authority is licensed to discharge up to 1.55×10^6 metric tons (t) of sludge per year [192 (dry) t d^{-1} in 1977] into a 6-km^2 designated disposal area with a depth of 70–80 m (Fig. 17.1). The Garroch Head sediments are said to be of the accumulating type (U.K. Department of the Environment, Standing Committee on the Disposal of Sewage Sludge, 1979). Sedimentary material settles out of the water column rapidly and is not widely distributed over the area. The fate of the deposited sludge and its effect on the benthic communities of the area is, therefore, of some concern, and previous studies of the area have been undertaken (Mackay et al., 1972; Halcrow et al., 1974; MacIntyre, 1977). Since 1977 sedimentary conditions and distribution of faunal communities in the area have been monitored. Some preliminary observations from these investigations are reported in this chapter and are used to assess the desirability of the disposal of carbon-rich wastes into this type of marine environment.

17.2. METHODS

In order to assess faunal distributions over the area, two transect lines of stations were occupied on each sampling occasion (Fig. 17.1). These lines, corresponding to Decca Navigator grid lines for subsequent ease of station location, intersect in the center of the disposal site at right angles to each other. Stations were spaced at ~1-km intervals for 6 km on either side of the center. Two 0.1-m^2 Van Veen grab samples of the sediments were obtained at each station. One grab was sieved through a 1-mm mesh screen, and the residue was retained for examination of the macrobenthic fauna. The other grab was used to provide samples for chemical analysis before being retained as a back-up faunal sample. Faunal change along the transects was, therefore, assessed by intercomparison of a line transect series of single samples taken at spaced intervals. Point variability at any one position was not assessed; however, the uniformity of the background sediments and the relatively close spacing of the sampling points suggests a low background variability.

In addition to the grab samples, core samples were taken at 56 sampling stations throughout the area. The redox potential (Eh) and acidity (pH) profiles for the upper sediment column were recorded using the methods described by Pearson and Stanley (1979). Values from 40 mm deep in the sediment were used for interstation comparison. Following pretreatment to remove carbonate carbon, organic carbon and nitrogen were measured for samples from the transect stations by using a Perkin-Elmer carbon–hydrogen–nitrogen analyzer (Perkin-Elmer Corp., Norwalk, Connecticut).

17.3. RESULTS

17.3.1. Sedimentary Conditions

The natural sediments throughout the area consist of a uniform fine brown silt. Beyond the area of sewage-sludge deposition the Eh and pH values for these sediments were high. These, however, decline rapidly toward the center of the dumpsite, where pH values of ~7.0 and Eh values <-150 mV were found (Fig. 17.2). High carbon values (>5%) were found in the sediments within 1 km of the center of the site, and moderately high values (>3.5%) were recorded up to 3 km from the center (Fig. 17.3). As might be expected, high carbon levels and low Eh and pH values coincided.

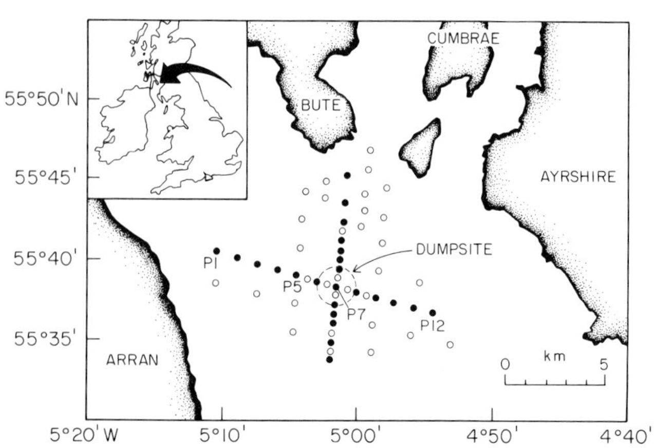

Figure 17.1. Map of the Garroch Head disposal area in the Firth of Clyde, Scotland, showing (●) transect sampling stations and (○) core sampling stations.

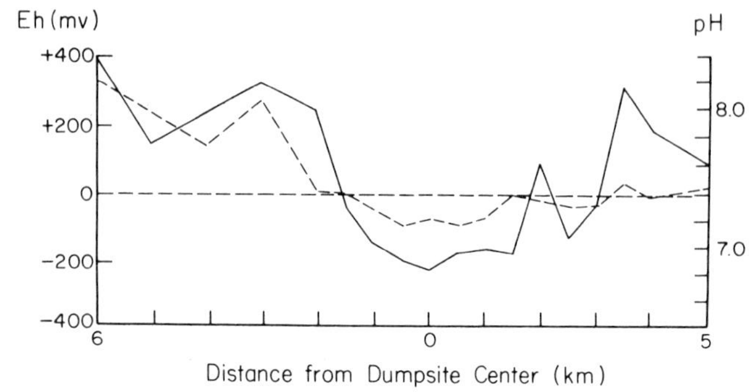

Figure 17.2. Variation in redox potential (Eh) (solid line) and acidity (pH) (dashed line) at a 40-mm depth in the sediments along the east–west transect across the center of the dumpsite.

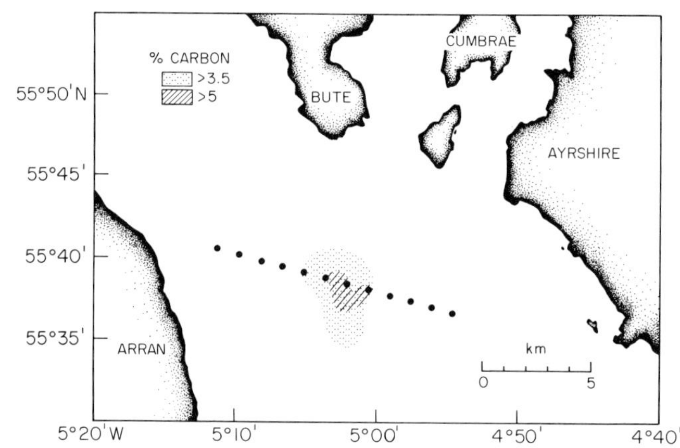

Figure 17.3. Areas of Garroch Head sediment showing enhanced carbon levels.

17.3.2. Hydrographic Conditions

Water conditions in the vicinity of the Garroch Head dumpsite have been described by Dooley (1978). Residual currents over the dumping area are subject to relatively unpredictable changes of speed and direction, but in general vary in speed between 5 and 15 cm s^{-1}. The water immediately above the sediment surface remained fully oxygenated at all times (values between 7 and 9 mg $liter^{-1}$ for water overlying sediment cores).

17.3.3. Sludge Deposit Depth

Long core samples were taken by gravity corer from the center of the dumpsite (station P7), the edge of the dumpsite (station P5), and from station P1, situated 7 km west of the center of the dumpsite (Fig. 17.1). In the center of the dumpsite the upper 15 cm of the sample consisted of a black deposit of sewage sludge. From 15 to 30 cm the core consisted of a mixture of sludge and silty sediment. Below 30 cm the core contained normal gray silt typical of the deeper sediment of the unaffected areas surrounding the dumpsite. No sludge was obvious in the cores from P5 to P1.

17.3.4. Benthic Faunal Distribution

The general distribution of benthic fauna on and in the vicinity of the designated dumpsite can best be described by a consideration of the faunal changes taking place along the transect of sampling stations. In our present study the general distribution of animals varied similarly along both transects; thus, for the purposes of illustration, changes along the east–west transect were considered. Figure 17.4 shows the variation of the principal population statistics along that transect. These varied in the characteristic manner of populations along a gradient of organic enrichment, as described by Pearson and Rosenberg (1978). In the center of the dumpsite species numbers were low, but biomass and abundance were high. One kilometer from the center, species numbers began to increase, whereas biomass and abundance declined. A secondary biomass peak was found between 1 and 2 km from the center; this peak was associated with the increasing variety of species present, whose developing abundance was offset by the decline in numbers of the species predominant in the center. Species numbers were highest 4 km from the center and declined toward the outer ends of the transect, where all three parameters reached the relatively low levels characteristic of the normal communities of the area.

A consideration of the changes in the abundances of the dominant species along the transect provides a more detailed

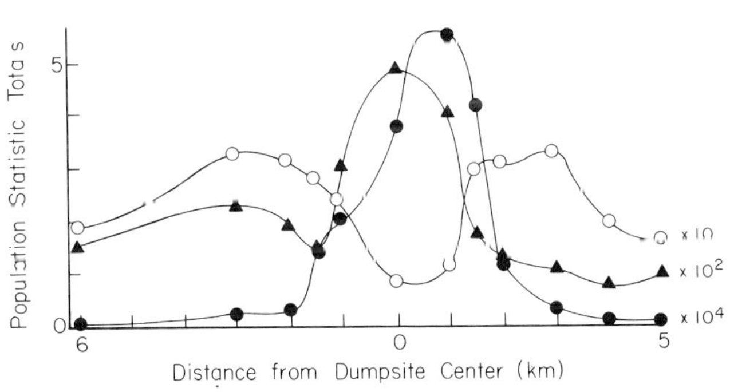

Figure 17.4. Change in the major population statistics along the east–west transect across the center of the dumpsite, based on single grab samples from each sampling station; (○) total number of species, (▲) total wet weight biomass per m^2, and (●) total abundance per m^2.

illustration of the effects of increased organic input on the benthic fauna. The center of the dumpsite was occupied by a large number of small annelid and nematode worms. The four principal species are illustrated in Fig. 17.5a. Nematodes (*Pontonema* sp.) dominated the station in the very center of the dumpsite (P7) but declined very rapidly on either side of the center. *Tubificoides benedeni* and *Capitella capitata* predominated over the whole central area, whereas *Pontonema* and *Scolelepis fuliginosa* were concentrated toward the central part of the transect. A number of other species that were either absent or present in very low numbers in the center of the dumpsite appeared on the edge of the central area, where they reached high densities. The principal species in this group are illustrated in Fig. 17.5b, and included the polychaetes *Mediomastus* sp., *Notomastus latericeus*, and

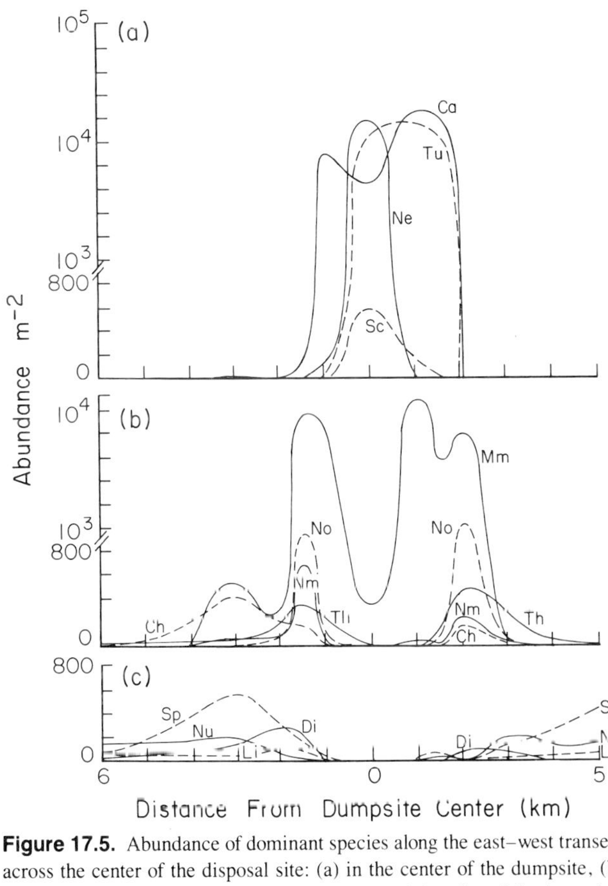

Figure 17.5. Abundance of dominant species along the east–west transect across the center of the disposal site: (a) in the center of the dumpsite, (b) around the center of the dumpsite, and (c) beyond the edge of the dumpsite. Abbreviations are as follows: Ca, *Capitella;* Ch, *Chetozone;* Di, *Diplocirrus;* Li, *Lipobranchius;* Mm, *Mediomastus;* Ne, *Pontonema;* Nm, *Nemertina* sp.; No, *Notomastus;* Nu, *Nucula;* Sc, *Scolelepsis;* Sp, *Spiophanes;* Th, *Thyasira;* and Tu, *Tubificoides*. Note the semi-logarithmic scale giving greater emphasis to the less abundant species.

Chaetozone setosa, and the mollusc *Thyasira flexuosa*. Beyond the central area, toward either end of the transect, the above-mentioned species declined in number, and another group characteristic of the unpolluted areas beyond the dumpsite became dominant. The abundances of some of these, including the molluscs *Nucula tenuis* and *Nucula sulcata* and the polychaetes *Spiophanes kroyeri, Lumbrinereis hibernica,* and *Lipobranchius jeffreysii,* are shown in Fig. 17.5c. These species were generally larger than those replacing them as the center of the site was approached. This general trend is illustrated in Fig. 17.6 by a comparison of the two composite statistics, the abundance ratio A:S (where A is the total abundance per m^2 and S is the total number of species per sample), which represents the mean number of individuals per species, and the size ratio B:A (where B is biomass), which represents the mean weight of all individuals (Pearson et al., 1982). The size ratio, B:A, increases with distance from the center of the dumpsite, whereas the abundance ratio, A:S, is greatest at the center of the dumpsite.

17.3.5. Comparisons with Previous Surveys

The data reported in the previous section are from the June 1981 survey. Comparable surveys were also undertaken during the two previous years. During this period the pattern of dumping at the site remained unaltered and the amount of material discharged was relatively constant. The results from each annual survey were remarkably similar. The distribution of species along the transects remained unchanged, and abundance and biomass fluctuations in relation to the center of the dumpsite showed a remarkable stability, as illustrated by the 3-y means of population statistics (Fig. 17.7).

17.3.6. Faunal Depth Distribution

The distribution of fauna with depth in the sediment was investigated at a series of stations along the transect. Gravity core samples 9.5 cm in diameter and 55 cm in length were taken at five stations varying in distance from 0 to 6 km from the center of the dumpsite. The cores were sectioned into 3-cm-long slices, each of which was sieved on a 0.5-mm mesh screen, and then the fauna was extracted, identified, and counted. Figure 17.8 illustrates the distribution of animals found in each core sample. The total number of animals in each sample increased markedly toward the center of the dumpsite, as was the case with the grab samples. Throughout the transect most of the animals were found in the upper 6 cm of the sediment. In the areas farthest from the point of deposition, 2–3% of all animals were found below 12 cm. Less than 1% were found below that depth on the edge of the dumpsite, but in the center of the site >5% were found below 12 cm in the anoxic areas of the sediment (Fig. 17.9). The maximum depth at which animals were found in each sample increased progressively toward the center of the dumpsite. The range of species found in these

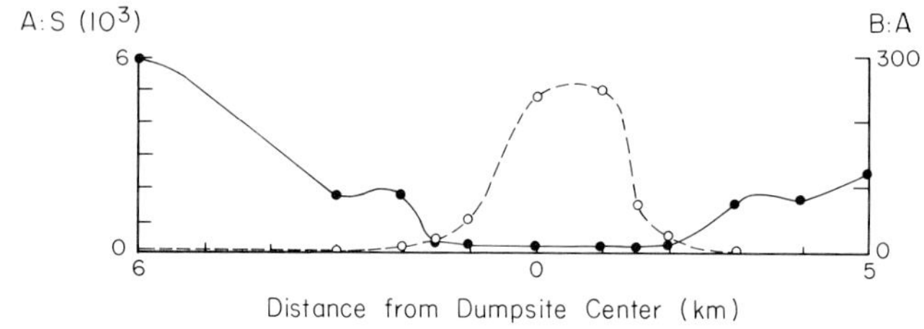

Figure 17.6. Along-transect changes in (○) abundance ratio, *A:S*, and (●) size ration, *B:A*.

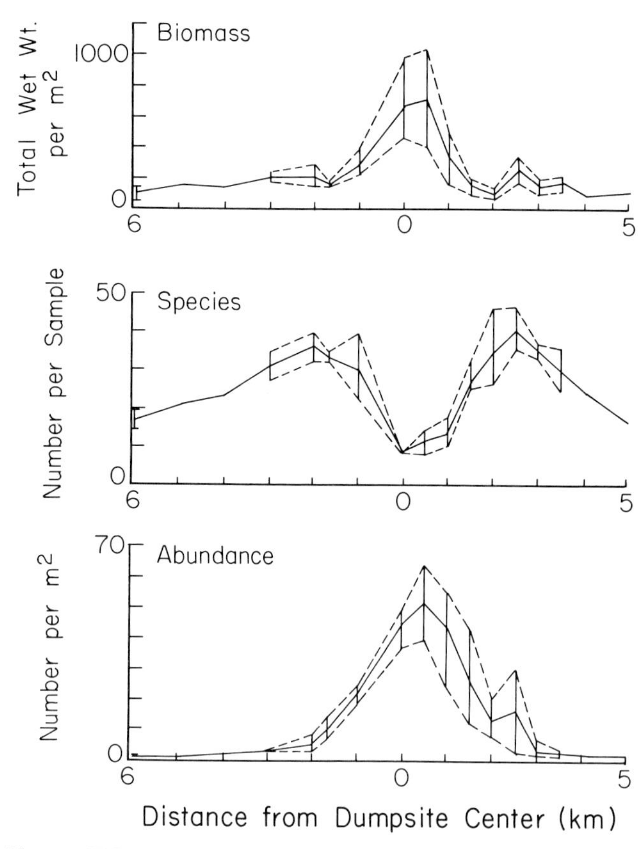

Figure 17.7. Changes in the mean values of the principal population statistics over 3 y. Means of single grab samples taken at each sampling station in each year, together with the range of values over the 3 y, are shown.

samples differed little from those recorded in the grab samples. The species at depth were predominantly polychaetes, particularly capitellids, in the dumpsite center.

17.4. DISCUSSION AND CONCLUSIONS

Faunal distribution in the areas subjected to sewage-sludge deposition allows some conclusions to be drawn as to the effect of that deposition. In addition, certain speculations may be made on the basis of these conclusions concerning the advisability of sludge disposal in this type of marine environment. The detrital input to the center of the disposal site is enormous when compared to areas outside of the dumpsite. At the Garroch Head dumpsite the chemical and biological conditions of the sediments in the vicinity of the dumpsite suggest that the sludge rapidly reached the sediments following discharge. If it is assumed that the area of low redox potential corresponds approximately to the area of deposition, then ~200 t (dry weight) of material per day were spread over 10 km^2. This is equivalent to an input of 20 g m^{-2} d^{-1}, a figure one to two orders of magnitude higher than the highest levels of natural carbon input to inshore sediments [0.1–0.3 g cm^{-2} d^{-1} (Parsons et al., 1977; Pearson, 1982; Seki, 1982)]. Despite the enormous input, there was no large accumulation of sludge in the center of the site. Active macrofaunal communities were present throughout the area, and animals were present in large numbers even in the anoxic sediments in the center of the dumpsite, where depth penetration was greatest. The sequential change in species along the increasing gradient of organic enrichment toward the center of the site closely followed the idealized pattern of all such successions (Pearson and Rosenberg, 1978), suggesting a highly evolved response in this type of benthic community for maximizing carbon utilization. Thus, the species distribution changed along the gradient from a diverse community of many trophic types, mostly breeding once a year [for example, *Nucula* (Allen, 1954) and *Melinna* (Guillou and Hily, 1983)], at the outer, lower end of the gradient, to a restricted community of deposit feeders breeding continuously [for example, *Scolelepis* (Guerin, 1973) and *Capitella* (Warren, 1977)] in the center of the dumpsite (Fig. 17.10). Biomass levels increased by one, and faunal densities by two orders of magnitude along the gradient. Thus, at the community level, adaptation to increasing carbon input was a progressive multiplex process involving continuous species replacement to maximize density, biomass, and reproductive effort in order to utilize all available carbon. It is apparent that total benthic productivity increases to accommodate increasing input, at least for the high levels recorded at the center of the Garroch Head dumpsite. The dissipative pathways whereby these high production levels are disseminated through the benthic ecosystem of the whole area are less easy to describe. Preliminary observations suggest that high levels of predation, which involve a variety of epifaunal invertebrate and demersal fish, occur over the cen-

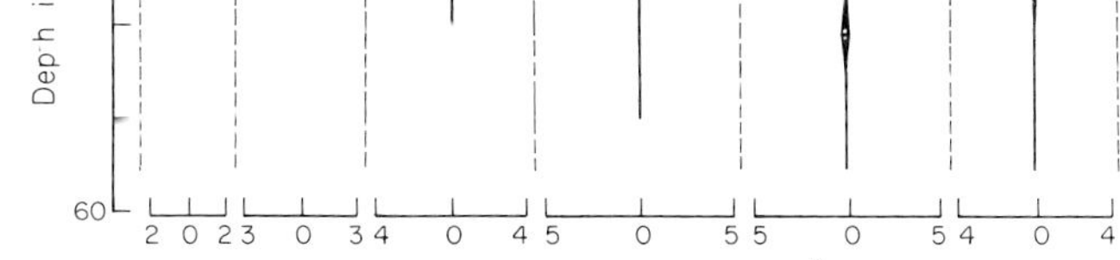

Figure 17.8. Distribution of fauna with depth in the sediment in core samples of 20-cm^2 cross-sectional area. Samples were taken at increasing distances from the center of the dumpsite.

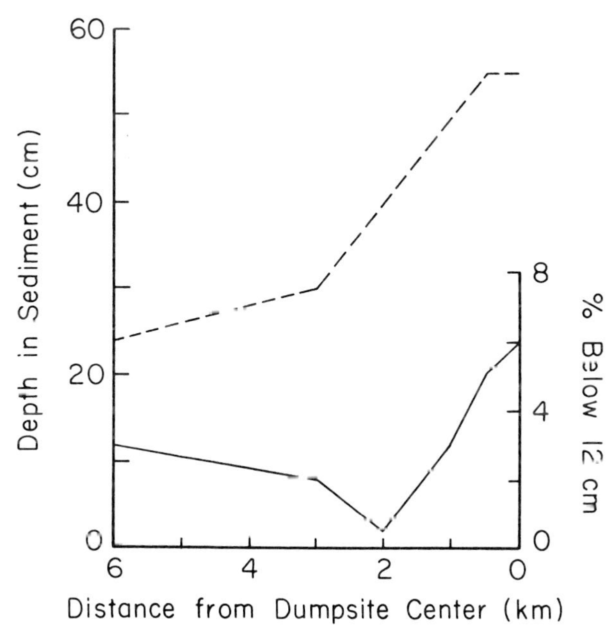

Figure 17.9. Maximum depth of penetration of animals (dashed line) and the percentage of the total number of animals found below 12 cm in the sediments (solid line). Numbers were gathered from core samples of 20-cm^2 cross-sectional areas taken at increasing distances from the center of the dumpsite.

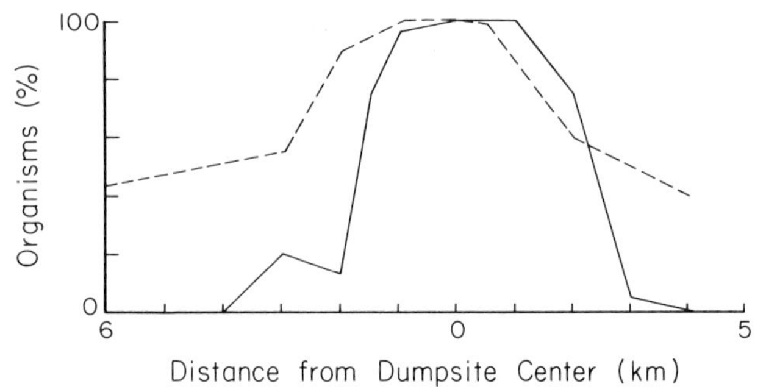

Figure 17.10. Changes in the percentages of deposit-feeding animals (dashed line) and animals having a continuous breeding cycle (solid line) along the east–west transect. The center of the disposal area is inhabited exclusively by despoit-feeding worms, which breed continuously.

ter of the dumpsite; however, these suggestions have yet to be quantified. Despite such gaps in understanding the whole system, it seems that complete adjustment to the high carbon input in the area has taken place.

A successional faunal system that maximizes the utilization of available carbon has been established over the Garroch Head dumpsite. Faunal density and productivity have increased progressively to accommodate increasing carbon enrichment of the sediment at the present input level of 20 g m^{-2} d^{-1}. The system has remained remarkably stable over a 3-y period, and carbon accumulation in the sediments, if discharges are maintained at their present level, is unlikely to present an environmental hazard to the area.

ACKNOWLEDGMENTS

The monitoring surveys at the Garroch Head sewage-sludge disposal site were carried out under contract to Strathclyde Regional Council on behalf of the Department of Sewerage. The dedicated assistance of K. Price and B. Archer is gratefully acknowledged.

REFERENCES

Allen, J. A. 1954. A comparative study of the British species of *Nucula* and *Nuculana*. *Journal of the Marine Biological Association of the United Kingdom,* **33,** 457–472.

Dooley, H. D. 1978. Factors influencing water movements in the Firth of Clyde. *Estuarine and Coastal Marine Science,* **9,** 631–641.

Guerin, J. P. 1973. Premières données sur la longévité, le rhythme de ponte et la fécondité de *Scolelepis* cf. *fuliginosa* (polychète spionidé) en élaevage. *Marine Biology,* **19,** 27–40.

Guillou, M., and C. Hily. 1983. Dynamics and biological cycle of a *Melinna palmata* (Ampharetidae) population during the recolonization of a dredged area in the vicinity of the harbour of Brest (France). *Marine Biology,* **73,** 43–50.

Halcrow, W., D. W. Mackay, and J. Bogan. 1974. PCB levels in Clyde marine sediments and fauna. *Marine Pollution Bulletin,* **5,** 134–136.

MacIntyre, A. D. 1977. Effects of pollution on inshore benthos. *In*: Ecology of Marine Benthos, B. C. Coull (Ed.). University of South Carolina, Belle W. Baruch Library of Marine Science, Columbia, pp. 301–318.

Mackay, D. W., W. Halcrow, and I. Thornton. 1972. Sludge dumping in the Firth of Clyde. *Marine Pollution Bulletin,* **3,** 7–10.

Parsons, T. R., M. Takahashi,and B. Hargrave. 1977. Biological Oceanographic Processes, 2nd edition. Pergamon Press, Oxford, 332 pp.

Pearson, T. H. 1982. The Loch Eil Project: assessment and synthesis with a discussion of certain biological questions arising from a study of the organic pollution of sediments. *Journal of Experimental Marine Biology and Ecology,* **57,** 93–124.

Pearson, T. H., and R. Rosenberg. 1978. Macrobenthic succession in relation to organic enrichment and pollution of the marine environment. *Oceanography and Marine Biology Annual Review,* **16,** 229–311.

Pearson, T. H., and S. O. Stanley. 1979. Comparative measurement of the redox potential of marine sediments as a rapid means of assessing the effect of organic pollution. *Marine Biology,* **53,** 371–379.

Pearson, T. H., G. Duncan, and J. Nuttall. 1982. The Loch Eil Project: population fluctuations in the macrobenthos. *Journal of Experimental Marine Biology and Ecology,* **56,** 305–321.

Seki, H. 1982. Organic Materials in Aquatic Ecosystems. CRC Press, Boca Raton, Florida, 201 pp.

U.K. Department of the Environment, Standing Committee on the Disposal of Sewage Sludge. 1979. Report of the Sub-committee on the Disposal of Sewage Sludge to Sea, 1975-1978. Standing Technical Committee Report No. 18, Department of the Environment–National Water Council, London, 65 pp.

Warren, L. M. 1977. The ecology of *Capitella capitata* in British waters. *Journal of the Marine Biological Association of the United Kingdom,* **57,** 151–159.

Chapter 18

Changes in the Benthos at a Sewage-Sludge Dumpsite in the Elbe Estuary

Hubert Caspers

Institut für Hydrobiologie und Fischereiwissenschaft
Federal Republic of Germany

ABSTRACT

Twenty years of investigations on the benthos in the outer part of the Elbe Estuary, which includes a sewage-sludge dumpsite, show that the community structure was highly unstable. Some changes were brought about by hydrological factors and the variable pattern of sedimentation. Most of the changes, however, resulted primarily from biological factors, such as the predominance of one or two species, the availability of physical space, predation, and the life span of the species. The continual arrival of larvae, belonging to various taxa, caused a more rapid fluctuation in the benthic community structure than is usually encountered in freshwater habitats. Communities in the coastal marine area of the Elbe Estuary might best be designated as astatic and characterized by seasonal and long-term fluctuations; they are relatively unstable but highly resilient.

18.1. INTRODUCTION

The mouth of the Elbe River is narrow due to extensive sandy mud flats on either side of the main channel. Only farther into its seaward progression does the estuary widen into the deeper zone of the German Bight. In the outer region of the estuary a wide variety of sediment types are found, including coarse materials in the eroded navigational channel, fine sands, mud, and a firm clay, which is found in the area of the mud-flat embankment (Caspers, 1952, 1954, 1968, 1978, 1980a, 1980b, 1981; Stripp, 1969a, 1969b; Ziegelmeier, 1970; Rachor, 1977; Rachor and Gerlach, 1978). Each type of sediment has a characteristic macrobenthos that fits well into classical classification systems of associations and biocoenoses. Such classification systems presuppose that variations in community structure are followed throughout the entire annual cycle.

A sewage-sludge dumpsite is located ~4.8 km north of the Lightship Elbe (Fig. 18.1), where water depths range from 14.5 to 19.5 m. Each year from 1961 to 1981, ~250,000 t of sludge was dumped; following the provisions of international conventions, sewage-sludge dumping was discontinued in the outer Elbe area in 1981 (Fuchs, 1982). Nauke (1974) examined heavy-metal distributions in the sediments of the area and provided data on the distribution of sediment types in the upper 10 cm. In the center of the dumpsite, silty sand forms the surface layer, and clay-like layers of silt occur 30–35 cm below the surface layer. At the sediment surface, there is usually a thin cover of very fine organic matter that generally has an oxidized layer. Below this layer the sediment ranges from grayish to black.

To estimate the effect of sewage-sludge dumping on benthic fauna in this region (or at any dumpsite), it is important to follow benthic community development over a time period long enough to observe both periodic and aperiodic fluctuations. Rachor (1980) described the inner German Bight as "an ecologically sensitive area." He considered the benthic

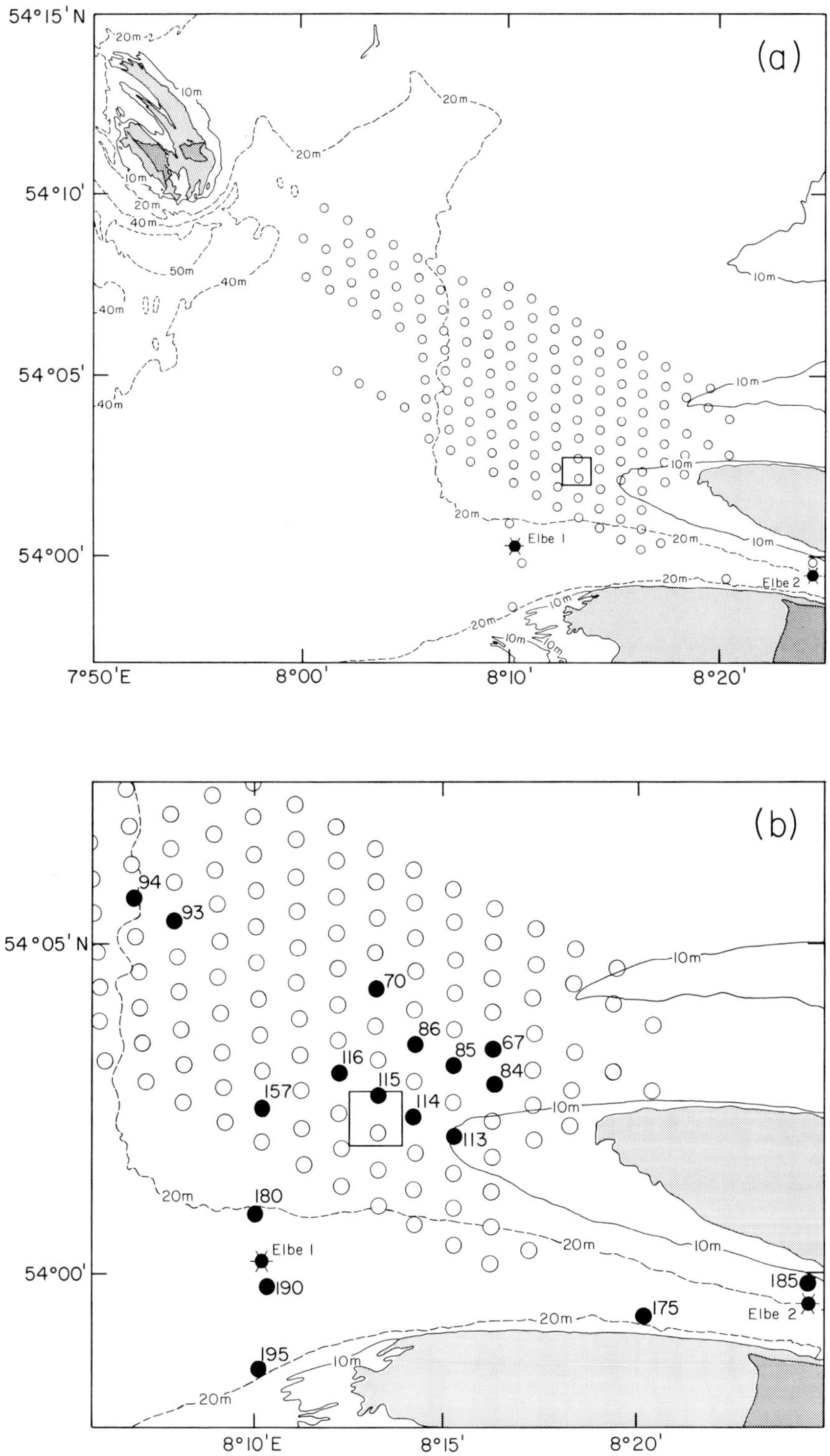

Figure 18.1. Map of (a) the network of 250 sampling stations between the mouth of the Elbe Estuary in the east and Helgoland in the northwest, and (b) the location of pertinent benthic sampling stations and the Lightships Elbe 1 and 2. The square denotes the sludge dumping area.

macrofauna of a muddy station to have a tendency toward gradual impoverishment interrupted by temporarily successful recolonization, which disturbs both the resistance and the resilience of the community. Such broad fluctuations in abundance, including sudden breakdowns, are accompanied by a steady impoverishment in species. Rachor attributes this to geological and hydrographic factors, especially oxygen depletion in specific water layers, and he concludes that such an ecologically sensitive area is not suited for the dumping of large amounts of wastes (Rachor, 1980).

Rachor's suggestions require proof that can only be obtained by long-term observations. A benthic community reflects both sediment conditions and hydrographic factors. Sewage-sludge dumping increases the nutrient supply for particle feeders in the benthic community, and, as long as the currents transport oxygen to the sediment surface, these animals will benefit.

A criterion for stability in a community is the richness in species and ecotypes. All faunal compositions change with time but physical changes in the environment may not be the primary cause. Faunal changes are more likely to result from biological factors such as life history, settlement patterns, competition, and food availability.

In this chapter, data on the benthic fauna at selected stations inside and outside the dumping area were analyzed with two objectives in mind: (1) to determine the degree of benthic stability and whether there is a higher degree of instability at areas influenced by sludge disposal, and (2) to demonstrate that changes in benthic fauna may result from inter- and intraspecific relationships, not primarily from changes in environmental conditions.

18.2. SAMPLING

In 1951, a network of ~250 stations was established throughout the outer Elbe to obtain a regional overview of faunal distribution. Benthic samples taken with a Van Veen grab (0.1-m^2) were quantitatively evaluated over a 3-y period (1951–1954). The results of this 3-y investigation show that settlement patterns were different in different parts of the region and appear to be related to the distribution of sediment types (Caspers, 1952, 1954). Sampling continued during the next 10 y, and some stations were sampled as many as 12 times. At each sampling, a minimum of two samples were taken to ensure that no atypical faunal aggregations were included. Long-term changes were determined on the basis of a network of several stations within an area. Therefore, both temporal changes and the relocation of regional borders could be recognized. The long-term effects of sewage-sludge dumping on benthic communities in the outer Elbe were further evaluated from 1970 to 1979.

18.3. RESULTS

18.3.1. Faunal Settlement Patterns

As a result of spatial heterogeneity and hydrographical factors, great differences were found in faunal settlement patterns at individual stations and in general regions. Both sediment structure and faunal composition changed rapidly.

Communities with relatively high species diversity are illustrated in Fig. 18.2. They were not dominated by any ecological type; this reflects a structural balance and gives the impression of stability. Sediment consumers, detritus feeders, and predators were represented. Station 180 (Fig. 18.2a) was located north of the Lightship Elbe 1, on the border of a shipping route (Fig. 18.1) at a depth of 20 m. The soft muddy sediment at this station had a thin aerobic surface layer. Station 185 (Fig.18.2b) was located in a polyhaline to mesohaline zone of the estuary. In September 1958, the fauna consisted of species characteristic of a fine muddy sand. Station 190 (Fig. 18.2c) was typical of the firm muddy sand in the outer part of the Elbe Estuary, where species diversity of polychaetes, bivalves, and echinoderms was high.

Communities such as those shown in Fig. 18.2, however, were atypical. More often it was found that the fauna was dominated by one or two species, as shown in Fig. 18.3. The dominant species from a fine-mud section of the dumpsite in October 1970 was the brittle star *Ophiura texturata* (Fig. 18.3a) (Caspers, 1980a). During subsequent years, brittle stars were displaced by a large population of the bivalve *Abra alba*. At another station the dominant species was the sedimentary polychaete, *Lanice conchilega* (Fig. 18.3b).

Settlement patterns at individual stations varied considerably over the sampling period (Fig. 18.4). Great care was taken to return to precise sampling locations during the sampling period, and individual samples at a specific station were not more than 20 m apart. At a station in the mesohaline region of the estuary, almost identical faunal settlement patterns were observed in 1965 and in 1968 (Fig. 18.4a). In contrast, at a station located farther seaward, considerable differences in fauna were observed in the 1952 and the 1958 samples (Fig. 18.4b). A similar comparison is depicted in Fig. 18.4e–f, representing faunal changes at a neighboring

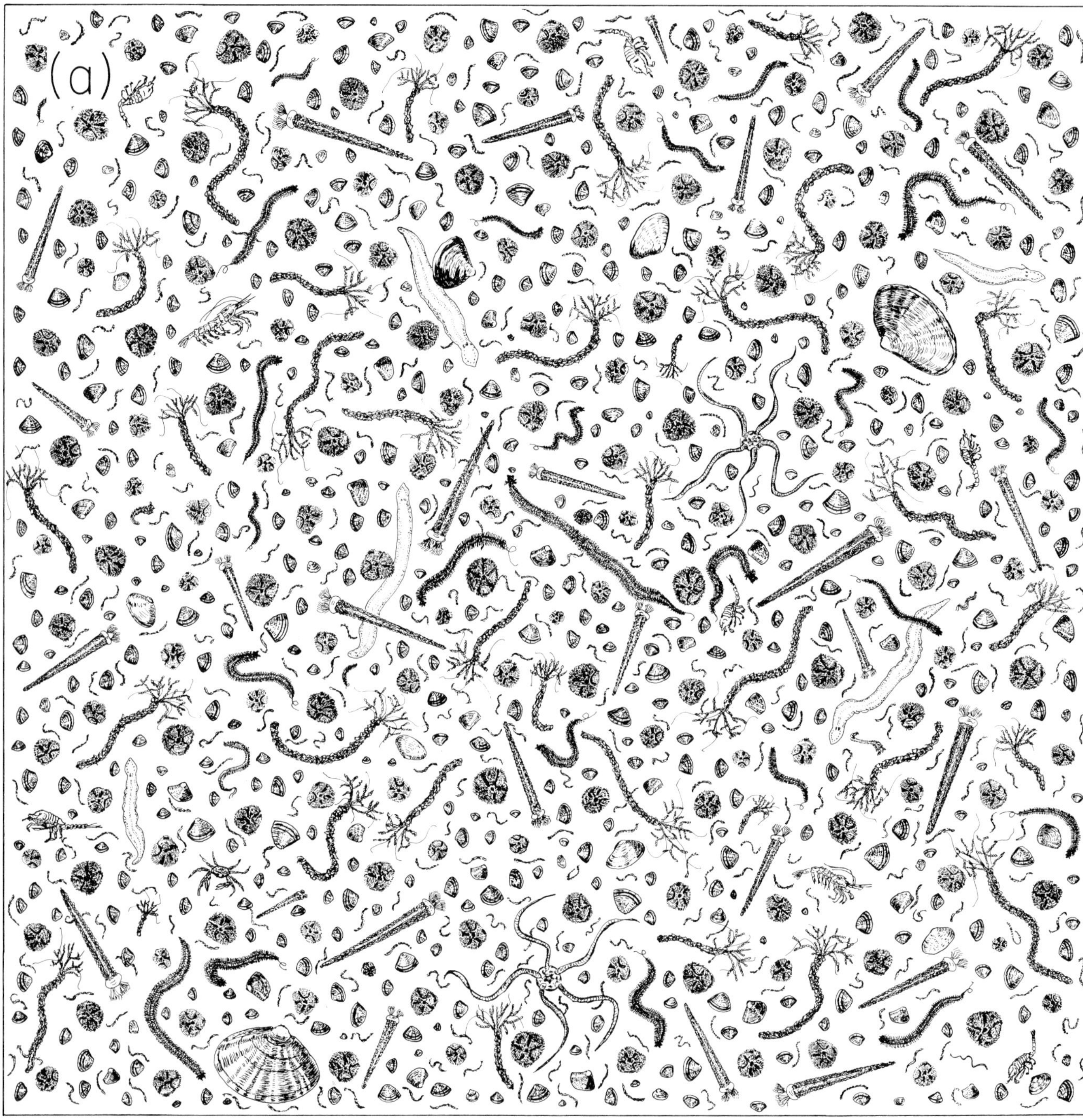

Figure 18.2. (a) A 0.1-m^2 benthic sample of a well-balanced community with a high species diversity north of Lightship Elbe 1 at station 180 in August 1951 (the sediment was very soft mud with a thin aerobic surface layer): 455 *Nucula turgida;* 360 *Pygospio elegans;* 158 *Echinocardium cordatum;* 36 *Lanice conchilega;* 35 *Aloidis gibba;* 30 *Mysella bidentata;* 26 *Pectinaria koreni;* 22 *Nephtys hombergii;* 6 *Diastylis rathkei;* 6 Nemertinea; 5 *Abra alba;* 5 *Angulus fabula;* 5 *Spiophanes bombyx;* 3 *Anaitides groenlandica;* 3 *Astarte sulcata;* 2 *Ampharete grubei;* 2 *Carcinus maenas;* 2 *Ophiura albida;* 2 *Scoloplos arminger;* 1 *Crangon crangon;* 1 *Mactra corallina;* 1 *Nephtys caeca;* 1 *Spisula solida;* and 1 *Spisula subtruncata.*

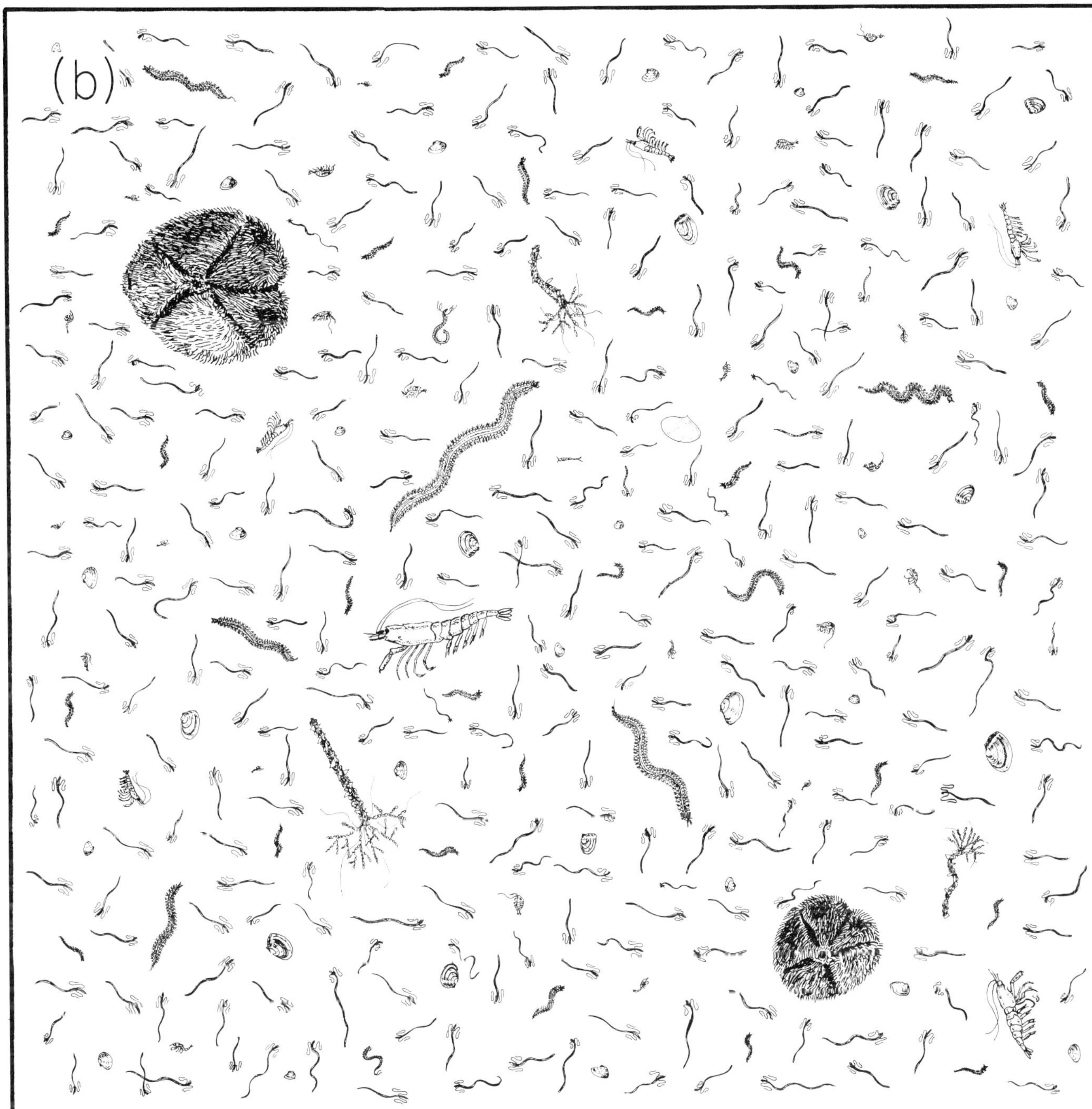

Figure 18.2. *(Continued)* (b) A 0.1-m^2 benthic sample of fine muddy sand in the mesohaline zone near Lightship Elbe 2 at station 185 in September 1958: 284 *Magelona papillicornis;* 33 *Nephtys longosetosa;* 22 *Angulus fabula;* 12 *Diastylis rathkei;* 10 *Angulus tenius;* 9 *Urothoe grimaldii* var. *poseidoni;* 7 *Crangon crangon;* 5 *Eteone longa;* 3 *Lanice conchilega;* 2 *Echinocardium cordatum;* 2 *Montacuta ferruginosa;* 1 *Abra alba;* and 1 *Scoloplos armiger.*

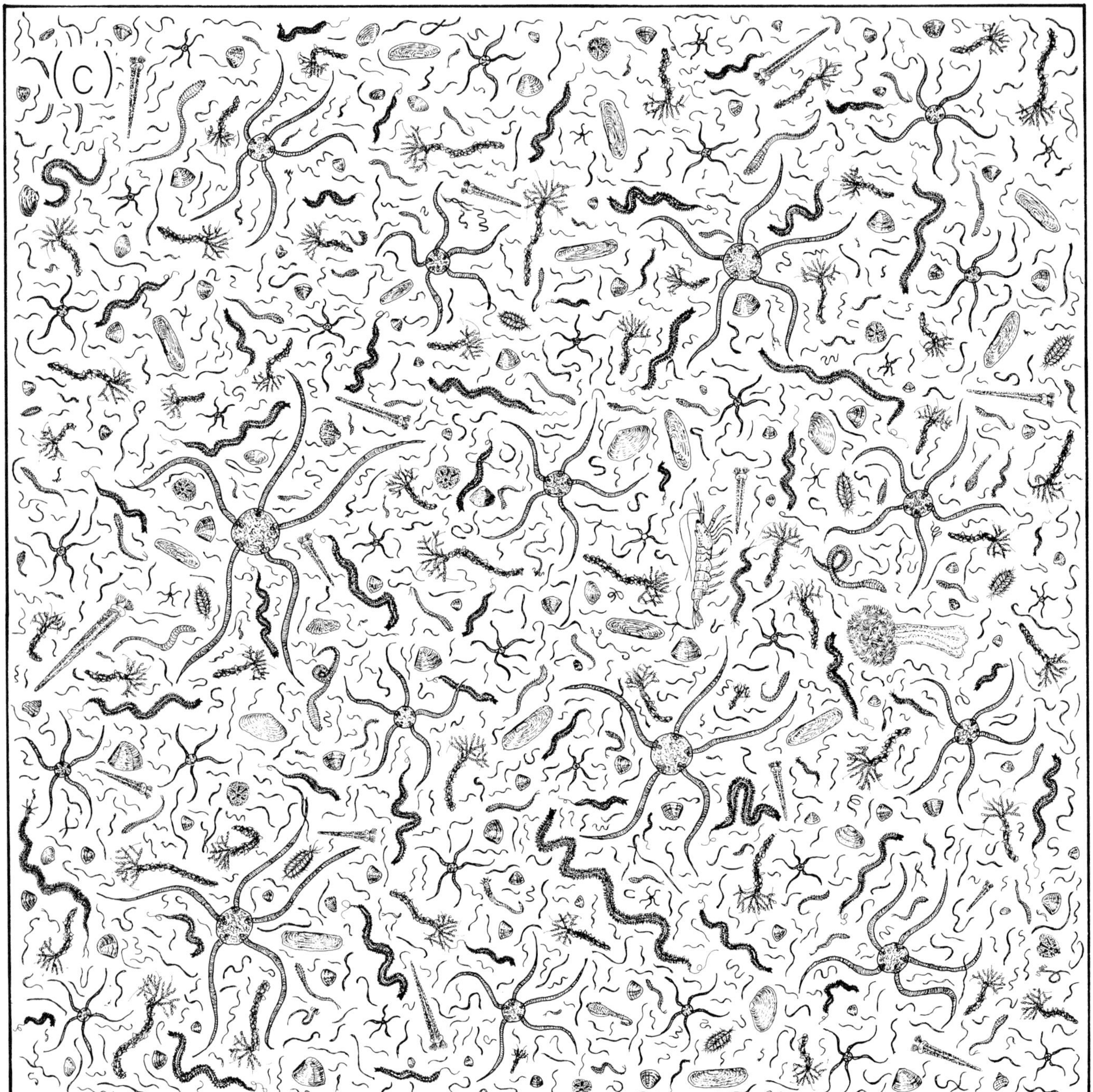

Figure 18.2. *(Continued)* (c) A 0.1-m^2 benthic sample of firm muddy sand in the outer part of the Elbe Estuary near Lightship Elbe 1 at station 190 in September 1957: 1235 *Spiophanes bombyx;* 67 *Scoloplos armiger;* 49 *Lanice conchilega;* 36 *Ophiura albida;* 18 *Nereis longispina;* 18 *Phaxas pellucida;* 17 *Eulalia sanguinea;* 17 *Nucula turgida;* 14 *Abra alba;* 14 *Pectinaria koreni;* 7 *Echinocardium cordatum;* 7 *Eteone longa;* 7 *Harmothoe impar;* 6 *Magelona papillicornis;* 6 *Mysella bidentata;* 5 *Goniada maculata;* 4 *Notomastus latericeus;* 4 *Polydora antennata;* 4 *Scalibregma inflatum;* 3 *Nephtys hombergii;* 3 *Nymphon grossipes;* 2 *Angulus fabula;* 2 *Natica poliana alderi;* 2 Nemertinea; 2 *Ophelina acuminata;* 2 *Pholoe minuta;* 1 *Ampelisca brevicornis;* 1 *Ampharete grubei;* 1 *Cardium* sp.; 1 *Crangon crangon;* 1 *Metridium senile;* 1 *Petricola pholadiformis;* 1 *Sphaerodorum flavum;* and 1 *Venus gallina.*

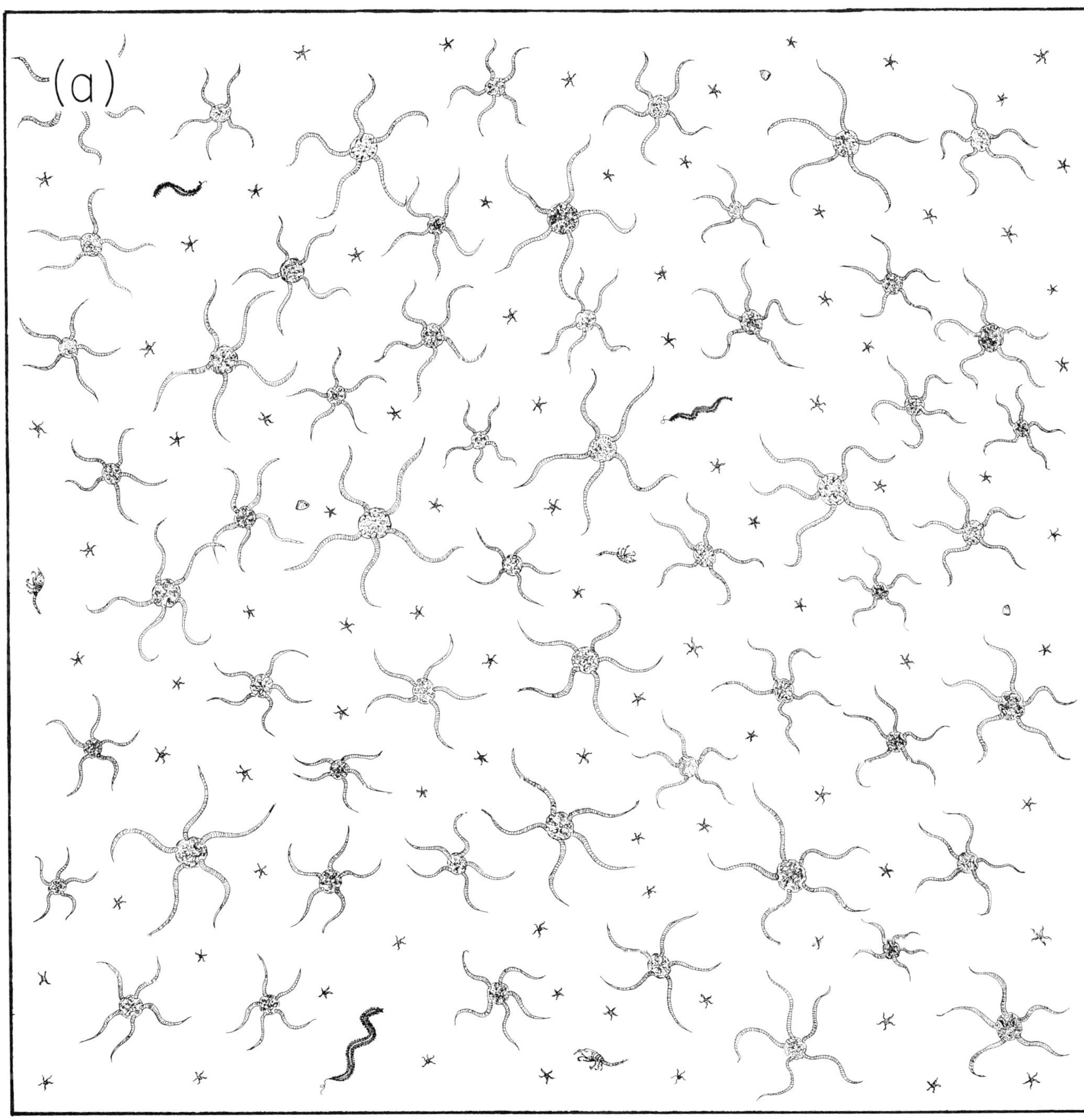

Figure 18.3. (a) A 0.1-m^2 benthic sample at station 116 in October 1970: 140 *Ophiura texturata* (57 adults and 83 juveniles).

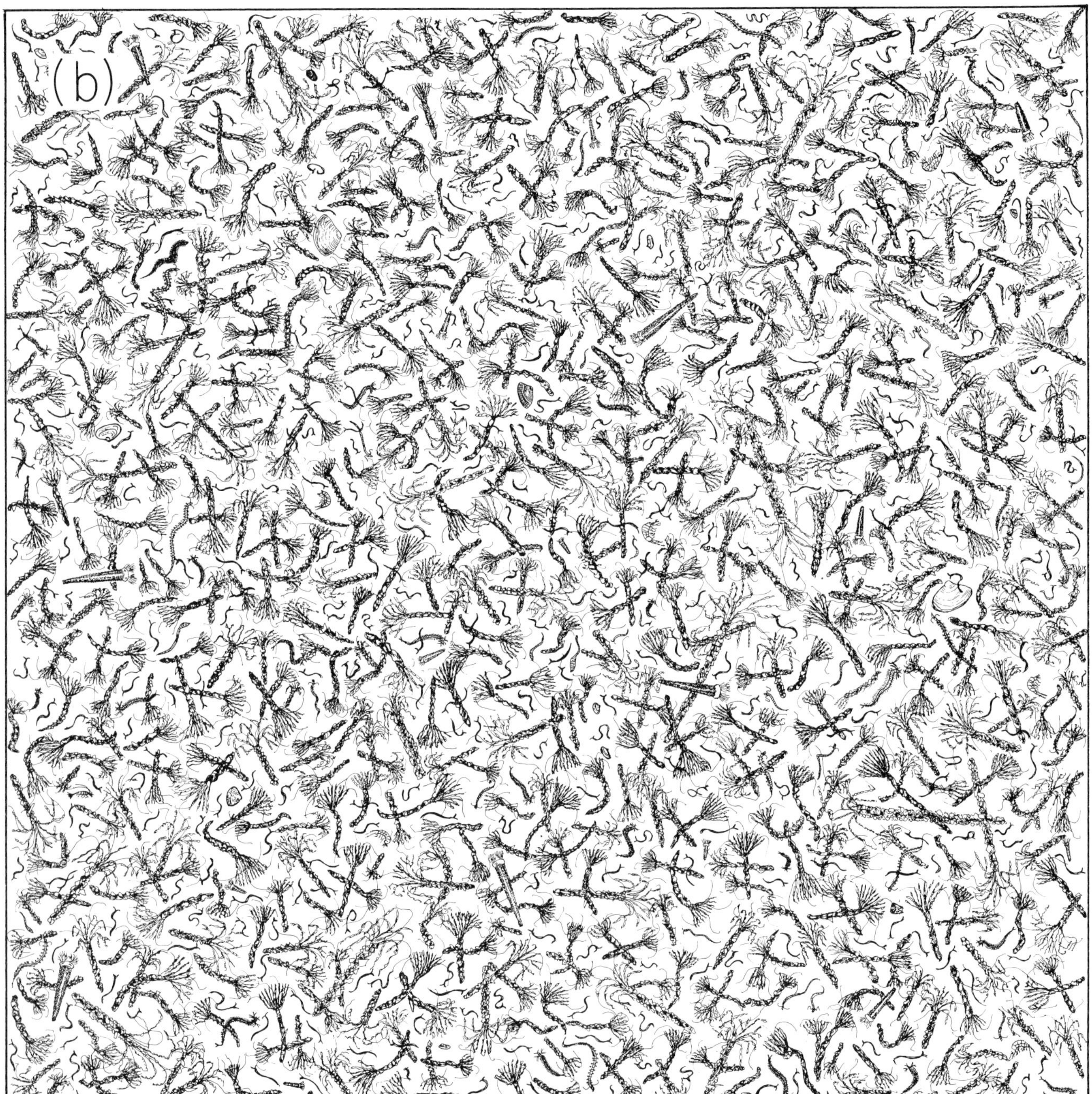

Figure 18.3. ***(Continued)*** (b) A 0.1-m^2 benthic sample at station 84 in July 1976 (for organism names and numbers, refer to Table 18.5).

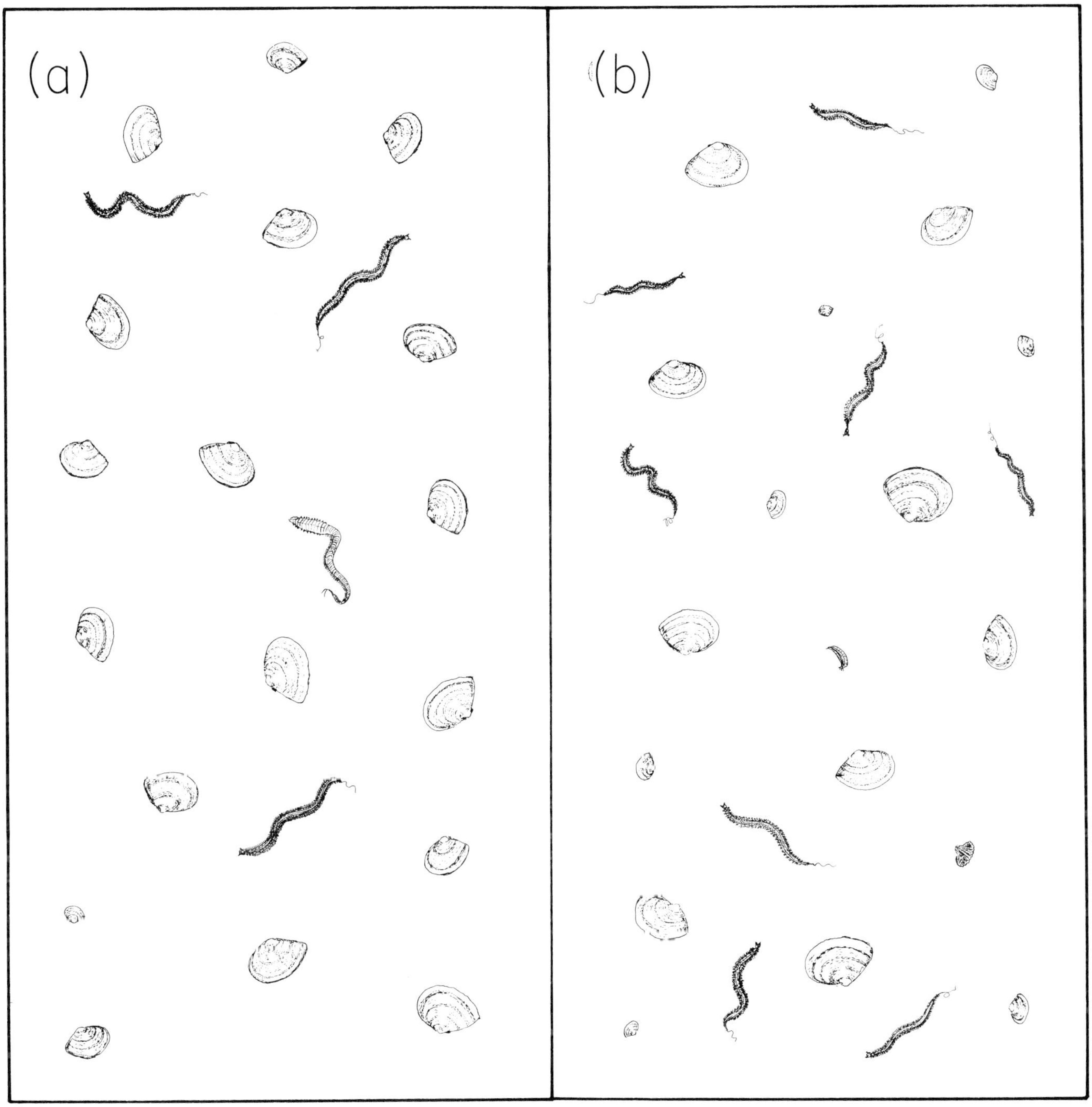

Figure 18.4. A 0.5-m^2 benthic sample at station 175 in the mesohaline zone of the Elbe Estuary in (a) 1965 and (b) 1968.

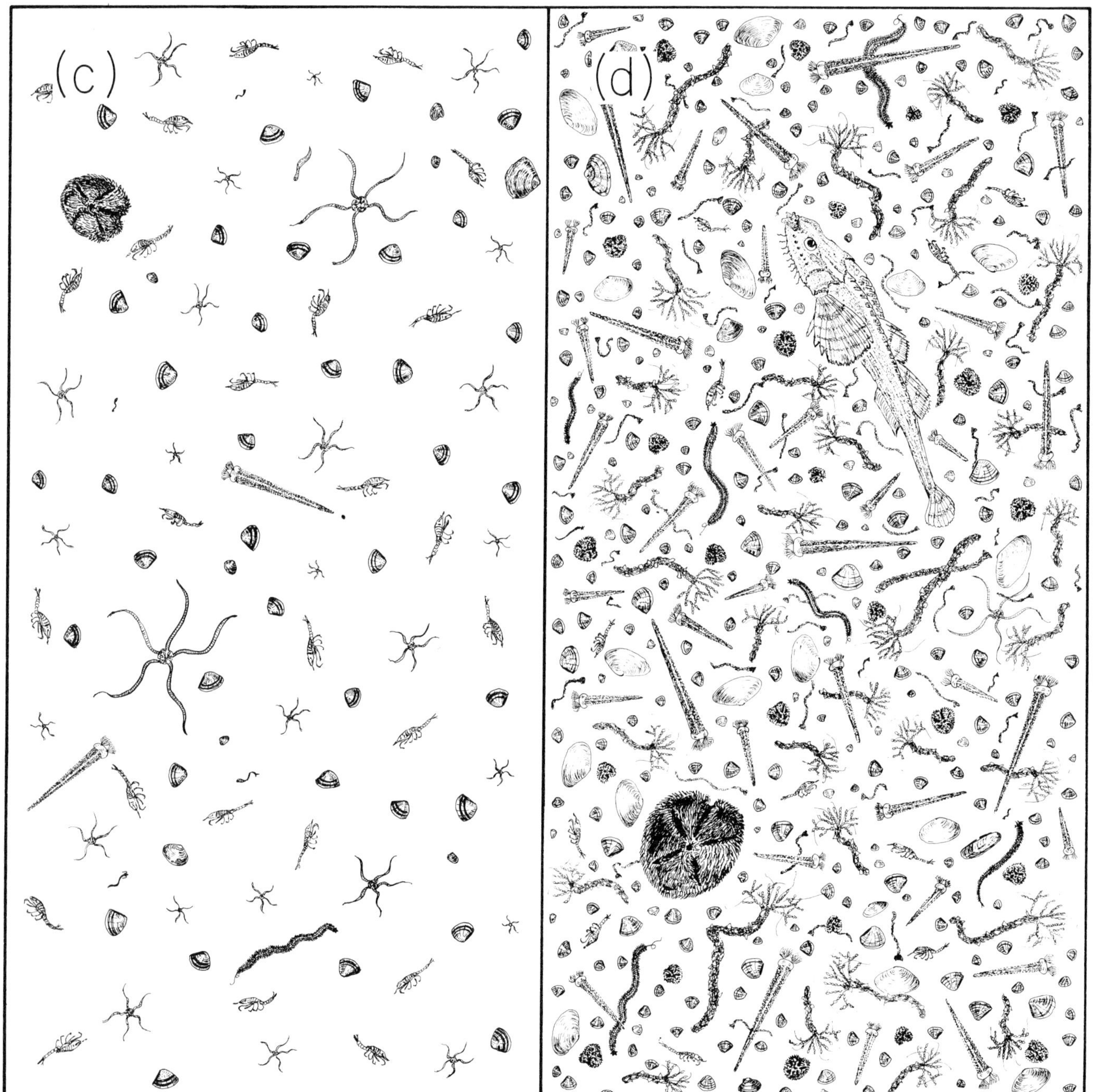

Figure 18.4. ***(Continued)*** A 0.5-m^2 benthic sample at station 157 in the polyhaline outer area of the Elbe Estuary in in September of (c) 1952 and (d) 1958.

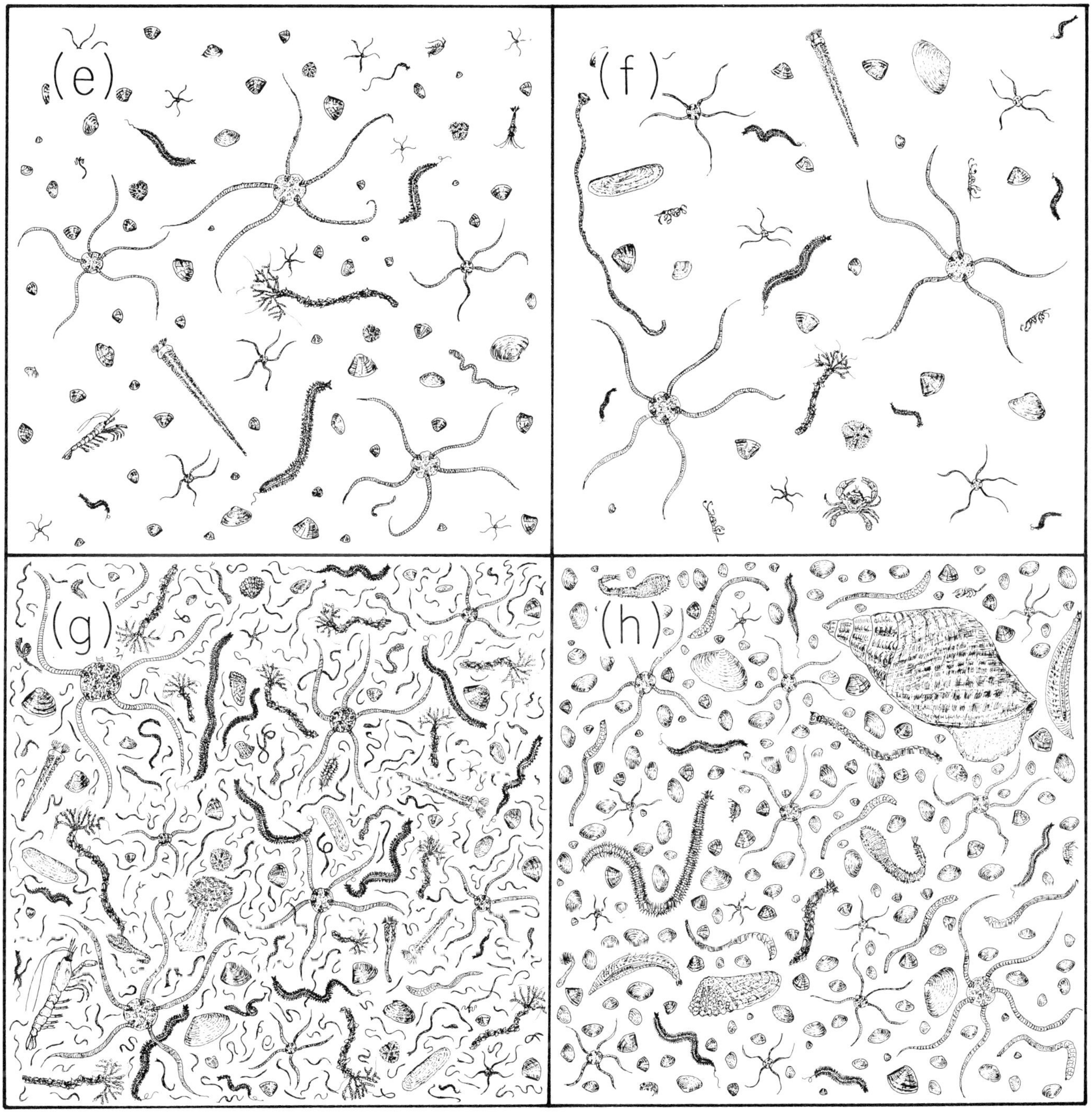

Figure 18.4. ***(Continued)*** A 0.5-m^2 benthic sample at station 190 in the summers of (e) 1951, (f) 1953, (g) 1957, and (h) 1965.

station in 1951, 1953, 1957, and 1965. These examples demonstrate the dramatic changes in the composition of benthic fauna. I infer that these community differences were related to the life span of the individuals, competition, and the chance arrival of planktonic larvae ready for settlement. No fundamental environmental changes, such as variations in sediment types, were observed.

18.3.2. Sludge Dumping

During sludge dumping, fine particles of sludge were distributed over a wide area by the currents (Nauke, 1974), and the larger particles sank relatively quickly to the bottom. The sedimentation region could be easily identified by the presence of tomato and apple seeds. The benthos in this region was dominated at times by massive populations of *Abra alba* (Caspers, 1978, 1980b). The settlement of the growing bivalves was occasionally so dense that further development was limited by the availability of physical space (Fig. 18.5). In the surrounding areas, also characterized by fine sediment, the bivalve *Nucula turgida* predominated. Figure 18.6 shows the densities of *Abra* and *Nucula* during 1973 and 1974. There appears to be an association of *Abra alba* with the polychaete *Pectinaria koreni*, and of *Nucula turgida* with the cumacean *Diastylis rathkei* (Fig. 18.7). Figure 18.8a depicts a faunal assemblage dominated by the *Abra–Pectinaria* association. The settlement of the central dumping area remained relatively constant during 1973 and 1974.

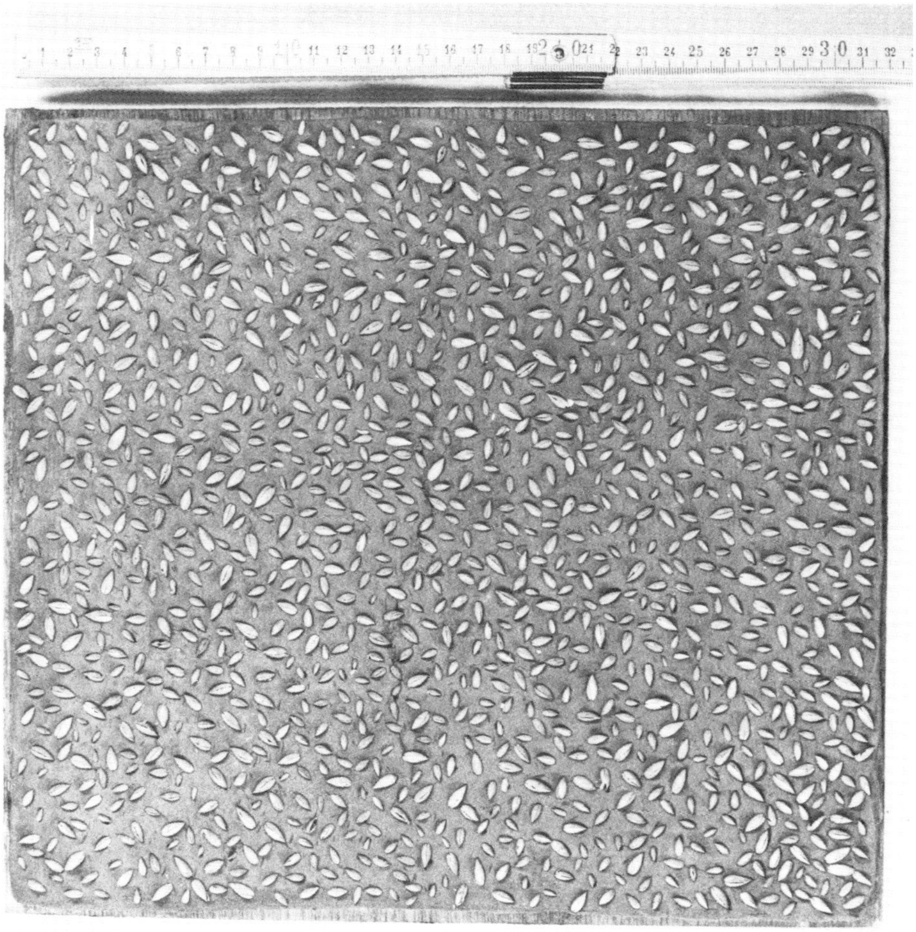

Figure 18.5. *Abra alba* (with a shell length of 4–13 mm) in the center of the sludge dumpsite during 1973 (depicted in their natural position at the surface of the mud).

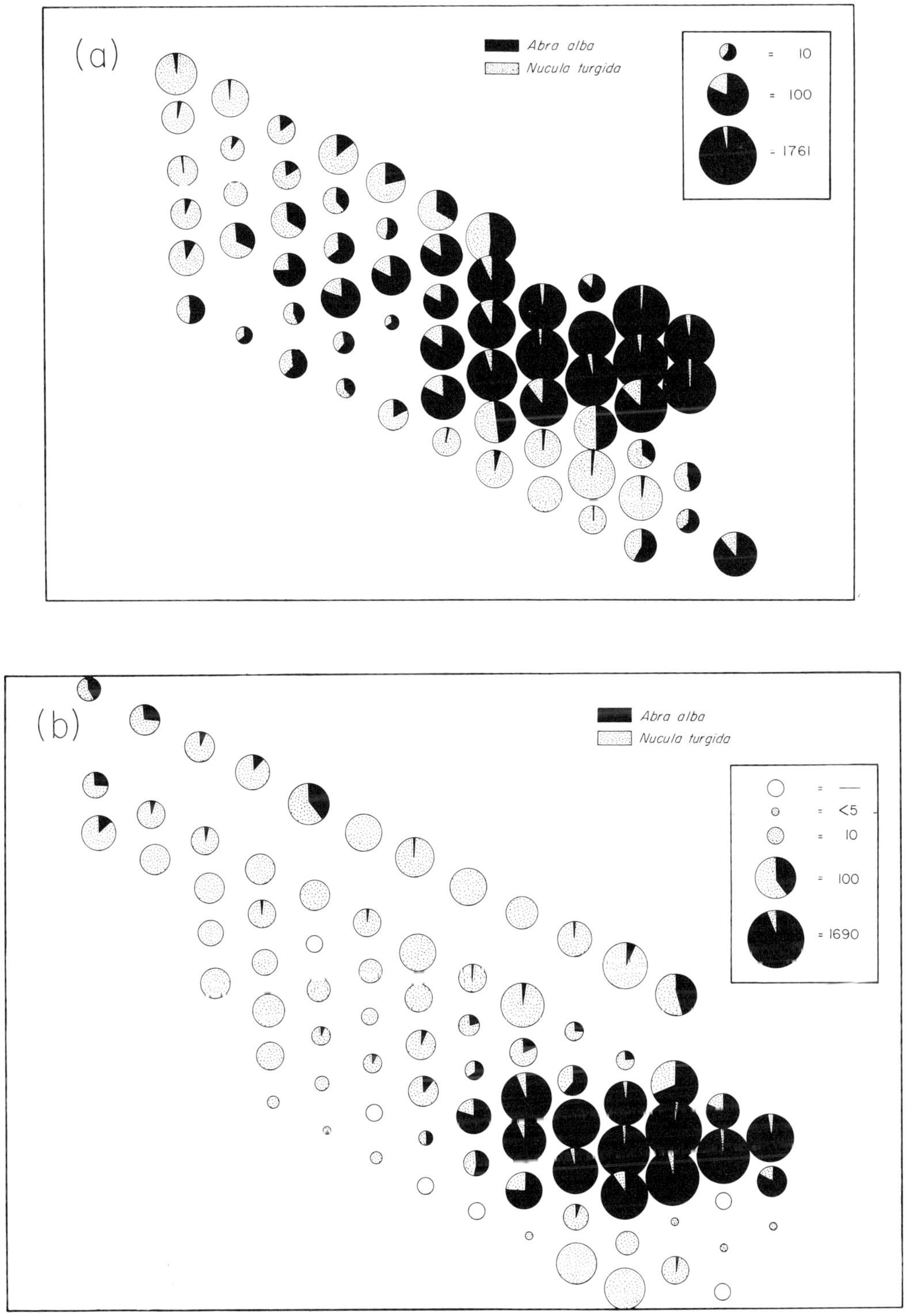

Figure 18.6. Numbers of *Abra alba* and *Nucula turgida* (logarithmic scale) per 0.1 m^2 at the sampling stations in the sewage-sludge area (Fig. 18.1) during (a) July 1973 and (b) April 1974.

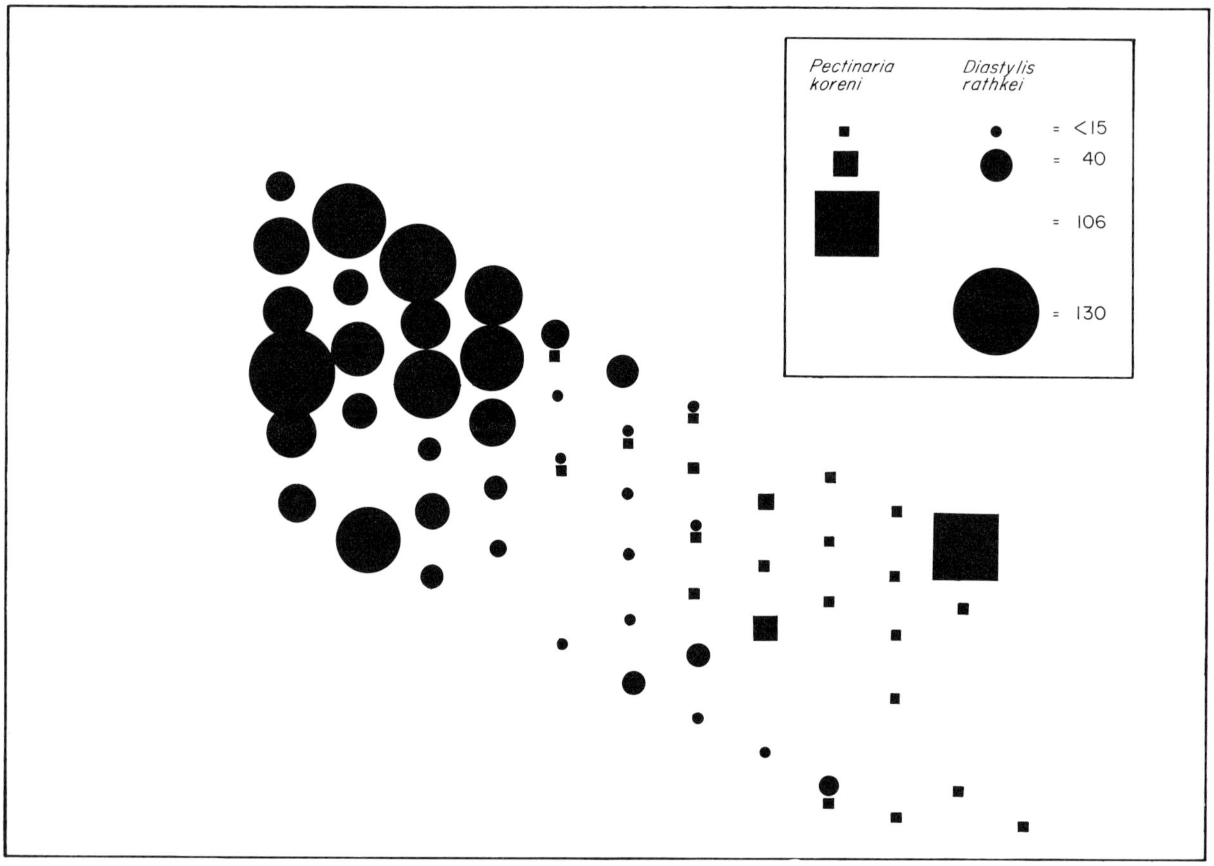

Figure 18.7. Numbers of *Pectinaria loreni* and *Diastylis rathkei* per 0.1 m^2 at the sampling stations in the sewage-sludge area (Fig. 18.1) in July 1973.

The instability of the benthic fauna was most pronounced at the sludge dumpsite itself. Species associations developed with a large proportion of *Abra alba* and, in addition to *Pectinaria,* a large number of polychaetes of different species and small brittle stars (Fig. 18.8b). Faunal patterns changed considerably, as seen in Fig.18.9, which represents the settlement patterns at one station from 1970 to 1977. In 1961, when sludge dumping began, the community contained few individuals. In 1973, a dense population of *Abra* was found. In 1974, the number of individuals in this species were reduced, but some of the growing bivalves from the previous year were still present. In the following years, some of the *Abra* were larger, but massive populations did not develop again. Yearly changes in species composition and abundance were not limited to areas influenced by sludge dumping but also occurred in regions free from such influences, as illustrated at a station northwest of the dumpsite (station 94) (Fig. 18.10). Fauna at this station was dominated by *Nucula turgida.* The abundances of each species at this station are listed in Table 18.1.

It must be determined whether the numbers found in single samples at one station were representative for the area and valid for the interpretation of the changes. Table 18.2 provides the numbers of each species found in samples taken at a neighboring station 1000 m away on the same days. Stations 94 and 93 showed identical increases and decreases in abundance and similar trends in species composition. For example, at both stations in July 1974, there were massive populations of the polychaete *Owenia fusiformis* and the phoronid *Phoronis mulleri.* The bivalve *Nucula turgida* and the cumacean *Diastylis rathkei* also showed parallel development at both stations. Therefore, the samples were believed to be representative, and the development of fauna in the area was reflected by the data.

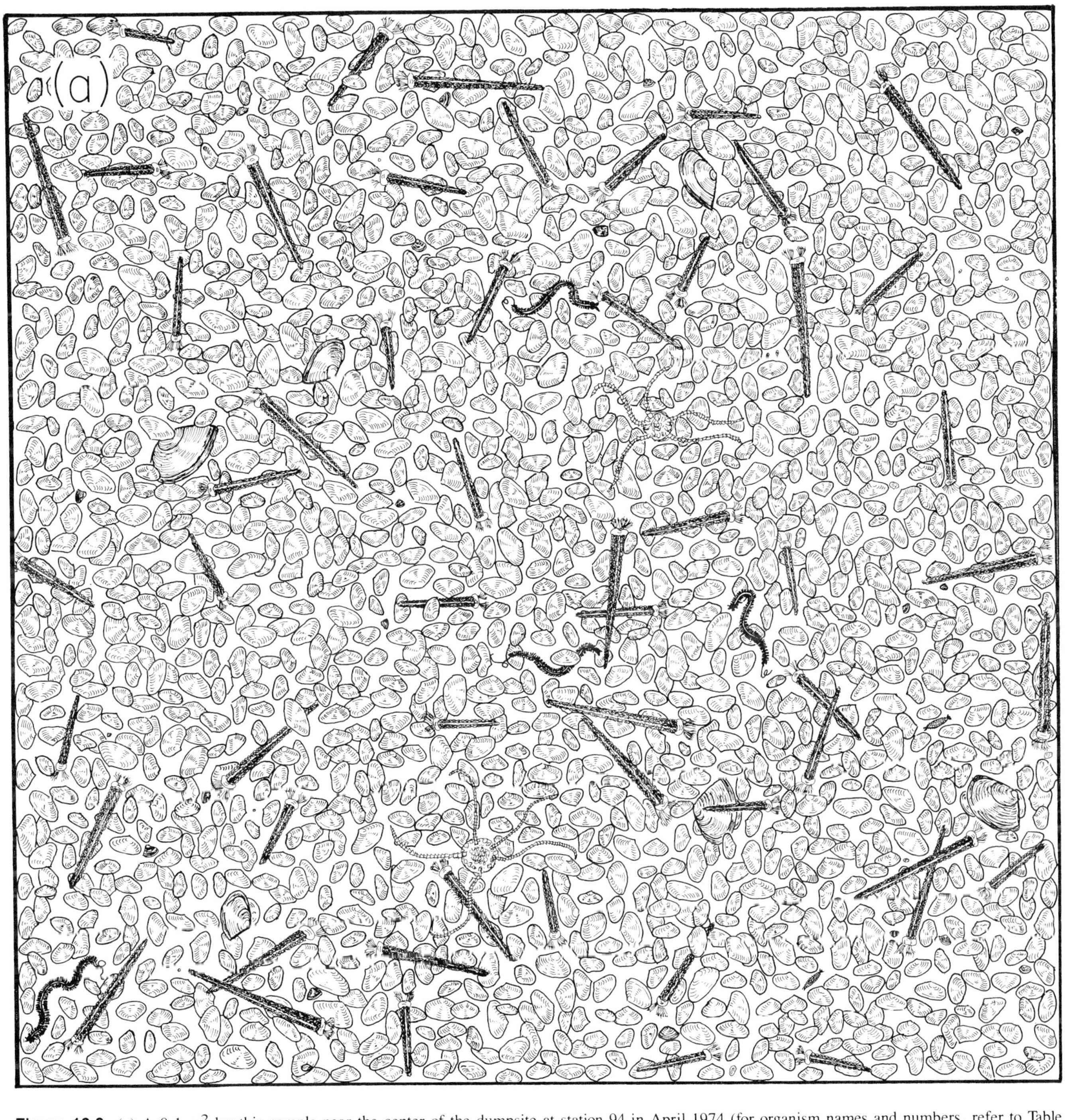

Figure 18.8. (a) A 0.1-m^2 benthic sample near the center of the dumpsite at station 94 in April 1974 (for organism names and numbers, refer to Table 18.1).

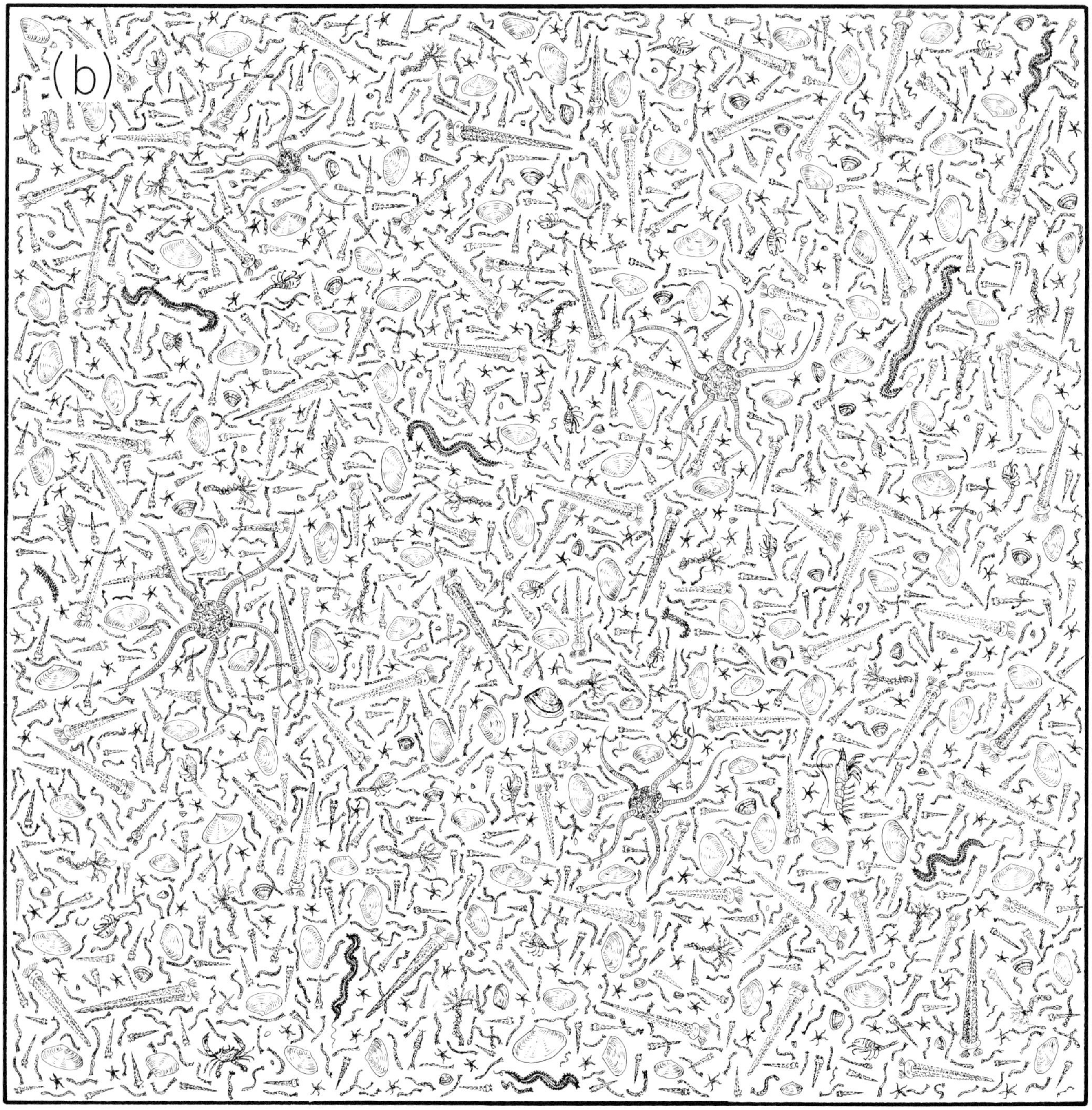

Figure 18.8. (Continued) (b) A 0.1-m^2 benthic sample at station 70: 1614 *Owenia fusiformis;* 296 *Pectinaria koreni;* 228 *Ophiura texturata;* 116 *Abra alba;* 34 *Spiophanes bombyx;* 24 *Diastylis rathkei;* 22 *Nucula turgida;* 20 *Lanice conchilega;* 4 *Angulus fabula;* 2 *Nepthys hombergii;* 1 *Ampharete acutifrons;* 1 *Crangon crangon;* 1 *Gattyana cirrosa;* 1 *Portunus holsatus;* and 1 *Spisula subtruncata.*

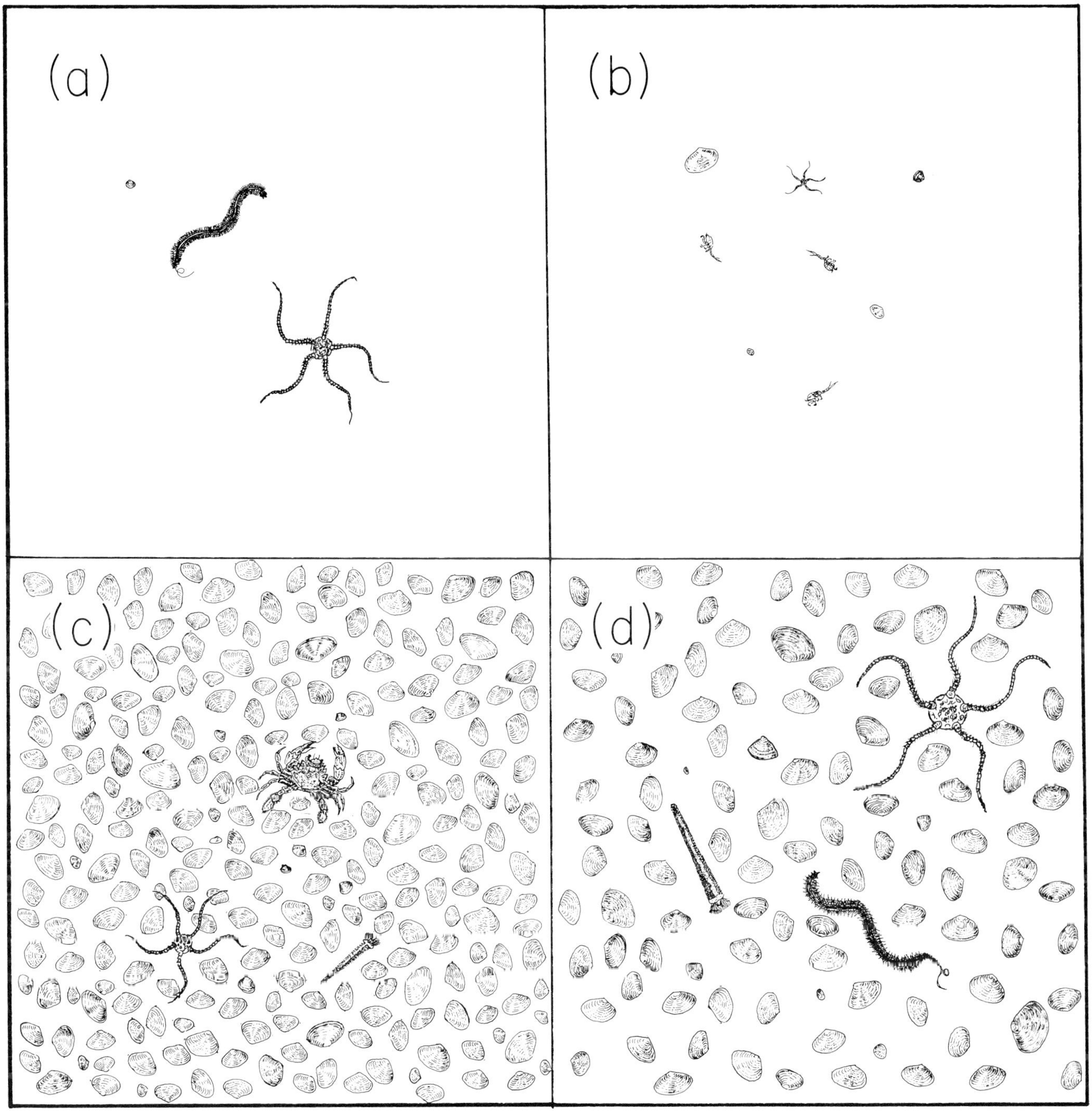

Figure 18.9. Changes in the benthic community (in a 0.02-m^2 area) at station 115 in the center of the sludge dumpsite from 1970 through 1977 (Table 18.7): (a) October 1970, (b) July 1972, (c) July 1973, (d) April 1974.

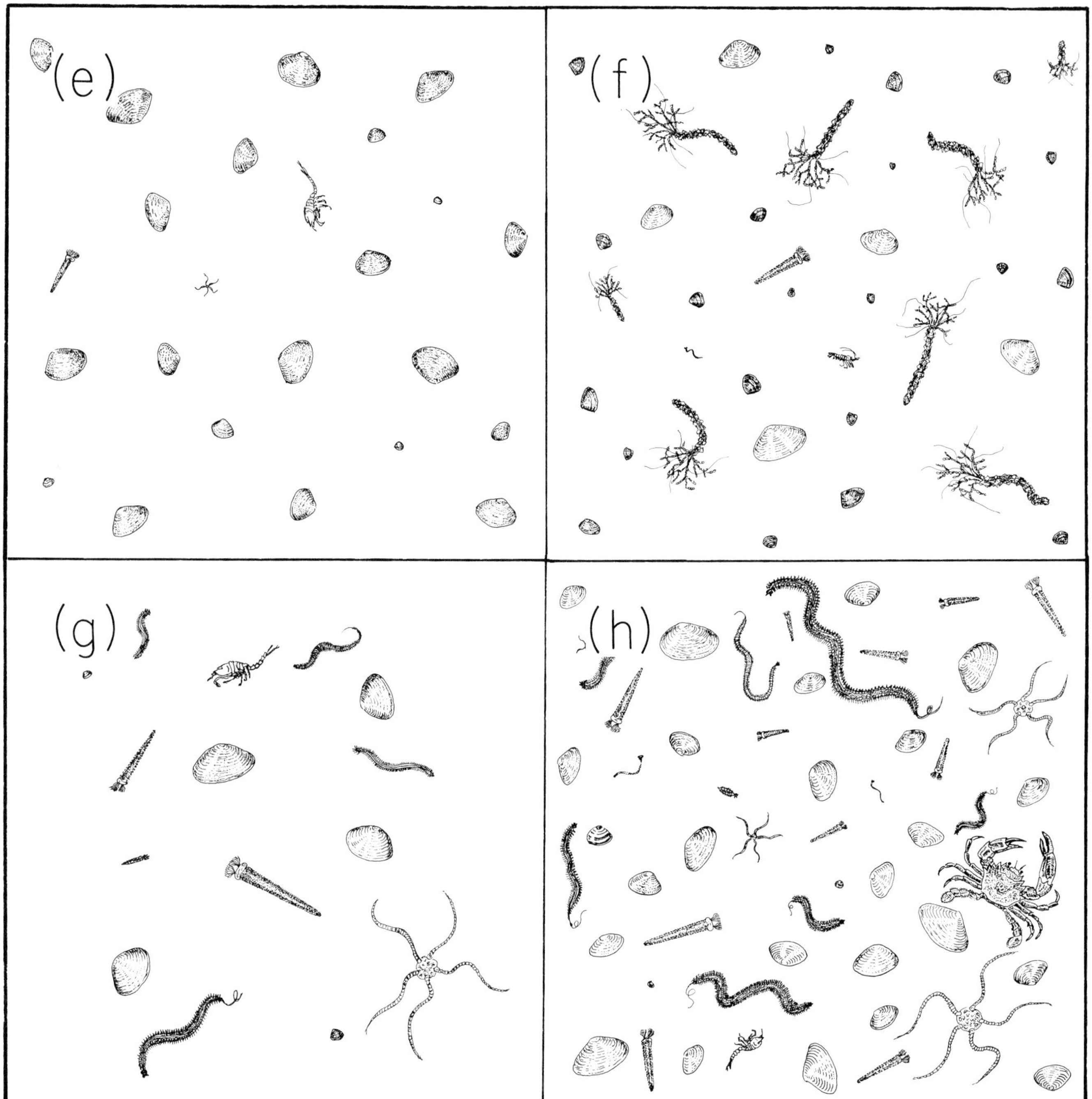

Figure 18.9 (Continued) Changes in the benthic community (in a 0.02-m^2 area) at station 115 in the center of the sewage sludge dumpsite from 1970 through 1977 (Table 18.7): (e) July 1975, (f) July 1976, (g) May 1977, and (h) July 1977.

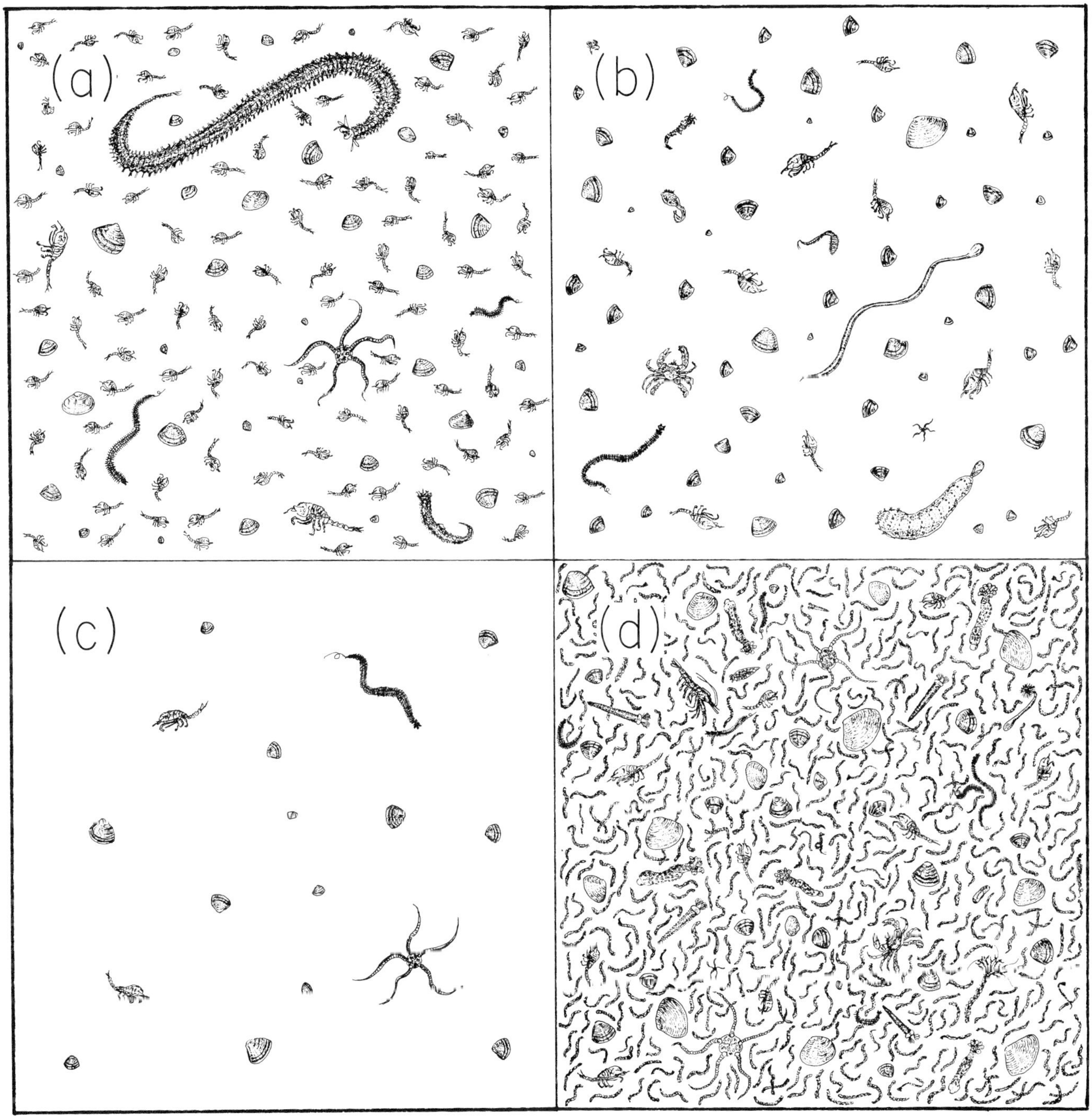

Figure 18.10. Changes in the benthic community (in a 0.02-m^2 area) at station 94 from 1972 through 1978 (Table 18.1): (a) August 1972, (b) August 1973, (c) April 1974, (d) July 1976.

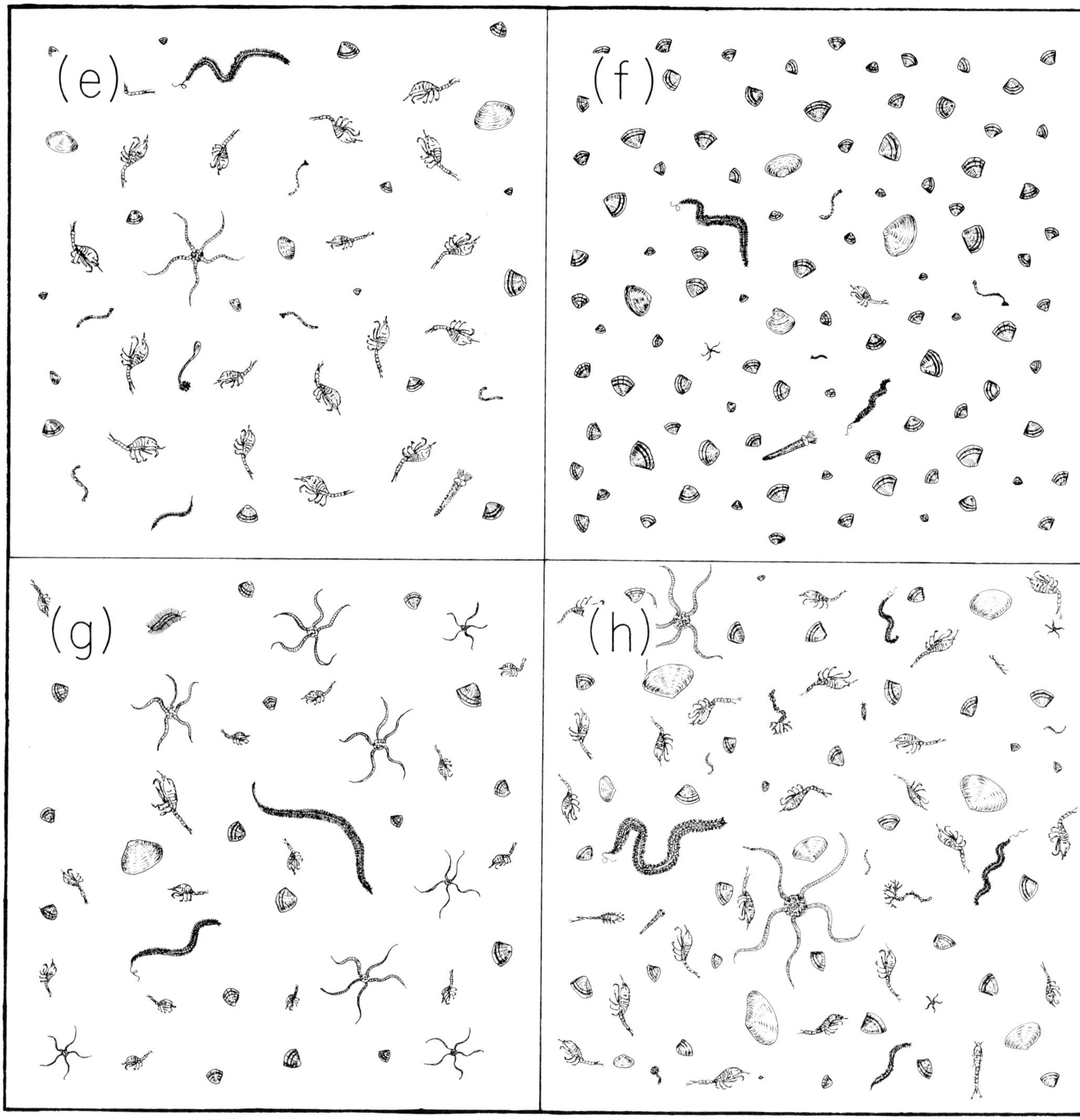

Figure 18.10. (Continued) Changes in the benthic community (in a 0.02-m^2 area) at station 94 from 1972 through 1978 (Table 18.1): (e) September 1976, (f) May 1977, (g) May 1978, and (h) July 1978.

Table 18.1. Station 94[a]: Numbers of Each Benthic Macrofaunal Taxon in a 0.1-m^2 Area

	Date								
	1972	1973	1974		1975	1976	1977	1978	1979
Organism	August	July	April	July	July	September	May	July	July
Edwardsia danica				22					
Cerianthus lloydii				1	1			1	2
Nemertini		1							
Aphrodite aculeata							1		
Harmothoe longisetis							1		3
Pholoe minuta							1		
Anaitides groenlandica				1		2		2	5
Nereis virens	1								
Nephtys hombergii	6	8	1	6	5	9	3	26	8
Scoloplos armiger		1		2					
Spiophanes bombyx				24				3	3
Scalibregma inflatum		1					1		47
Polydora pulchra				4					
Ophelina acuminata		1							3
Owenia fusiformis				2074	13	3			
Notomastus latericeus							1		
Pectinaria koreni		1		19				5	6
Ampharete acutifrons	1	1							
Lanice conchilega				8				7	
Echiurus echiurus	1								
Lora turricula							1		
Lunatia nitida		1							
Nucula turgida	123	196	53	77	55	354	65	159	78
Mysella bidentata	3		1	2	2			1	
Venus gallina						1			
Spisula subtuncata					1	1			
Abra alba and *Abra nitida*	17	4		45	10	13	3	40	2
Mya subtruncata				4					
Cultellus pellucidus				1					
Corbula gibba	1	2		3				1	
Pariambus typicus									1
Ampelisca brevicornis									1
Diastylis rathkei	394	45	8	36	74	3	59	123	
Crangon crangon				1					
Portunus holsatus				1					
Ophiura texturata	2	1	3	6	5	2	19	7	3
Ophiura albida				1				2	
Phoronis mülleri				48	7	2			
Total									
Species	9	14	5	22	10	10	11	13	13
Individuals	550	265	66	2366	2182	391	45	167	183

[a] See Fig. 18.1 for station location.

Table 18.2. Station 93[a]: Numbers of Each Benthic Macrofaunal Taxon in a 0.1-m² Area

	Date								
	1972	1973	1974		1976	1977	1978		1979
Organism	**August**	**July**	**April**	**July**	**September**	**May**	**May**	**July**	**July**
Edwardsia danica				13					
Nemertini						1			1
Harmothoe longisetis						1			1
Harmothoe sarsi	1								
Pholoe minuta						1			
Anaitides groenlandica				1			1	1	
Anaitides maculata									1
Nereis virens	1								
Nephyts hombergii	9	7	5	13	4	9	14	12	2
Scoloplos armiger	1	3							
Spiophanes bombyx				14					
Scalibregma inflatum	1								4
Polydora pulchra				1					
Ophelina acuminata	2								
Owenia fusiformis				2120	9				1
Pectinaria koreni				57	3			2	5
Ampharete acutifrons		2				1		1	
Lanice conchilega				8					
Echiurus echiurus									1
Nucula turgida	79	135	49	81	198	124	156	41	9
Mysella bidentata				6					
Cardium edule				1					
Abra alba and *Abra nitida*	5	5	1	42	10	15	26	23	1
Macoma balthica					1				
Spisula subtruncata						1			
Mya truncata, juvenile				14					
Corbula gibba		2	1	1		1			1
Nototropis swammerdami				1					
Ampelisca brevicornis		1						1	
Gammarus locusta							1		
Diastylis rathkei	371	115		54	8	86	98	58	37
Portunus holsatus				1					
Ophiura texturata	4	41	33	35	3	30	22	2	1
Ophiura albida				1					
Phoronis mülleri				71	2				
Total									
Species	10	9	5	20	10	11	7	9	13
Individuals	477	315	98	2527	240	281	318	141	27

[a] See Fig. 18.2 for station location.

Another example of similar trends is listed in Table 18.3 (station 67); it reveals an increase in the abundance of the sedentary polychaete *Lanice conchilega* between July 1974 and July 1975, a decline in 1976, and low abundance during the following years. Other polychaete populations (*Anaitides, Eumida,* and *Spiophanes*) also showed such yearly changes. At neighboring stations, a similar composition of the benthos was observed. It was considered that there might be patchiness of a species within very small areas; however, the narrow net of stations took such chance variations into account. The data, therefore, provided insights into the dynamics of the benthos. The settlement of larvae, the competition between the growing species, and their disappearance at the end of their life span could be observed. Frequently, the arrival of new larvae of the same species stabilized the colonization, but in many cases one species was displaced by another due to changing patterns of larval drift and settlement. This was a natural phenomenon not confined to the sludge dumpsite.

Table 18.3. Station 67[a]: Numbers of Each Benthic Macrofaunal Taxon in a 0.1-m^2 Area

	Date						
	1974		1975	1976	1977	1978	
Organism	April	July	July	July	May	May	July
Halcampa chrysanthellum			1				
Edwardsia danica		1	1				
Sagartia troglodytes		81	1				
Cerianthus lloydii			2				
Nemertini			2				
Harmothoe longisetis				1			
Gattyana cirrosa		8					
Pholoe minuta		1	9		1	4	
Eteone longa			7	2		1	
Anaitidis maculata and *Anaitidis mucosa*		200	318				1
Anaitidis groenlandica		6	2				
Anaitidis subulifera			1				
Eumida sanguinea		53	128	16			
Nereis longissima		1	8	4			
Nephtys hombergii	2	6	8	3	21		2
Nephtys caeca		1	2				2
Scoloplos armiger	1			2	1		
Polydora pulchra		1	12				
Spiophanes bombyx		9	463	7	4	121	1
Magelona papillicornis			2				1
Scalibregma inflatum			1				
Ophelia limacina				1			
Spio filicornis	1						
Heteromastus filiformis					1		
Capitella capitata	1						
Owenia fusiformis		10					
Pectinaria koreni		38	5	19	6	3	11
Lanice conchilega		1233	3379	434	3		94
Hydrobia ulvae				1			
Nucula turgida	17	24	3	2	4		
Mytilus edulis		1					
Mysella bidentata		5	1				
Mactra corallina			1				
Spisula subtruncata			2				
Donax vittatus			1				
Abra alba	72	96	67	9	2	15	10
Macoma balthica		2	2	2		5	21
Angulus fabula		1	10			1	
Nototropis swammerdami		1					
Caprella linearis			1		1		
Diastylis rathkei			1		1	23	18
Crangon crangon		2	2				
Portunus holsatus		3	2				1
Eupagurus bernhardus			3	2			
Ophiura texturata		1	4	5	1	1	3
Echinocardium cordatum			1				
Total							
Species	6	25	35	16	12	9	12
Individuals	94	1907	4452	491	46	174	155

[a] See Fig. 18.1 for station location.

To show the yearly changes in populations of the bivalve *Abra alba,* the number of specimens at six stations in the central dumping area from 1970 to 1979 are listed in Table 18.4. In 1970 and 1972, there were very few specimens, but in 1973 and 1974, a large mass of newly settled individuals was observed. Between July 1973 and April 1974, their numbers decreased due to grazing and space limitation at some stations. On the other hand, the number of *Abra* continued to increase due to new settlement by young individuals at stations 84, 85, and 86 during the same period; however, this stock began to decrease between April and July 1974 due to biological factors. Only at station 86 was the population in April 1974 not limited by space. In July 1975, the numbers of *Abra* diminished because those animals had reached the end of their life spans. Some young *Abra* continued to settle, but they never reached the density observed in 1973 and 1974. Polychaetes and crustaceans occupied the space, and some of these species developed large populations. In some cases, the number of species increased, but there was generally no fundamental change in species diversity. At some stations in the dumpsite as many as 15 species of molluscs, polychaetes, crustaceans, and echinoderms were found in one grab sample.

In Tables 18.5, 18.6, and 18.7 (stations 84, 114, and 115), the species composition between 1970 and 1979 at the three stations compared in Table 18.4 demonstrated the degree of change. The species diversity was lowest at the beginning of the study (1970 and 1972) and then increased when *Abra* developed large populations. These stations showed parallel increases and decreases in species diversity and number of individuals present. Species diversity was lower in the dumpsite than in regions with muddy sand, such as station 67 (Table 18.3), which had many ecotypes. Fewer species were also found in some muddy regions not influenced by sludge dumping.

Table 18.4. The Number of *Abra alba* in a 0.1-m^2 Area at Six Stations in the Center of the Sewage-Sludge Dumpsite

	1970	1972	1973	1974		1975	1976	1977		1978		1979
Station	**October**	**August**	**July**	**April**	**July**	**July**	**July**	**May**	**July**	**May**	**July**	**July**
84		2	872	1669	391	336	3	13		6	41	121
85			1431	1559	608	148	9	16		6	0	90
86		4	43	212	217	102	0	11		0	47	103
113	0		1035	0		128	117	25	11	3	71	
114	0	5	1289	504		136	58	23	57	2	65	
115	0	9	1005	416		68	18	89	60	0	75	107
Mean												
84–86		3	782	1447	403	195	4	13		4	29	105
113–115	0	7	111	307		111	64	46	43	2	70	

Table 18.5 Station 84[a]: Numbers of Each Benthic Macrofaunal Taxon in a 0.1-m^2 Area

	Date								
	1973	1974		1975	1976	1977	1978		1979
Organism	July	April	July	July	July	May	May	July	July
Cerianthus lloydii								1	
Nemertini			1						
Pholoe minuta		2	4	24		3		4	
Anaitides groenlandica					1				
Anaitides maculata	70			2					2
Nephtys hombergii	9	4	5	8	4	12	14	11	4
Eumida sanguinea					29				
Scoloplos armiger						1	1		
Spiophanes bombyx		1	1	4	391	15		1	1
Polydora pulchra					23				
Magelona papillicornis					6			1	
Capitella capitata					10				
Owenia fusiformis			1						
Pectinaria koreni	106	55	13	66	19			1	66
Lanice conchilega			4		654			11	
Nucula turgida	51	21	41	38	2				4
Mysella bidentata	12	14	17	25		7	13	6	1
Cardium edule									44
Mactra corallina				1			1	1	
Macoma balthica	30	7	6	59		3	1	12	70
Abra alba	872	1669	391	336	3	236	7	52	204
Angulus fabula			2	8	7		1	9	6
Cultellus pellucidus									48
Mya arenaria							1		
Balanus crenatus			2						
Nototropis swammerdami					1				
Diastylis rathkei			1	2		4	45	10	6
Crangon crangon	1				2				
Portunus holsatus			3						
Ophiura texturata	4	2	3	1		1	1		14
Total									
Species	9	9	16	13	14	9	10	13	12
Individuals	1463	1777	495	574	1152	282	85	120	470

[a] See Fig. 18.1. for station location.

Table 18.6. Station 115[a]: Numbers of Each Benthic Macrofaunal Taxon in a 0.1-m^2 Area

	Date										
	1970	1972	1973	1974	1975	1976	1977		1978		1979
Organism	October	August	July	April	July	July	May	July	May	July	July
Segartia troglodytes							1				
Cerianthus lloydii								2			
Pholoe minuta							4	4		1	
Anaitides groenlandica							1				
Anaitides maculata and *Anaitides mucosa*										1	2
Nereis succinea							6				
Nereis longissima							1				
Nephtys hombergii	2			4			5	23	2	5	5
Eumida sanguinea										1	
Spiophanes bombyx						2	1	1		2	
Pectinaria koreni			2	1	4	2	11	43		56	27
Lanice conchilega						30			318	370	
Nucula turgida	1	2	15	11	1	84	3				
Mysella bidentata				10	1		5	6		1	1
Cardium edule						1					
Abra alba		9	1005	416	68	18	16	87		79	76
Macoma balthica			1					2		23	14
Angulus fabula			1		6			2		23	
Cultellus pellucidus											1
Diastylis rathkei		11			1	1	2	2	2	11	2
Crangon crangon					1	1				1	
Portunus holsatus			1					1		1	
Ophiura texturata	3	1	4	2	1		5	16	1	5	2
Phoronis mülleri								1			
Total											
Species	3	4	7	6	8	8	13	13	4	15	9
Individuals	9	27	1036	450	910	147	139	203	327	595	130

[a] See Fig. 18.1 for station location.

Table 18.7. Station 114[a]: Numbers of Each Benthic Macrofaunal Taxon in a 0.1-m² Area

	Date										
	1970	1972	1973	1974	1975	1976	1977		1978		1979
Organism	October	August	July	April	July	July	May	July	May	July	July
Cerianthus lloydii										3	1
Pholoe minuta								2			
Eumida sanguinea										2	
Anaitides maculata			75			1	1				10
Anaitides mucosa and *Anaitides groenlandica*										1	
Nereis virens										1	
Nephtys hombergii	16		3	2	2		2		4	7	1
Nephtys caeca							1	3			
Scoloplos armiger							3	1			
Spiophanes bombyx							2	4		11	1
Scalibregma inflatum			2					2			
Polydora pulchra						1		1			
Owenia fusiformis										2	
Pectinaria koreni			6		12	6	3	183		10	112
Lanice conchilega						77		2		112	5
Nucula turgida	5		35	49	3					1	1
Mysella bidentata							2	3			
Cardium edule						8	3	1			3
Mactra corallina					1		3	1	1		
Mytilus edulis											1
Macoma balthica	1		1	4	4	6	1	3		22	17
Abra alba		5	1289	504	136	58	17	73	2	68	11
Angulus fabula					1			1		4	6
Cultellus pellucidus											6
Diastylis rathkei		7					6	3	16	10	10
Corystes cassivelaunus										1	
Crangon crangon					1	2					
Portunus holsatus								1		1	
Carcinus maenas						1					
Ophiura texturata	1		3		1		7	5		1	2
Total											
Species	4	2	8	4	9	9	13	17	5	15	15
Individuals	23	12	1414	559	162	161	64	295	294	262	187

[a] See Fig. 18.1 for station location.

18.4. DISCUSSION AND CONCLUSIONS

Pearson and Rosenberg (1978) reviewed reports on macrobenthic succession in relation to organic enrichment and described methods to measure diversity by involving similarity indexes and ordering techniques. We applied similar methods for data from the Elbe Estuary. Individual species or groups can be used to characterize the degree of organic pollution. The development of saprobic classification system, which is controversial among limnologists (Caspers, 1977), is even more problematic when applied to marine systems. Indicator species react most rapidly to the oxygen concentration, depending on their trophic level in the habitat. The large increase in numbers of the bivalve, *Abra alba,* as well as of some polychaetes, crustaceans, and echinoderms, can be interpreted as a reaction of preadapted organisms to the disposal of additional waste matter (Rachor, 1980). Such species take advantage of organic nutrients; however, they also require a sufficient oxygen supply. At the center of the dumpsite, as well as at distant locations, the mud was anaerobic, but an aerobic surface layer supplied most of the benthic organisms with oxygen.

There was some correlation between massive population development and hydrographic factors. New sludge was introduced into the dumping area almost every day, thus producing a considerable input of detritus. The sludge could have had negative effects if anaerobic conditions were produced in the sediment. Rachor (1977) stated: "The long-term trend of impoverishment of the fauna is attributed to the development of anaerobic conditions in the muddy sediment." However, this situation is counteracted by the continuous import of oxygen by the water as it moves with tidal currents. Thus, the sediment surface is aerobic, as evidenced by the color of the surface layer. Most of the benthic animals utilize this oxidized surface layer as a source of oxygen and nutrients. The bivalve *Abra alba* lives just below the surface. Its siphon reaches the sediment–water interface, where it obtains organic detritus. *Abra* occur in such densities that they are limited by physical space (Caspers, 1978). Similar conditions were encountered in the neighboring zones of the outer Elbe Estuary. The surface of the sediment was dark and exhibited anaerobic conditions in only a few areas of the dumpsite. In such locations, there was a rapid decrease in the number of individuals and species.

Among the biotic factors influencing the variability in the community structure, four seem to be of particular importance: (1) the life span of individuals, (2) competition for space, (3) predation, and (4) settlement by larvae. It does not necessarily signify a change in habitat conditions when a bivalve population becomes senescent and dies off. *Abra alba* can live for a maximum of three years, but most individuals die after two. The area is then available for resettlement when, by chance, larvae of the same species of polychaete or bivalve that had previously inhabited the area arrive at a stage sufficiently advanced to permit their transition to a benthic existence. After these organisms have found suitable places to settle, predators may enter the habitat and decimate the newly established populations.

The seasonal arrival of larvae belonging to various taxa brings about a more rapid fluctuation in the benthic community structure than is encountered in freshwater habitats. This suggests that such marine associations are relatively unstable. It may be, however, that a dynamic stability is exhibited over a period of many years; more likely, there is a rapid and irregular succession of species. The community might best be designated as astatic, and characterized by seasonal and long-term fluctuations. Holling (1973) provides a detailed discussion of stability and resilience in ecological systems and, according to his definitions, the associations in the outer Elbe Estuary may be characterized as relatively unstable but highly resilient.

Buchanan and Warwick (1974), in their study of offshore macrofaunal production, pointed out that, after the pioneer investigations of Petersen, only a few surveys have been repeated for consecutive years. Such surveys include those by Boysen Jensen (1919) in the Limfjord from 1909 to 1917, Segerstråle (1960) in the Baltic, and Muus (1966), who investigated the Oresund for three years. These studies reveal a considerable degree of fluctuation in population density from year to year. Buchanan (1964) also observed these fluctuations during his study of the echinoderm fauna of the Northumberland coast.

The reasons for the changes in diversity of marine benthic species have long been discussed (Sanders, 1956, 1968, 1969). In his review of experiments on the exclusion of predators, Peterson (1979) pointed out that many marine infaunal systems include a large proportion of short-lived, seasonally abundant organisms and that instability is due mainly to predation (Thorson, 1966; Naqvi, 1968; Mills, 1969; May, 1973; Woodin, 1974, 1976, 1978; Menge and Sutherland, 1976; Levinton, 1977; Steimle, 1977; Steimle and Radosh, 1979; Moore, 1978; Pearce et al., 1981; Hulbert et al., 1982; Alaftheriou et al., 1982; Reid, 1982; Emmerson et al., 1983).

As this short review of the literature shows, the instability of the marine benthos has been observed by quite a few investigators; however, most have not continued their studies over a period of several years. The long-term study of the benthic populations in the outer Elbe region has provided examples of population fluctuations caused by factors that are primarily biotic, such as life cycle and settlement patterns, predation, and crowding (Caspers, 1978, 1980a, 1980b, 1981). Rachor and Gerlach (1978), who sampled the benthic macrofauna in a sandy region of the German Bight from 1967 to 1975, suggest that the variations are caused by wave erosion. Ziegelmeier (1964, 1970) also discusses this situation.

The sludge-dumping area is also exposed to wave action, which may have an effect on benthic populations and bring about yearly differences. However, direct observations of sudden catastrophic changes in benthic populations produced by such environmental factors has not been made in this area. Even if such influences could be demonstrated, community structure may still depend primarily on biotic factors. A dominant population of adult bivalves or polychaetes disappears after the individuals die at the end of their natural life span, which is a maximum of three years for *Abra alba*. New colonization by the same or another species is then possible; this depends mainly on the transport of larvae that are ready to settle (Caspers, 1980a, 1980b).

Biotic instability makes prediction more difficult than it is when environmental factors only, such as grain size, oxygen supply, temperature, or nutrient supply, must be considered. A marine benthic community does not develop or disappear according to a stable pattern, because each population consists of individuals with specific spawning periods and life spans. Intra- and interspecific competition for food and space are important. Each species reacts as an element of a complicated, multifaceted ecosystem. This makes the study of the effects of sewage-sludge dumping more complicated; however, the dumping itself can be viewed as a large-scale, *in situ* experiment providing the chance to investigate natural processes in the sea in considerable detail. Such investigations require extensive monitoring. It takes several years to understand the abiotic and biotic factors involved in the reaction of the communities. This may be considered basic research that cannot always be completed quickly enough to provide answers concerning the effects of a proposed project. However, for a study of anthropogenic effects, such as sludge dumping, a minimum of five years to observe any effects is necessary. During this period, biotic factors could be expected to produce changes in the composition of the fauna, depending on the life span of the individuals that must resettle as larvae. To record long-term changes, additional years are necessary. Currently there is not sufficient basic information about benthic-community changes caused by oceanic influences. To observe these trends, long-term monitoring is necessary. This is expensive, but it would increase what is known about the biological processes that take place in the sea.

Investigations of benthic fauna based on data gathered over a period of only a few years can lead to false conclusions. If the interpretations of the faunal changes are too hasty, environmental or anthropogenic influences can be blamed for short-term events that are actually part of a normal randomness of the community structure. If the statistical methods chosen presuppose temporal randomness in a system, then the indexes or probabilities calculated may show that there is a relationship between the temporal or spatial settlement patterns and some abiotic factors (Elliot,1971). Although the relationship seems probable, it is not possible to prove definitely that the relationship is not due simply to coincidence; therefore, later investigations in the same area often fail to support earlier assumptions. This statement is not meant to criticize the statistical methods or the reporting of preliminary findings, but, rather, to stress the importance of continuing studies over a long term to preclude errors from premature interpretations.

REFERENCES

Alaftheriou, A., D. C. Moore, D. J. Basford, and M. R. Robertson. 1982. Underwater experiments on the effects of sewage sludge on a marine ecosystem. *Netherlands Journal of Sea Research,* **16,** 463–473.

Boysen Jensen, P. 1919. Valuation of the Limfjord. I: Studies on the fish-food in the Limfjord 1909–1917, its quantity, variation and animal production. *Report of the Danish Biological Station to the Board of Agriculture,* **26,** 1–44.

Buchanan, J. B. 1964. A comparative study of some features of the biology of *Amphiura filiformis* and *Amphiura chiajei* (Ophiuroidea) considered in relation to their distribution. *Journal of the Marine Biological Association of the United Kingdom,* **44,** 565–576.

Buchanan, J. B., and R. M. Warwick. 1974. An estimate of benthic macrofaunal production in the offshore mud of the Northumberland coast. *Journal of the Marine Biological Association of the United Kingdom,* **54,** 197–222.

Caspers, H. 1952. Bodengreiferuntersuchungen über die Tierwelt in der Fahrrinne der Unterelbe und im Vormündungsgebiet der Nordsee. *Zoologischer Anzeiger,* **16**(Supplement), 404–418.

Caspers, H. 1954. Biologische Untersuchungen über die Lebensraüme der Unterelbe und des Vormündungsgebietes der Nordsee. *Mitteilungen aus dem Geologischen Staatsinstitüt in Hamburg,* **23,** 78–85.

Caspers, H. 1968. Der Einfluss der Elbe auf die Verunreinigung der Nordsee. *Helgoländer Wissenschaftliche Meeresuntersuchungen,* **17,** 422–443.

Caspers, H. 1977. Qualität des Wassers — Qualität der Gewasser. Die Problematik der Saprobiensysteme. *Archiv für Hydrobiologie,* **9,** 3–14.

Caspers, H. 1978. Ecological effects of sewage sludge on benthic fauna off the German North Sea coast. *In:* Proceedings of the 8th International Conference on Water Pollution Research, 1976, S. H. Jenkins (Ed.). Pergamon Press, Oxford, pp. 951–956.

Caspers, H. 1980a. Adaptation and biocoenotic associations of echinoderms in a sewage sludge dumping area of the southern North Sea. *In:* Proceedings of the European Colloquium on Echinoderms, Brussels, 1979, M. Jangoux (Ed.). A. A. Balkema, Rotterdam, pp. 189–198.

Caspers, H. 1980b. Long-term changes in benthic fauna resulting from sewage sludge dumping into the North Sea. *Progress in Water Technology,* **12,** 461–479.

Caspers, H. 1981. Ten year studies on marine pollution as a part of a priority program sponsored by the German Research Society. *In:* Some Approaches to Saprobiological Problems, M. Sudzuki (Ed.). Tokyo, pp. 12–28.

Elliott, J. M. 1971. Some Methods for the Statistical Analysis of Samples of Benthic Invertebrates. Scientific Publication No. 25, Freshwater Biological Association, Ambleside, Westmoreland, U. K., 144 pp.

Emmerson, W. D., A. McLachlan, H. R. Watling, and R. J. Watling. 1983. Some ecological effects of two sewage outfalls in Algoa Bay. *Water S.A.,* **9,** 23–30.

Fuchs, D. 1982. Sludge disposal to the sea in the Federal Republic of Germany. *In:* Disposal of Sludge to the Sea, Vol. 14, M. C. Dart and S. H. Jenkins (Eds.). Water Technology, London, pp. 163–168.

Holling, C. S. 1973. Resilience and stability of ecological systems. *Annual Review of Ecology and Systematics,* **4,** 1–23.

Hulbert, A. W., K. J. Pecci, J. D. Witman, L. G. Harris, J. R. Sears, and R. A. Cooper. 1982. Ecosystem definition and community structure of the macrobenthos of the NEMP monitoring station at Pigeon Hill in the Gulf of Maine. Technical Memorandum NMFS-F/NEC-14, National Oceanic and Atmospheric Administration, Boulder, Colorado, 143 pp.

Levinton, J. S. 1977. Ecology of shallow water deposit-feeding communities Quisset Harbor, Massachusetts. *In:* Ecology of Marine Benthos, B. C. Coull (Ed.). University of South Carolina Press, Columbia, pp. 191–228.

May, R. M. (Ed.). 1973. Stability and Complexity in Model Ecosystems. Princeton University Press, Princeton, New Jersey, 233 pp.

Menge, B. A., and J. P. Sutherland. 1976. Species diversity gradients: synthesis of the roles of predation, competition, and temporal heterogeneity. *American Naturalist,* **110,** 351–369.

Mills, E. L. 1969. The community concept in marine zoology, with comments on continua and instability in some marine communities. *Journal of the Fisheries Research Board of Canada,* **26,** 1415–1428.

Moore, J. W. 1978. Some factors influencing the density of invertebrates near a sewage outfall. *Hydrobiologia,* **61,** 81–93.

Muus, K. 1966. A quantitative 3-year survey on the meiofauna of known macrofauna communities in the Öresund. *Veröffentlichungen des Instituts für Meeresforschung in Bremerhaven,* **2,** 289–292.

Naqvi, S. M. Z. 1968. Effects of predation on infaunal invertebrates of Alligator Harbor, Florida. *Gulf Research Reports,* **2,** 313–321.

Nauke, M. 1974. Die Schwermetallgehalte der Sedimente im Klärschlamm–Verklappungsgebiet vor der Elbmündung. *Deutsche Hydrographische Zeitschrift,* **27,** 313–321.

Pearce, J. B., D. J. Radosh, J. V. Caracciolo, F. W. Steimle, Jr. 1981. Benthic Fauna. New York Bight Atlas Monograph 14, Marine Ecosystems Analysis Program, New York Sea Grant Institute, Albany, New York, 79 pp.

Pearson, T. H., and R. Rosenberg. 1978. Macrobenthic succession in relation to organic enrichment and pollution of the marine environment. *Oceanography and Marine Biology: An Annual Review,* **16,** 229–311.

Peterson, C. H. 1979. Predation, competitive exclusion, and diversity in the soft-sediment benthic communities of estuaries and lagoons. *In:* Ecological Processes in Coastal and Marine Systems, J. Livingston (Ed.). Plenum Press, New York, pp. 233–264.

Rachor, E. 1977. Faunenverarmung in einem Schlickgebiet in der Nähe Helgolands. *Helgoländer Meeresuntersuchungen,* **30,** 633–651.

Rachor, E. 1980. The inner German Bight — an ecologically sensitive area as indicated by the bottom fauna. *Helgoländer Meeresuntersuchungen,* **33,** 522–530.

Rachor, E., and S. A. Gerlach. 1978. Changes of macrobenthos in a sublittoral sand area of the German Bight, 1967 to 1975. *Rapports et Procès-Verbaux des Réunions, Conseil International pour l'Exploration de la Mer,* **172,** 418–431.

Reid, R. 1982. Benthic macrofauna community structure. *In:* Contaminants in New York Bight and Long Island Sound. Sediments and Demersal Species, and Contaminant Effects on Benthos, Summer 1980, J. E. O'Reilly and V. S. Zdanowicz (Eds.). Technical Memorandum NMFS-F/NEC-16, National Oceanic and Atmospheric Administration, Boulder, Colorado, pp. 63–76.

Sanders, H. L. 1956. The biology of marine bottom communities. *Bulletin of the Bingham Oceanographic Collection,* **15,** 344–414.

Sanders, H. L. 1968. Marine benthic diversity: a comparative study. *American Naturalist,* **102,** 243–282.

Sanders, H. L. 1969. Benthic marine diversity and the stability–time hypothesis. *Brookhaven Symposia in Biology,* **12,** 71–80.

Segerstråle, S. G. 1960. Fluctuations in the abundance of benthic animals in the Baltic area. *Commentationes Biologicae, Societas Scientiarum Fennica,* **23,** 1–19.

Steimle, F. W. 1977. A preliminary assessment of the 1976 New York Bight oxygen depletion phenomenon on the benthic invertebrate megafauna. *In:* Oxygen Depletion and Associated Environmental Disturbance in the Middle Atlantic Bight in 1976. Technical Series Report No. 3, Northeast Fishery Center, Woods Hole, Massachusetts, pp. 127–175.

Steimle, F. W., Jr., and D. J. Radosh. 1979. Effects on the benthic invertebrate megafauna. *In:* Oxygen Depletion and Associated Benthic Mortalities in New York Bight, 1976, D. L. Swanson and C. J. Sindermann (Eds.). Technical Memorandum NMFS-F/NEC-14, National Oceanic and Atmospheric Administration, Boulder, Colorado, pp. 281–291.

Stripp, K. 1969a. Jahreszeitliche Fluktuationen von Makrofauna und Meiofauna in der Helgoländer Bucht. *Veröffentlichungen des Instituts für Meeresforschung in Bremerhaven,* **12,** 65–94.

Stripp, K. 1969b. Die Assoziationen des Benthos in der Helgoländer Bucht. *Veröffentlichungen des Instituts für Meeresforschung in Bremerhaven,* **12,** 95–142.

Thorson, G. 1966. Some factors influencing the recruitment and establishment of marine benthic communities. *Netherlands Journal of Sea Research,* **3,** 267–293.

Woodin, S. A. 1974. Polychaete abundance patterns in a marine soft-sediment environment: The importance of biological interactions. *Ecological Monographs,* **44,** 171–187.

Woodin, S. A. 1976. Adult–larval interactions in dense infaunal assemblages: patterns of abundance. *Journal of Marine Research,* **34,** 25–41.

Woodin, S. A. 1978. Refuges, disturbance and community structure: a marine soft-bottom example. *Ecology,* **59,** 274–284.

Ziegelmeier, E. 1964. Einwirkungen des kalten Winters 1962/63 auf das Makrobenthos im Ostteil der Deutschen Bucht. *Helgoländer Wissenschaftliche Meeresuntersuchungen,* **10,** 276–282.

Ziegelmeier, E. 1970. Über Massenvorkommen verschiedener makrobenthaler Wirbelloser während der Wiederbesiedlungsphase nach Schädigungen durch "katastrophale" Umwelteinflüsse. *Helgoländer Wissenschaftliche Meeresuntersuchungen,* **21,** 9–20.

Chapter 19

Pollution Studies in False Bay, South Africa: Chemical Versus Meiofaunal Indicators

Paul D. Bartlett and Helmke F-K.O. Hennig

National Research Institute for Oceanology
Council for Scientific and Industrial Research
Stellenbosch, South Africa

ABSTRACT

False Bay lies to the east of the Cape Peninsula at the south-western tip of Africa and adjacent to metropolitan Cape Town. In view of the increasing problems of effluent disposal, a multi-disciplinary study has been started in order to assess the present and future pollution status of False Bay. This chapter outlines the various studies being undertaken and describes the chemical and biological program. Sediments were examined for metals, organohalogens, fecal steroids, organic content, particle size, and meiofaunal populations. Meiofauna studies were undertaken to assess population densities of nematodes, harpacticoid copepods, and flatworms. These studies revealed areas of perturbation that were not shown by chemical studies. However, apart from these areas most of the results revealed False Bay to be essentially unpolluted.

19.1. INTRODUCTION

False Bay is east of the Cape Peninsula at the southwestern tip of Africa (Fig. 19.1), covers approximately 1000 km^2, and is enclosed except on the southern side. It is adjacent to the growing urban area associated with Cape Town and its satellite towns. In view of the increasing problem of effluent disposal and fears of pollution, the Department of Environment Affairs, in response to local pressure, initiated a study of the present pollution condition and possible future trends in False Bay. The National Research Institute for Oceanology (NRIO), Council for Scientific and Industrial Research, was contracted for this study, which involved both the marine science and the coastal engineering divisions of the institute.

Little information is available about False Bay and any previous pollution problems. The physical oceanographic and meteorological conditions have been described by Atkins (1970a, 1970b). Cram (1970) further described the currents and found an area of low surface temperature in the center of the bay. Sedimentological studies have been performed by

Bowie et al. (1970) and included some organic carbon analyses. Glass (1980) and the NRIO (1982) further described sediment distributions. Some chemical data on the rivers in False Bay catchment (Cape Town City Engineer) are included in reports by Morant and Grindley (1982) and Bickerton (1982).

The Marine Chemistry and Marine Biology Divisions have undertaken the largest part of the present field program. Initially, studies concentrated on the sediments, which were examined for heavy metals, organohalogens, organic content, particle size, and meiofaunal populations. Coprostanol (a specific fecal steroid) analyses have also been undertaken to assess the extent of sewage pollution in the bay. This chapter deals mainly with the chemical and biological studies.

19.2. CHEMICAL AND PHYSICAL FEATURES OF FALSE BAY

False Bay is almost square in shape and is open only to the Atlantic Ocean in the south. The mouth is 29 km across, the northern shore is about 35 km in extent, and the north–south axis is 27 km long (Mallory, 1970). The bottom topography shoals (at a gradient of 1 in 400) from about 80 m deep at the mouth to the northern shoreline (Fig. 19.2). The wind and current patterns, general hydrology, and temperature conditions affect sedimentary deposition, which collectively controls benthic community structure.

19.2.1. Chemical Features

The population of the Cape Metropolitan area has a direct influence on False Bay by its varying use of land on the margins of the bay. Treated sewage effluent and industrial wastes are discharged directly into the bay. Presently, most waste is directed to the northern shore (Fig. 19.2), where effluents from five sewage works, fertilizer and explosive plants, and food factories are discharged (Eagle, 1980). Most future development is expected to occur along the northern shore, and the greatest impact from increases in effluent loads may occur in this area.

Currents and thermocline structure have a definite influence on the dispersion of effluent in the bay. The water has an average residence time of 4–6 days (Atkins, 1970a, 1970b). It was decided therefore, that effects in the water column

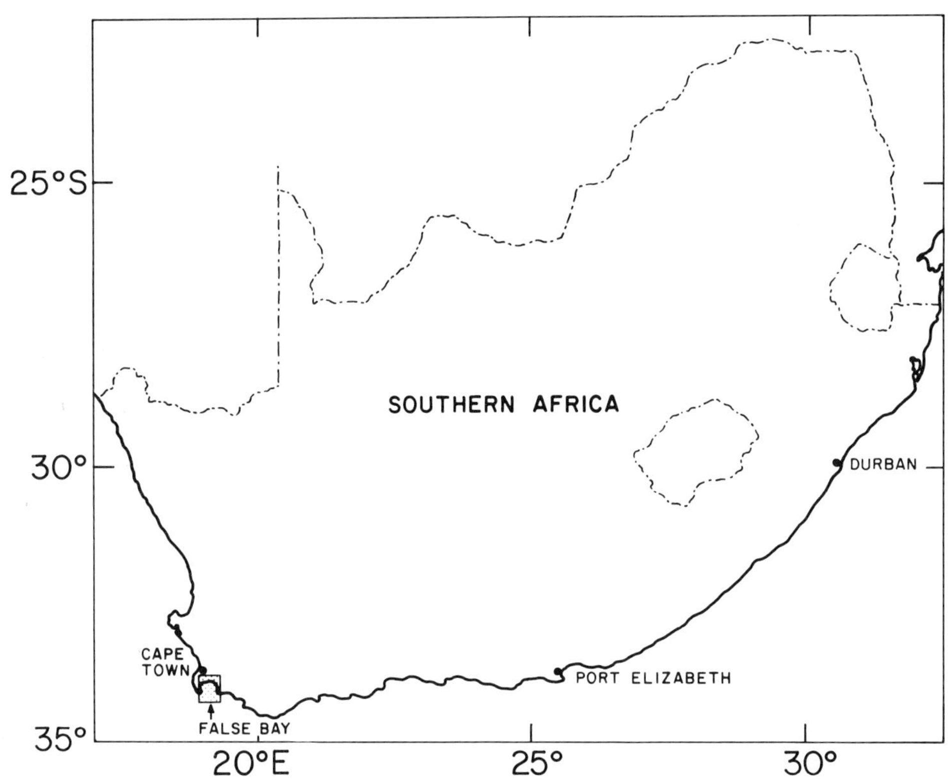

Figure 19.1. Location of False Bay in Southern Africa.

are too transient to exhibit any pollution accumulation over a long period of time.

19.2.2. Sedimentology

Sediment distribution maps have been produced not only by the present study, but also by Bowie et al. (1970), Glass (1980), and the NRIO (1982). Several features can be identified. Very coarse sands composed of shell fragments occur along the western and eastern margins of the bay, where high-energy conditions prevail and the sediment supply is low. A large patch of coarse material is found in the northeastern corner of the bay on all three surveys. Another prominent feature that appears in all of the surveys is the large cresent-shaped area of fine sediments extending across the center of the bay. In this area, the floor of the bay is covered with very fine sediment in deep, relatively low-energy areas. Another small area of fine sediments is found immediately to the south of Gordon's Bay.

19.3. SAMPLING AND METHODS

Sediment sampling was carried out in May 1982 by using a Shipek grab operated from a fishing vessel. Station positions were based on the sediment-distribution map produced by Glass (1980). Samples were successfully taken from 65 stations (Fig. 19.3).

At each station, two subsamples were placed in glass jars for organohalogen and coprostanol determinations. The jars had been previously cleaned with pesticide-grade acetone and hexane. A futher subsample was collected for determinations of trace metals, organic compounds, and particle size. All samples were frozen as soon as possible after collection. An additional subsample taken for meiofaunal determinations was fixed with 4% formalin.

Concentrations of the trace metals Cu, Mn, Fe, Ni, Zn, Pb, and Cd were assayed by leaching the whole-sediment samples with 4:1 nitric:perchloric acid (Watling and Watling, 1982), followed by standard flame atomic absorption analysis. Separate analyses were carried out for Hg by using permanganate oxidation followed by cold vapor atomic absorption analysis.

Checks on the accuracy of the methods are maintained. Over several years, intercalibration exercises have been undertaken by laboratories in South Africa. Additionally,

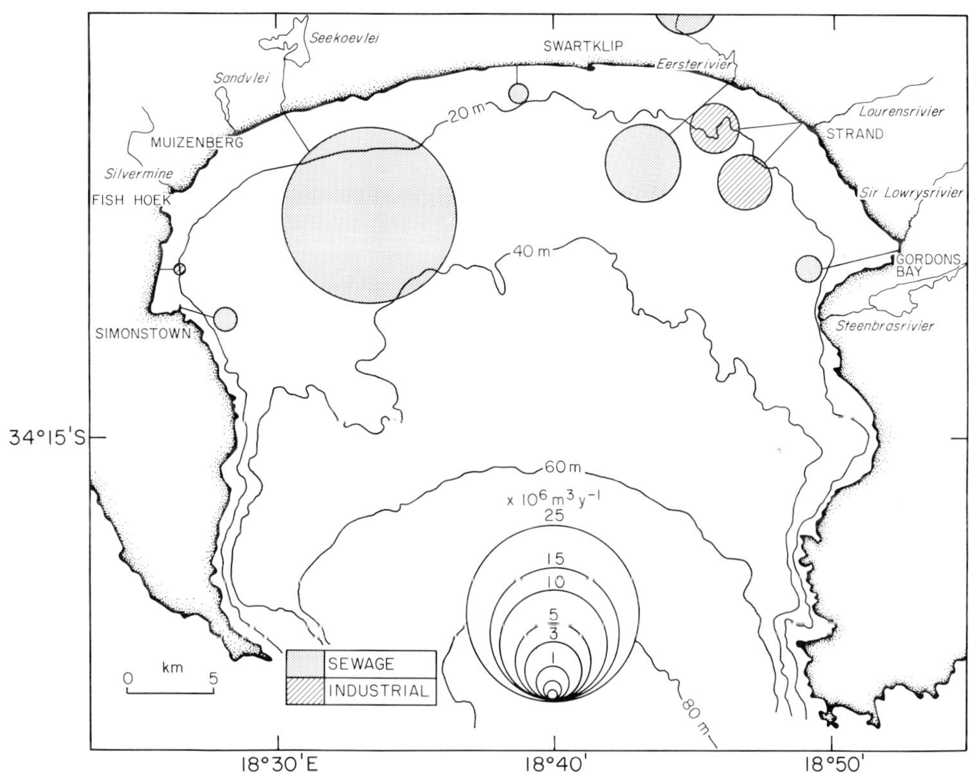

Figure 19.2. False Bay depth contours in meters and discharged effluents.

these laboratories have used standards from the International Atomic Energy Agency (IAEA) and the National Bureau of Standards and have participated in intercalibration exercises.

Results from local intercalibration exercises on marine sediments are shown in Table 19.1. The results obtained by using the nitric:perchloric acid digestion were satisfactory. An analysis of the intercomparison with the IAEA water sample (W-3/1) also yielded satisfactory results, with the exception of iron, which had a value of 70% of the accepted figure.

The total organic content of the samples was measured by loss of weight on ignition to 570° C for 4 h. The maximum clay-sized content of these sediments was 5%, which was shown by X-ray diffraction to be composed mostly of illite (Rosental, 1984). Illite contains 8% water, which gives a maximum possible error of 0.4% due to dehydration of the clay. Since completion of these analyses, a carbon analyzer has become available, and tests have shown that errors due to carbonate decomposition were small for these sediments, so the loss-on-ignition value is considered satisfactory for this study.

Particle-size analysis was performed on the sand fraction (0.063–2.0 mm) by using a hydraulic settling tube (Flemming and Thum, 1978). Pesticide determinations were performed, under subcontract, at the University of Port Elizabeth according to the method described by Kerkhoff et al. (1982). Samples for coprostanol analyses were extracted with hexane:propan-2-ol and cleaned on alumina columns. Analysis was by gas chromatography on an OV-1 capillary column. Finally, meiofauna were separated by using a modified Oostenbrink apparatus (Fricke, 1979) and then counted into classes under a dissecting microscope.

19.4. RESULTS AND DISCUSSION

19.4.1. Sediment Description

The sediments were not analyzed for mineralogy. However, by visual inspection they fell into roughly three categories: (1) those consisting largely of shell fragments and, hence, with a high carbonate content and a small percentage of quartz; (2) those consisting largely of sand (quartz grains); and (3) finer-grained sediments consisting of fine quartz and fine shell fragments together with some mud and organic material.

The organic carbon content of the sediments was determined by the loss of weight on ignition; the distribution is illustrated in Fig. 19.3. Again, some prominent features can be noticed. One feature is the large, crescent-shaped area of highly (2–5%) organic, fine sediment in the middle of the bay, and another is the area with a low (<0.8%) organic content near Fish Hoek. Both of these features have been recorded before (Bowie et al., 1970; Glass, 1980). Bowie found organic contents of 4–11% in the middle of the bay; however, these analyses were by wet oxidation, which probably accounts for the higher absolute values.

19.4.2. Metals in the Sediments

The sediments can provide an integrated record even of irregular pollutant inputs. Sediment samples were therefore analyzed for nine metals. Ranges and mean values are shown in Table 19.2.

The concentrations of Cd, Cu, Zn, Ni, and Al had a similar distribution to that of organic carbon and were highest in the south–central area of the bay. This was the area in which the finest sediments lay and also the area in which organic content of the sediments was the highest. Mercury concentrations were <0.144 $\mu g\ g^{-1}$ and were detectable only in three areas close to shore.

Iron and Mn often behave similarly (Dayal et al., 1983). In this case, the areas of highest concentrations were in the northeastern corner of the bay, and a strong land-derived source of Mn seems likely. Without further investigation, however, this cannot be assumed to be pollution, since the Table Mountain Sandstone series contains manganese mineralization, and this Mn may be transported to the bay by the Lourens River.

Table 19.1. Results of National Intercalibration of Metal Concentrations ($\mu g\ g^{-1}$) in Three Marine Sediments

Sediment	Cu	Mn	Fe	Ni	Zn	Pb	Cd
A (Mean)	3.2	24	3200	2.2	14.2	6.3	0.04
A (NRIO)	2.9	21	3088	1.7	12.1	6.3	0.04
B (Mean)	2.8	51	3368	2.0	11.5	5.3	0.04
B (NRIO)	1.6	47	3115	1.7	10.4	5.2	<0.04
C (Mean)	2.1	28	2463	0.4	5.5	2.6	0.05
C (NRIO)	1.5	29	2407	<0.4	5.8	3.2	0.05

Table 19.2. Metal Concentrations ($\mu g\ g^{-1}$, dry wt) in False Bay Sediments

Concentration	Al	Cu	Mn	Fe	Ni	Zn	Pb	Cd	Hg
Minimum	152.4	<0.3	1.0	110	<0.4	<0.8	<0.2	0.1	<0.002
Maximum	2349.1	2.1	25.6	3456	4.6	9.0	5.0	0.8	0.144
Mean	631.6	0.7	6.3	833	0.8	2.0	1.7	0.1	0.006

The distribution pattern for Pb was completely different from those of the other metals. Definite inputs from the shore were seen at Simonstown, Muizenberg, Swartklip, and Strand; the lowest concentrations of lead were in the center of the bay.

The distribution of metals in the sediments of an area like False Bay is usually controlled by three major factors: (1) inputs, (2) sediment grain-size distribution, and (3) organic content of sediments. To test these control mechanisms, various correlations were tested.

Most of the metal concentrations correlated well with organic content at the 95% level, decreasing in the following order: Cu → Zn → Ni → Al → Cd → Fe → Mn. This provides a means of identifying point-source inputs. The metal concentrations can be normalized to a constant organic content, and any significant metal inputs will show up as high concentrations. In this way, most of the metals showed a fairly even distribution over the entire bay. The principal exceptions were Mn and Pb. Manganese showed a similar distribution when both the actual and the normalized levels were plotted, confirming a source of Mn in the Strand area. Lead presented a similar example and confirmed the source as being principally urban run-off. Some of the other metals showed slightly elevated concentrations near the coast, especially near Simonstown.

Metals are generally associated with fine sediments such as clays and associated organic particles. The metal concentrations in this study, however, did not correlate with particle size. This lack of concentration implies that sediment size is not a suitable tool to use in determining metal associations, and that the mud fraction (<0.063 mm) should be more closely examined.

A guide to the significance of the metal concentrations can be obtained by comparisons with other sites. In Table 19.3, mean values for False Bay are compared to those in other areas in the Western Cape (Table Bay, Saldanha Bay, Noordwesbaai, and Camps Bay) and some values from British sites, all of which have a predominantly sandy sediment. Table Bay is located adjacent to urban Cape Town and is subject to a pollution load; Saldanha Bay (120 km north of Cape Town) is subject to a small pollutant load from ore-loading operations; Noordwesbaai (160 km north of Cape

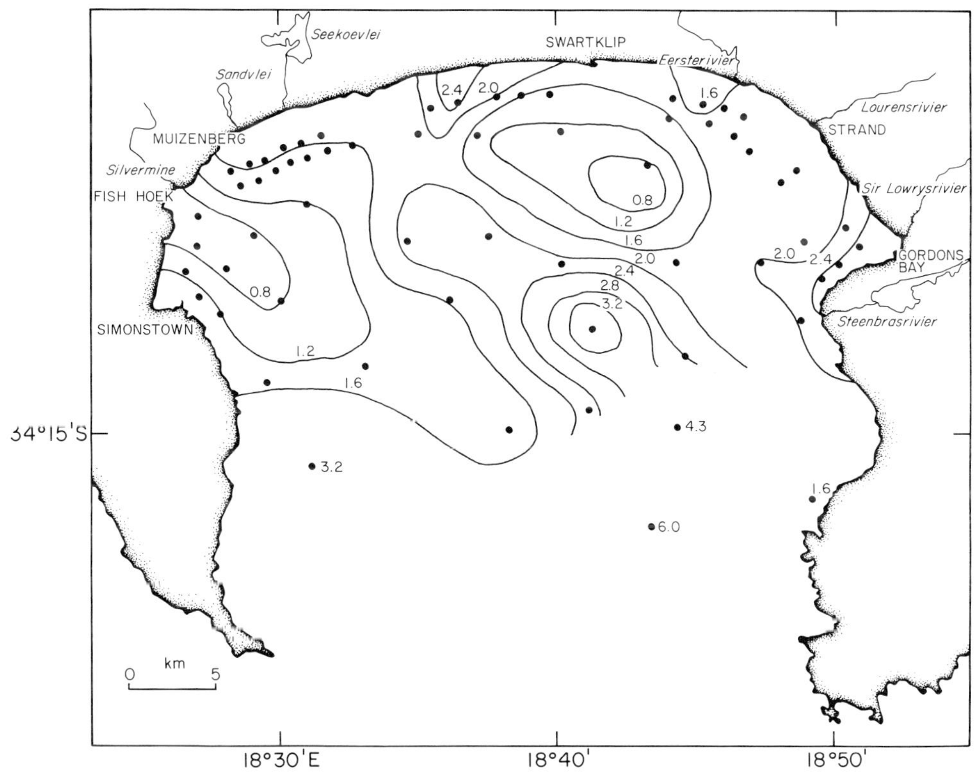

Figure 19.3. Percent of organic content of False Bay sediments.

Town) is presently pollution-free; and Camps Bay, although located within the urbanized part of the Cape Peninsula, is considered to be pollution-free. The concentrations of metals in the sediments of Tor Bay and Solway Firth approached those found in the Western Cape sites. Tor Bay is a semi-urban area similar to False Bay. A comparison of these mean values gives an approximate indication that the Western Cape sites, including False Bay, have no serious metal pollution problem.

19.4.3. Organohalogens and Fecal Steroids in Sediments

The sediment samples were analyzed for polychlorinated biphenyls (PCBs) and a variety of pesticide residues. Only PCBs and lindane were found in a substantial number of the samples. Other residues found were endosulphan cyclic sulfate; β-BHC; p,p-DDE; dieldrin; and hexachlorobenzene. All of the concentrations were extremely low (Table 19.4) and their distribution was essentially random. The randomness of the distribution indicates that there are no point sources of organohalogens in False Bay. The low concentrations show that there is, at present, no pollution problem involving these organic compounds in False Bay.

To test for sewage pollution, coprostanol analyses were also undertaken with the sediments. The levels found were also low (<10 to 117 ng g^{-1}) and exhibited a random distribution, illustrating the effective treatment of sewage effluents, which generally undergo secondary treatment.

19.5. MEIOFAUNA

A large amount of background work on meiofauna distribution and population densities was carried out by the NRIO (Eagle et al., 1979; Fricke et al., 1979; Orren et al., 1979, 1981a, 1981b; Hennig et al., 1982). However, all of the previous work applied to intertidal sandy beaches, and there has been no published work on subtidal meiofauna that can be used for direct comparison. Some unpublished data exist for subtidal meiofauna in Table Bay, and these can be used for comparison. Nonetheless, the meiofauna distribution patterns showed some extremely interesting features that were not revealed by chemical analysis. Hennig et al. (1983) showed that the number of nematodes and harpacticoid copepods in the intertidal environment depended on sand grain size and factors such as organic enrichment or high oxygen levels due to mechanical perturbation. These relationships were used to compare the results obtained from False Bay.

Table 19.3. Mean Concentrations (μg g^{-1}, dry wt) of Metals in False Bay, Western Cape Sites, and British Sites

Location	Cu	Mn	Fe	Ni	Zn	Pb	Cd	Hg
False Bay	0.7	6.3	833	0.8	2.0	1.7	0.12	0.006
Table Bay	4.8	17.4	1986	1.0	11.1	9.0	0.06	0.012
Camps Bay	0.3	12.1	490	<0.4	11.5	1.0	<0.04	
Noordwesbaai	0.7	9.9	1040	0.9	3.6	1.6	0.48	0.003
	1.0	9.9	1680	1.0	8.0	1.5	0.23	
Tor Bay[a]	4.2	12.8		7.2	24.8	31.2	0.37	0.07
River Clyde (control)[b]	16	355		30	85	42	1.6	0.1
Solway Firth[c]	10	341			63	36	1	

[a] Taylor (1974).
[b] Halcrow et al. (1973).
[c] Perkins et al. (1973).

Table 19.4. Organohalogen Concentrations (μg g^{-1}, dry wt) in False Bay Sediments

Organohalogen	Minimum	Maximum	Mean (n = 65)	Number of Positive Samples
PCB	<0.01	0.30	0.02	22
Lindane	<0.001	0.210	0.01	32
Endosulphan cyclic sulfate	<0.001	0.01		5
β-BHC	<0.001	0.003		2
p,p-DDE		0.002		1
Dieldrin		0.001		1
Hexachlorobenzene		0.001		1

Table 19.5. Nematode Density per 100 cm^3 in Relation to Grain Size in False Bay and Other Locations

Location	Sediment Size (μm)		
	0–180	181–300	301–700
False Bay	1506	943	379
Table Bay	644	93	507
Southern Bight[a]	200	615	407
Reference beaches[b]	64	189	90
Beaches polluted by organic material[b]	299	607	
Oil-polluted beaches[b]		84	76

[a] Willems et al. (1982).
[b] Hennig et al. (1983).

19.5.1. Nematodes

Table 19.5 compares nematode numbers in False Bay with those from other locations. It is well known that discrete faunal communities do occur where distinct physical boundaries (for example, pore size) exist in the environment (Wieser, 1953; Gray, 1976; Bennington, 1977; Hennig et al., 1983). The meiofauna of different grain-size categories (fine, medium, and coarse) consists of different populations (Lorenzen, 1974; Hennig et al., 1983); therefore, our data has been separated into these three categories (Table 19.5). The nematode fauna of False Bay, Table Bay, and the Southern Bight of the North Sea off Belgium (Willems et al., 1982), are all sublittoral. The data from the South African reference beaches, those disturbed by organic matter and those disturbed by oil are taken from a report by Hennig et al. (1983).

The number of nematodes found in the False Bay sediments was very high and indicated a disturbed condition, particularly in the western half of the bay (Fig. 19.4). The nematode numbers were some of the highest recorded by this laboratory, and were about twice those found in subtidal sediment known to be enriched by organic material in Table Bay and the Southern Bight (Table 19.5). A comparison with nematode densities in beach samples emphasized the high numbers of nematodes and the similarity in population densities of beaches enriched by organic material (Table 19.5).

The nematode distribution illustrated some fine features that were not indicated by the chemical analysis of the sediments (Fig. 19.4). For instance, the numbers of nematodes to the west of the Eerste River mouth were high, which supports the observation by Bartlett and Hennig (1981) that the discharge from the river flows predominantly to the east. The nematodes are likely to avoid the freshwater emanating from the river, but at the same time they would be attracted by the organic material brought in by the river. They are therefore likely to congregate near the edge of the plume. In the same way, another patch containing high numbers of nematodes was found on the west side of the Lourens River mouth.

19.5.2. Harpacticoid Copepods

The densitites of harpacticoid copepods in the sediments of False Bay and in various other locations are given in Table 19.6, and the copepod distribution in False Bay is given in Fig. 19.5. There is a highly significant relationship between sediment size [$\phi(\mu m)$] and harpacticoid numbers (*Harp*):

$$Harp = 70 + 0.5424\ \phi \quad (r = 0.3658,\ n = 51) \qquad 1$$

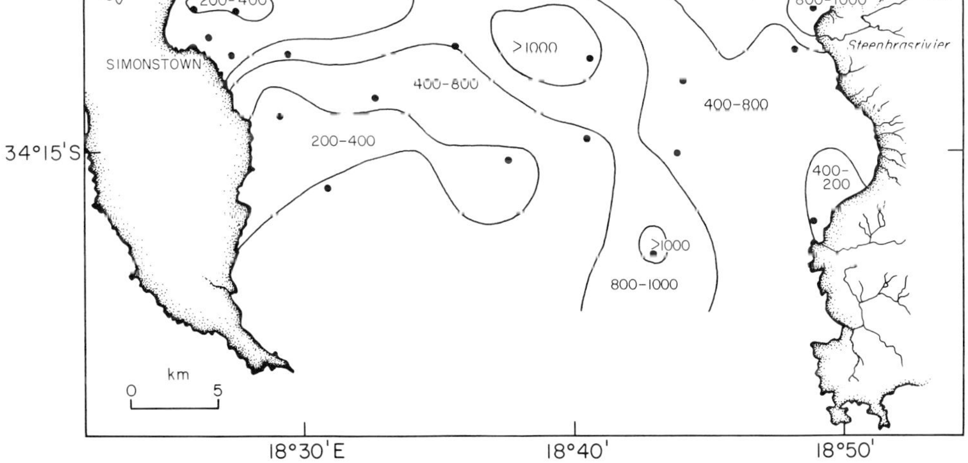

Figure 19.4. Distribution of nematodes per 100 cm^3 of sediment in False Bay.

This correlation between sediment size and copepods is probably related to oxygen content (Fricke et al., 1979; Orren et al., 1981a) because higher oxygen content is usually found in coarse sediments. Furthermore, the organic content was less in the coarse sediment (Table 19.7).

The population density in the False Bay sediment was very similar to that of the Southern Bight, the reference beaches and those polluted by organic material, and it appears to be normal as defined by Hennig et al. (1982) and Hennig et al. (1983). The harpacticoid numbers in False Bay could not be compared to those of Table Bay because it appeared that Table Bay was partly polluted by oil. This resulted in lower copepod numbers (Table 19.6) as seen by the oil-polluted beaches.

19.5.3. Flatworms

Flatworms usually occur in very low numbers in sandy beaches and therefore have not been studied in detail in the past. Known exceptions to this are the beach around the food-canning factory outfall near Mossel Bay (Eagle et al., 1979) and near the mouth of the Eerste River (Bartlett and Hennig, 1981).

The Table Bay studies have also indicated high numbers of flatworms in the vicinity of the Green Point sewage outfall (Table 19.8). Thus the presence of large numbers of

Table 19.6. Harpacticoid Copepod Density per 100 cm³ in Relation to Grain Size in False Bay and Other Locations

Location	Sediment Size (μm)		
	0–180	181–300	301–700
False Bay	77	179	439
Table Bay	9	16	382
Southern Bight[a]			184
Reference beaches[b]	43	194	73
Beaches polluted by organic material[b]	34	118	
Oil-polluted beaches[b]		15	34

[a] Willems et al. (1982).
[b] Hennig et al. (1983).

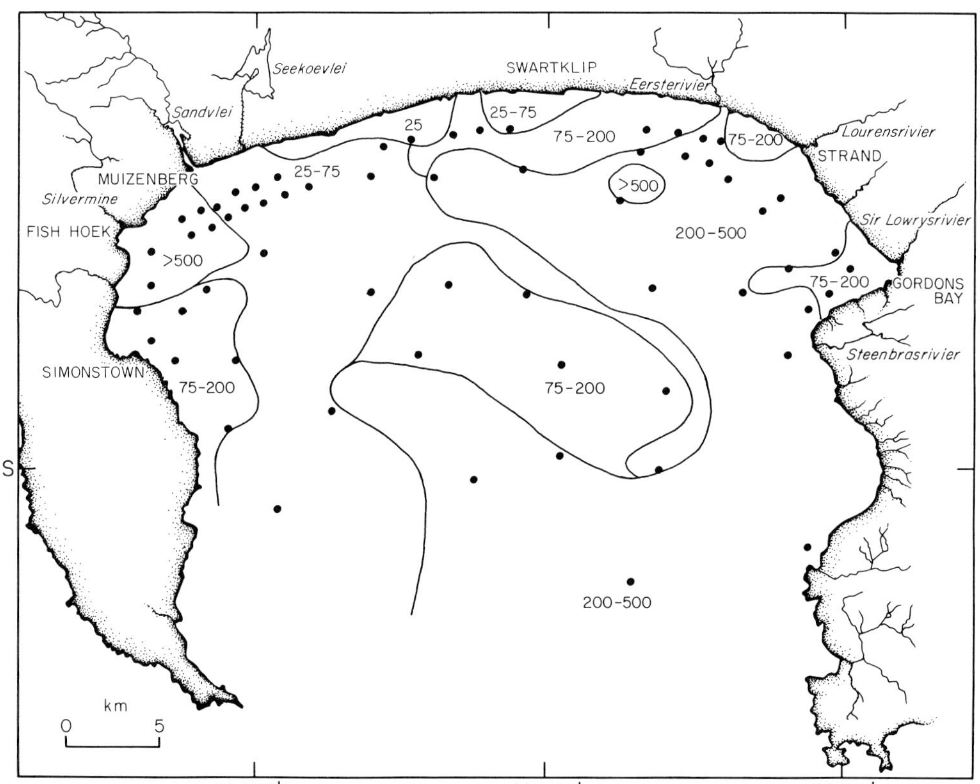

Figure 19.5. Distribution of harpacticoid copepods per 100 cm³ of sediment in False Bay.

Table 19.7. Organic Content (%) of Subtidal Sediments in False Bay and Other Locations in South Africa

Location	Sediment Size (μm)		
	0–180	181–300	301–700
False Bay	1.9	2.0	1.4
Table Bay	2.1	1.9	1.6
South coast (mean)[a]		0.15	
West coast (mean)[a]		0.11	

[a] Hennig et al. (1983).

Table 19.8. Flatworm Density per 100 cm^3 in Relation to Grain Size in False Bay and in Table Bay

Location	Sediment Size (μm)		
	0–180	180–300	301–700
False Bay	54	66	59
Table Bay	10	23	23

flatworms could be an indication of pollution hot-spots, which could be due either to anthropogenic inputs or to natural organic enrichment.

19.6. CONCLUSIONS

From the chemical and biological aspects of this pollution study, various conclusions can be drawn. The organic content of the sediments was generally a magnitude higher than that found in typical South African beaches; however, the organic content was higher in the fine sediments. The concentrations of most metals correlated well with the organic content of the sediments.

The concentrations of most metals measured were low throughout the bay, although they were slightly higher in fine sediments. A few hot-spots were indicated, notably for lead and for manganese. Lead had a definite input from the northern shore of the bay and there was a source of manganese in the Lourens River area. Organohalogen concentrations were also low throughout the bay.

The meiofaunal distributions indicated some anomalies that were not apparent from the chemical analyses. These anomalies were found between Simonstown and Fish Hoek, Muizenberg, west of Swartklip, the mouth of the Eerste River, and in Gordon's Bay.

By comparison with other countries in Europe and North America, the levels of pollution in False Bay are low. It is clear, however, that the False Bay system is sensitive to human influence and that the effects are likely to increase with time. The meiofaunal densities imply that the perturbation of False Bay by organic material is twice that of Table Bay. One reason for this could be the faster turnover rate of the water mass in Table Bay, while the water exchange between False Bay and the adjacent ocean is much less (Harris, 1978).

The complex hydrology and geology and the occurrence of discrete benthic communities in different regions of the bay emphasize the heterogeneity of the Bay. This indicates that isolated studies are not likely to be of help in understanding the bay's system or in conserving False Bay in a healthy state. The chemical and biological studies discussed in this chapter form only part of the project and are only in the preliminary stages.

This type of study provides a good example of interdisciplinary cooperation essential in the management of environmental problems.

ACKNOWLEDGMENTS

This study was carried out under contract to the Department of Environment Affairs. Thanks are due to all the National Research Institute for Oceanology and University of Cape Town personnel who participated in this study, especially Dr. G. A. Eagle.

REFERENCES

Atkins, G. R. 1970a. Thermal structure and salinity of False Bay. *Transactions of the Royal Society of South Africa*, **34**, 117–128.

Atkins, G. R. 1970b. Winds and current patterns in False Bay. *Transactions of the Royal Society of South Africa*, **34**, 139–148.

Bartlett, P. D., and H. F-K. O. Hennig. 1981. Pollution Monitoring Surveys of the Eerste River Estuary. Council for Scientific and Industrial Research Report T/SEA 8209, Stellenbosch, South Africa, 11 pp.

Bennington, S. L. 1977. Benthic meiofauna substrate relationships in an unpolluted and a polluted stream environment. *Mauri Ora*, **5**, 21–37.

Bickerton, I. B. 1982. Estuaries of the Cape. Part II: Synopsis of Available Information on Individual Systems. *In:* Report No. 15, Zeekoe (CSW5), A. E. F. Heydorn and J. R. Grindley (Eds.). Research Report 414, Research Council for Scientific and Industrial Research, Stellenbosch, South Africa, 54 pp.

Bowie, D. K., A. O. Fuller, and W. G Siesser. 1970. The marine sediments of False Bay. *Transactions of the Royal Society of South Africa*, **34**, 149–162.

Cram, D. L. 1970. A suggested origin for the cold surface water in central False Bay. *Transactions of the Royal Society of South Africa*, **34**, 129–138.

Dayal, R., M. G. Heaton, M. Fuhrmann, and I. W. Duedall. 1983. A geochemical study of the dredged-material deposit in the New York

Bight. *In:* Wastes in the Ocean, Vol. II: Dredged-Material Disposal in the Ocean, D. R. Kester, B. H. Ketchum, I. W. Duedall, and P. K. Park (Eds.). Wiley-Interscience, New York, pp. 123–149.

Eagle, G. 1980. Waste disposal into the Bay. *In:* The Future Management of False Bay, B. Gasson (Ed.). False Bay Conservation Society, Cape Town, pp. 93–96.

Eagle, G. A., P. J. Greenwood, H. F-K. O. Hennig, and M. J. Orren. 1979. Preliminary pollution surveys around the south-western Cape coast. Part 1: Mossel Bay. *South African Journal of Science,* **75,** 453–456.

Flemming, B. W., and A. B. Thum. 1978. The settling tube—an hydraulic method for the grain size analysis of sands. *Kieler Meeresforschung,* **4,** 82–95.

Fricke, A. H. 1979. Meiofauna extraction efficiency by a modified Oostenbrink apparatus. *Helgoländer Wissenschaftliche Meeresuntersuchungen,* **32,** 436–443.

Fricke, A. H., G. A. Eagle, W. J. Gledhill, P. J. Greenwood, and M. J. Orren. 1979. Preliminary pollution surveys around the south-western Cape coast. Part 3: Hout Bay. *South African Journal of Science,* **75,** 459–461.

Glass, J. 1980. Geology, morphology, sediment cover and movement. *In:* The Future Management of False Bay, B. Gasson (Ed.). False Bay Conservation Society, Cape Town, pp. 15–25.

Gray, J. 1976. Are marine base-line surveys worthwhile? *New Scientist,* **70**(998), 219–220.

Halcrow, W., D. W. Mackay, and I. Thornton. 1973. The distribution of trace metals and fauna in the Firth of Clyde in relation to the disposal of sewage sludge. *Journal of the Marine Biological Association of the United Kingdom,* **53,** 721–739.

Harris, T. F. W. 1978. Review of Coastal Currents in Southern African Waters. Report No. 30, South African National Scientific Programmes, Council for Scientific and Industrial Research, Pretoria, 103 pp.

Hennig, H. F-K. O., A. H. Fricke, P. J. Greenwood, and G. A. Eagle. 1982. Relationships between meiofaunal population densities and physico-chemical properties of unpolluted sandy beaches. *Environmental Monitoring and Assessment,* **1,** 337–344.

Hennig, H. F-K. O., G. A. Eagle, L. Fielder, A. H. Fricke, W. J. Gledhill, P. J. Greenwood, and M. J. Orren. 1983. Ratio and population density of psammolittoral meiofauna as a perturbation indicator of sandy beaches in South Africa. *Environmental Monitoring and Assessment,* **3,** 45–60.

Kerkhoff, M. A. T., A. de Vries, R. C. C. Wegman, and A. W. N. Hofspee. 1982. Analysis of PCB's in sediments by glass-capillary gas chromotography. *Chemosphere,* **11,** 165–174.

Lorenzen, S. 1974. Die Nematoden Fauna der subliteralen Region der Deutschen Bucht, insbesondere im Titan Abwassergebiet bei Helgoland. *Veröffentlichungen des Instituts für Meeresforschung Bremerhaven,* **14,** 305–327.

Mallory, J. K. 1970. The bathymetry and microrelief of False Bay. *Transactions of the Royal Society of South Africa,* **34,** 103–112.

Morant, P. D., and J. R. Grindley. 1982. Estuaries of the Cape. Part II: Synopsis of Available Information on Individual Systems. *In:* Report No. 14, Sand (CSW 5), A. E. F. Heydorn and J. R. Grindley (Eds.). Research Report 413, Council for Scientific and Industrial Research, Stellenbosch, South Africa, 70 pp.

National Research Institute for Oceanology. 1982. The Geology of False Bay with Special Empahsis on Modern Sediments. Report C/SEA 8253, Council for Scientific and Industrial Research, Stellenbosch, South Africa, 20 pp.

Orren, M. J., A. H. Fricke, G. A. Eagle, P. J. Greenwood, and W. J. Gledhill. 1979. Preliminary pollution surveys around the south-western Cape coast. Part 2: Green Point sewage outfall. *South African Journal of Science,* **75,** 456–459.

Orren, M. J., G. A. Eagle, A. H. Fricke, W. J. Gledhill, P. J. Greenwood, and H. F-K. O. Hennig. 1981a. The chemistry and meiofauna of some unpolluted sandy beaches in South Africa. *Water S. A.,* **7,** 203–210.

Orren, M. J., G. A. Eagle, A. H. Fricke, P. J. Greenwood, H. F-K. O. Hennig, and P. D. Bartlett. 1981b. Preliminary pollution surveys around the south-western Cape coast. Part 4: Salt River mouth. Table Bay. *South African Journal of Science,* **77,** 183–188.

Perkins, E. J., J. R. S. Gilchrist, O. J. Abbot, and W. Halcrow. 1973. Trace metals in Solway Firth sediments. *Marine Pollution Bulletin,* **4,** 59–61.

Rosental, R. 1984. Trace Element Distribution in Different Chemical Fractions of False Bay Sediments. M.Sc. Thesis, University of Cape Town, 75 pp.

Taylor, D. 1974. The natural distribution of trace metals in sediments from a coastal environment, Tor Bay, England. *Estuarine and Coastal Marine Science,* **2,** 259–264.

Watling, R. J., and H. R. Watling. 1982. Metal surveys in South Africa estuaries. I: Swartkops River. *Water S. A.,* **8,** 26–35.

Wieser, W. 1953. Die Beziehung zwischen Mundhöhlengestalt, Ernahrungsweise und Vorkommen bei freilebenden marinen Nematoden. *Arkiv für Zoologie,* **4,** 439–484.

Willems, K .A., M. Vincx, D. Claeys, S. Vanosmael, and C. Heip. 1982. Meiobenthos of a sublittoral sandbank in the Southern Bight of the North Sea. *Journal of the Marine Biological Association of the United Kingdom,* **62,** 535–548.

SUBJECT INDEX

AUTHOR INDEX